Fourier Modal Method and Its Applications in Computational Nanophotonics

Fourier Modal Method and Its Applications in Computational Nanophotonics

Hwi Kim • Junghyun Park
Byoungho Lee

CRC Press
Taylor & Francis Group
Boca Raton London New York

CRC Press is an imprint of the
Taylor & Francis Group, an **informa** business

MATLAB® is a trademark of The MathWorks, Inc. and is used with permission. The MathWorks does not warrant the accuracy of the text or exercises in this book. This book's use or discussion of MATLAB® software or related products does not constitute endorsement or sponsorship by The MathWorks of a particular pedagogical approach or particular use of the MATLAB® software.

CRC Press
Taylor & Francis Group
6000 Broken Sound Parkway NW, Suite 300
Boca Raton, FL 33487-2742

First issued in paperback 2017

© 2012 by Taylor & Francis Group, LLC
CRC Press is an imprint of Taylor & Francis Group, an Informa business

No claim to original U.S. Government works

Version Date: 20120127

ISBN 13: 978-1-4200-8838-0 (hbk)
ISBN 13: 978-1-138-07430-9 (pbk)

Library of Congress Cataloging-in-Publication Data

Kim, Hwi.
Fourier modal method and its applications in computational nanophotonics / Hwi Kim, Junghyun Park, Byoungho Lee.
  p. cm.
"A CRC title."
Includes bibliographical references and index.
ISBN 978-1-4200-8838-0 (hardcover : alk. paper)
1. Nanophotonics--Mathematics. 2. Nanoelectronics--Mathematics. 3. Fourier analysis. I. Park, Junghyun. II. Lee, Byoungho. III. Title.

TA1530.K56 2012
515'.2433--dc23   2011046920

Visit the Taylor & Francis Web site at
http://www.taylorandfrancis.com

and the CRC Press Web site at
http://www.crcpress.com

Contents

Preface

Recent great advances in nanophotonics have opened new possibilities in this field. The fusion of nanophotonics and micro- and nanoelectronics has stimulated new research and development trends for overcoming the microelectronics of the present, which are facing technological limitations. Nanophotonics is also perceived as a necessary prerequisite for achieving breakthroughs in the fields of bio- and nanotechnologies. The origin of these potentially new developments is the fascinating area of the physics of the interactions between optical fields and nanoscale structures.

The objective of this book is to provide researchers as well as graduate students with an in-depth mathematical understanding, a framework for developing a sound numerical analysis of nanophotonics phenomena, and practical skills required for the MATLAB® implementation of the Fourier modal method (FMM).

The development of rigorous mathematical modeling in nanophotonics research is one of the most important issues, from two viewpoints. First, accurate optical field calculation in nanoscale structures is a fundamental requirement for nanophotonics research. Various physical interpretations of nanoscale optical phenomena are necessarily obtained by the post-data processing of optical field distribution data obtained from rigorous mathematical models. Second, the large-scale system-level integration of nanophotonic devices, i.e., a nanophotonic network, is considered to be an ultimate point for advances in nanophotonics. From the viewpoint of this application, a mathematical model of such a large-scale nanophotonic network is an important issue. Although electromagnetic methods for analyzing individual devices and physical phenomena on small local regions have already been established, there are no definitive discussions of mathematical models of large-scale nanophotonic networks until now. The use of conventional methods for analyzing nanophotonic networks is almost impossible without some specific innovations, because of the limitation of computing resources. This book presents a perspective on the mathematical modeling of nanophotonic networks with the full exploitation of the linear system properties in FMM. As is well known, the core feature in the linear system theory is modal analysis. The principle of modal analysis states that a linear optical structure has its own electromagnetic eigenmodes, and thus the arbitrary possible optical fields are actually the coherent superposition of eigenmodes with appropriate coupling coefficients. The modal analysis of a linear optical structure is intuitive, informative, efficient, and flexible. These benefits of linear modal analysis make large-scale nanophotonic network modeling feasible.

The main objective of this book is to establish a stable and efficient numeri-cal framework for the rigorous modal analysis of very general linear optical structures on both micro- and nanoscales. The modal method is made up of two key subtheories.

FMM and the scattering matrix method (S-matrix method). In a general sense, the S-matrix method is a linear system algorithm for interconnecting separately characterized system blocks in a collinear manner. FMM is an algorithm for finding the Bloch eigenmodes of system blocks and represents the Bloch eigenmodes with a pseudo-Fourier series.

The S-matrix method and FMM can be extended to the local Fourier modal method (LFMM). LFMM is a method for the representation of Bloch eigen-modes of multiblock systems by means of a pseudo-Fourier series. Based on LFMM, the framework for modeling a nanophotonic network is addressed for advanced research on this topic.

This book is organized as follows:

Chapter 1: An introduction to nanophotonics and FMM is provided. The meaning and importance of FMM in nanophotonics are dis-cussed and computational nanophotonics is defined as a subfield of nanophotonics. A brief overview of several examples of the simu-lation of the nanophotonics of photonic crystals and plasmonics is included in this book; they constitute good references for training related to FMM.

Chapter 2: As a prerequisite to the study of FMM, the mathematical modeling of one-dimensional multiblock structures is elucidated using the S-matrix method. It is essential that the basic scheme and properties of the S-matrix method be understood in the simplified one-dimensional case. After reading this chapter, the reader will have become familiar with the commonly used mathematical frame-work and notations used in this book.

Chapter 3: The fundamentals of FMM are developed. The Fourier modal representation of electromagnetic fields and the S-matrix analysis of single-block structures and multiblock structures are described step by step. The concepts illustrated in this chapter provide the founda-tion for LFMM in the following chapters. Examples of modeling an extraordinary transmission phenomenon are presented with work-ing MATLAB codes.

Chapter 4: FMM can be used in investigations of aperiodic structures such as a single slit or waveguides that use perfect matched layers (PMLs). The PML is an artificial layer that fully absorbs incoming electromagenetic waves but does not reflect. In FMM, the presence of a PML between neighboring computation cells prevents any inter-ference from electromagnetic fields between them, resulting in com-pletely isolated computation space. The PML can be achieved either by the absorbing boundary layer or by the stretched nonlinear coordinate transformation. Both PML theories for FMM are described with detailed mathematics and working MATLAB codes. As examples of the aperiodic problems in FMM via the PML, the on- and off-axis plasmonic beamings, the generation of plasmonic hot spots, and the plasmonic vortex are provided.

Chapter 5: LFMM is developed based on FMM and the S-matrix method. The pseudo-Fourier representation of Bloch eigenmodes and the S-matrix analysis of single-super-block and multi-super-block structures are derived. The LFMM modeling of tapered photonic crystal resonators is presented with working MATLAB codes.

Chapter 6: Advanced theory and the direction of future development of LFMM for nanophotonic network modeling is discussed. General nanophotonic networks are decomposed by two-port blocks and four-port cross-blocks. The 4×4 S-matrix for a four-port cross-block is derived and a generalized S-matrix method describing the interconnection algorithm of the S-matrices of two-port blocks and four-port blocks is constructed. Finally, concluding remarks and perspectives regarding FMM are discussed.

This book provides FMM MATLAB source codes as well as key principles, concepts, and detailed mathematics for practical use and self-training, and further research, which have been developed in the OEQELAB (Optical Engineering and Quantum Electronics Lab led by Professor Byoungho Lee), School of Electrical Engineering, Seoul National University, Korea. Since the FMM package was developed for educational purposes, performance factors such as speed, compactness, and memory usage may be sacrificed. However, the MATLAB codes provided in the package directly match the mathematics described in the book for easy interpretation of the MATLAB codes. We believe that this approach is more helpful for readers than providing optimized codes. After achieving a complete understanding of the theory, one can further engineer the MATLAB source code to enhance the speed and memory usage.

The authors are indebted to colleagues and students in selecting and revising the content and presentation, and particularly appreciate the help and inspiration of the Taylor & Francis book editors, L.-M. Leong, K. A. Budyk, and A. Shih. The authors also acknowledge support by the National Research Foundation and the Ministry of Education, Science, and Technology of Korea through the Creative Research Initiative Program (Active Plasmonics Application Systems).

Hwi Kim, Junghyun Park, and Byounho Lee

MATLAB® is a registered trademark of The MathWorks, Inc. For product information, please contact:

The MathWorks, Inc.
3 Apple Hill Drive
Natick, MA 01760-2098 USA
Tel: 508-647-7000
Fax: 508-647-7001
E-mail: info@mathworks.com
Web: www.mathworks.com

1

Introduction

1.1 Nanophotonics and Fourier Modal Methods

Nanophotonics has attracted a substantial amout of interest, given the belief that nanophotonics can provide technological breakthroughs for modern technologies facing their limitations and eventually contribute to building a future knowledge base and sustainable society. Nanophotonics is a multidisciplinary field that encompasses the fields of the science and technology of light and interactions of matter on a nanoscale. Technological advances are the result of new achievements in nanophotonics constantly being reported [1–4]. Essentially the use of interactions between optical near fields and nanoscale matters is expected to guide revolutionary changes in modern technologies.

Nanophotonics with multidisciplinary characteristics provides numerous opportunities for multidisciplinary research [1,2]. For physicists, nanophotonics opens novel optical phenomena in nanocavities, single photon sources, nanoscale nonlinear optical processes, the nanocontrol of interactions between electrons, phonons, and photons, and time-resolved and spectrally resolved studies of nanoscopic excitation and dynamics. For engineers, nanophotonics encompasses nanolithography for the nanofabrication of emitters, detectors, and couplers, photonic crystal circuits and micro-cavity-based devices, combinations of photonic crystals and plasmonics to enhance various linear and nonlinear optical functions, quantum dot and quantum wire lasers, plasmonic lasers, and highly efficient broadband and light-weight solar panels. For biologists and biomedical researchers, nanophotonics includes the genetic manipulation of biomaterials for photonics, biological principles to guide the development of bioinspired photonic materials, novel biocolloids and biotemplates for photonic structures, bacterial synthesis of photonic materials, and novel optical nanoprobes for diagnostics, targeted therapy using light-guided nanomedicine, and nanotechnology for biosensors. Although the classification criteria are dependent on various viewpoints, such as physical concepts, physical quantities, and academic interest, conceptually, the interplay of nanoscale confined light with nanoscale confined matter is a critically important issue.

Focusing on the fundamental objects of nanophotonics, the area of nanophotonics can tentatively be divided into three categories: plasmonics, photonic crystals, and photonic metamaterials [1,2]. Plasmonics is a research field that deals with metal-dielectric nanostructures that feature a number of peculiar properties originating from surface plasmons (SPs) [5–8]. SPs are the coupled resonant oscillation of photons and electrons at a metal-dielectric interface. Representative properties of plasmonics include a high confinement of electromagnetic fields on a nanoscale, highly enhanced light intensity of an electromagnetic field mediated by SP, and dispersion properties that provide the basis for various photonic metamaterials. A photonic crystal is a general periodic structure on a nanoscale, which is an important subfield of nanophotonics. The goal of photonic crystal research is the perception of the analogy between photons in photonic crystals and electrons in solid, semiconductor crystals. The control of photonic light propagation in photonic dielectric structures through photonic band structure engineering produces novel concepts and applications of photonic devices [9–16]. With the logical similarity from well-known electron semiconductor devices, active photonic devices as well as passive photonic devices such as ultracompact photonic crystal waveguide, cavities, lasers, and optical memories are subjects of active investigation. The important properties of photonic crystals are ultra-compact photonic circuitry based on high confinement of light and low bending loss in photonic crystal waveguides, omnidirectional photonic bandgap, and a peculiar dispersion property, such as photonic interband transitions, effective negative refraction, superprism effects, and slow light.

A photonic metamaterial is one of the newest subfields in nanophotonics. Metamaterials [17–24], especially photonic metamaterials, refer to any artificial materials with optical properties that cannot be observed in natural materials. Metamaterials have fine optical structures at a deep subwavelength scale, which show various resonant responses to electromagnetic waves and result in unprecedented macroscopic optical properties, such as negative refraction, a strong magnetic response, and nonlinear coordinate transformation. Optical waves propagating in a metamaterial create an effective homogeneous medium with special dispersion or resonance properties. The difference between a photonic crystal and a photonic metamaterial can be understood with respect to critical size and a functional analogy to electronic solids. A photonic crystal is analogous to a natural material or a composite material with periodicity. Periodicity is the most critical point in a photonic crystal. However, the use of a photonic metamaterial to control optical wave propagation is a completely different approach. The periodicity of a metamaterial is on a deep subwavelength scale much smaller than the critical feature size of a photonic crystal. Using an analogy to an electron, we are able to understand that photonic crystals with periodic dielectric potential and a photonic band structure correspond to solid crystals with periodic electronic potential and an electronic bandgap structure. The photonic bandgap structure is fundamentally derived from the periodicity of the dielectric potential. The point of comparison is that the periodicity (including topology, size, and permittivity distribution in a unit primitive cell) is the key factor in creating a photonic band structure and managing the optical properties of a photonic crystal, while for a photonic metamaterial, individual scatter, i.e., photonic atoms, needs to be recreated with special resonance characteristics, which is analogous to creating a new atom with a special orbital structure. Plasmonic nanostructures are usually employed to make up individual photonic atoms of a photonic metamaterial due to its strong resonance properties on a deep subwavelength scale. Such collections of newly created atoms result in an artificial photonic metamaterial with unnatural optical properties such as negative refraction, invisibility, superlensing, and deep subwavelength nondiffracting ray propagation.

In addition to experimental nanophotonics research, nanophotonics research centered on mathematical modeling and computational simulations, that is, computational nanophotonics, is an important branch in nanophotonics and has already been considered as an independent special academic research area. The aim of computational nanophotonics involves the theoretical discovery and unveiling of new physical phenomena within the realm of nanophotonics and providing theoretical analyses through rigorous mathematical modeling, and it also covers the design and analysis of devices and systems as part of the applications. In computational nanophotonics, a rigorous and efficient computational method for optical field analysis is the most important and basic element. Over the last 30 years, computational methods in electrodynamics have been intensively researched. Today, the literature is replete with reports in this area, and many books on computational methods in electrodynamics have been published. The recent exponential advances in computational methods for electrodynamics owe a great deal to the great advances in computer science. Nanophotonics computation requires large-scale 3D modeling and the parallel Maxwell equation solver.

Among the well-known and popular computational methods in electromagnetic analysis, FMM was relatively limited to classical optics such as grating modeling and holography because of its inherent periodicity in the past. Thus few attempts have been made to publish a concrete textbook that deals with FMM, although there are abundant textbooks dealing with other computational electromagnetic methods, such as the finite-difference time-domain (FDTD) [25], the finite integration method (FIT) [26], and the finite-element method (FEM) [27–29]. However, recent advances in FMM have successfully established FMM as one of the major computational electromagnetic methods of choice, which is especially adapted for nanophotonics.

This book is intended to offer a complete guide to understanding the principles and detailed mathematics of the up-to-date FMM and to implement working MATLAB codes for the practical modeling of nanophotonic structures: plasmonics, photonic crystals, and metamaterials. To help the reader understand concepts and applicability of FMM and to apply it in practical research areas, this book offers various basic and important examples of

recent nanophotonics research that can be analyzed by FMM. How to prac-
tically model nanophotonic structures with FMM and how to analyze and
interpret the obtained data are stressed. FMM is applied to specific categories of problems with the constraints of linear and nonquantum phenomena,
i.e., only areas not beyond Maxwell's equations in linear media.

1.2 Elements of the Fourier Modal Method

The numerical schemes for solving Maxwell's equations can be classified into
two types of space domain methods and spatial frequency domain methods.
The space domain methods represent Maxwell's equations as partial differential equations in a space domain and analyze the numerical values of the
field distributions at spatial sampling points. Representative methods are
FDTD and FEM.

On the other hand, in the spatial frequency domain methods, Maxwell's
equations are transformed into the algebraic Maxwell eigenvalue equations
in the spatial frequency domain and the electromagnetic field distributions are represented by a Bloch eigenmode expansion. In FMM, the Bloch
eigenmode is represented by a pseudo-Fourier series. The relation between
a spatial domain field function and a spatial frequency domain field representation is similar to the relation between a time domain function and its
temporal Fourier transform. An important feature of FMM is that, by solving
the algebraic Maxwell eigenvalue equation in the spatial frequency domain,
it is possible to mathematically construct a complete set of eigenmodes for an
optical structure spanning all possible optical fields within the finite dimen-
sional numerical framework.

FMM is a representative spatial frequency domain method with the
pseudo-Fourier series representation of internal Bloch eigenmodes. During
the last decade, FMM [30–32] has been intensively researched and complex
problems related to the foundation of FMM have been solved. At present,
FMM is considered to be one of the most efficient and accurate electromagnetic analysis frameworks for optics and photonics. Among recent advances
in FMM, Li's Fourier factorization rule [33–35], the fast Fourier factorization rule by Popov and Neviére [36,37] for proper convergence in transverse
magnetic (TM) polarization, the scattering matrix (S-matrix) method [38,39],
and overcoming the staircase approximation [40,41] are particularly notable.
FMM has achieved mathematical completeness and numerical stability.

The great innovation of FMM can be attributed to the recent invention of
the PML formulation [42]. The absorbing boundary layer and nonlinear coordinate transform-based PML technique are important. Because of this invention, FMM has become one of the most powerful electromagnetic analytical
tools for linear photonic structures.

The modal analysis feature of the spatial frequency domain method can be easily combined with the linear system theory. The refined linear system theory can be built in the spatial frequency domain method. The S-matrix method was developed as a stable wave-propagating algorithm in multiblock structures. In a general concept, the S-matrix method interconnects several linear system blocks that are separately characterized. The S-matrix method is a fascinating algorithm derived from linear system theory. With this notion, it is possible to develop an elegant view of any linear electromagnetic structure as a linear system having its own S-matrix.

The large-scale system-level integration of nanophotonic devices [43–48], i.e., a nanophotonic network, is considered to be an ultimate point in advances in nanophotonics. Global structures such as networks and circuits will become a main issue based on the previous and present active research on local structures, such as devices and individual elements. However, the conventional methods cannot be directly applied to the analysis of nanophotonic networks without some specific innovations because of the limitations in computing resources. From this point of view, FMM itself is not yet a mature theory. As a perspective on the advances of FMM, in this book, we present a systematic linear system theory of complex nanophotonic network structures with LFMM and a generalized S-matrix method. In addition, the Bloch eigenmodes are analyzed within the framework of LFMM and with the form of a pseudo-Fourier series when the structure is collinear. This nanophotonic network modeling is introduced as an advanced topic of FMM in Chapter 6.

2

Scattering Matrix Method for Multiblock Structures

In this chapter, a scattering matrix (S-matrix) method is used to investigate optical wave propagation through a one-dimensional (1D) multiblock structure. The optical wave propagation algorithm for the classical transfer matrix method (TMM) suffers from numerical divergence when lossy dielectric or metallic materials are contained in the multiblock structure. Enhanced TMM (ETMM) is a refined algorithm of TMM that successfully overcomes this instability problem [1–3]. The S-matrix method has the same numerical stability as ETMM and additional advantages of computational parallelism and the fact that it can be extended to a more general mathematical framework that will be discussed in this book. Thus the S-matrix method is typically employed as our standard wave propagation algorithm. The S-matrix method established in this chapter will be applied to two-dimensional (2D) and three-dimensional (3D) multiblock optical structures in a straightforward manner.

The concept and principle of the S-matrix method are elucidated for a 1D multiblock structure with detailed mathematics and working MATLAB® codes. Since the same numerical framework of the S-matrix method described in this chapter is used for FMM and LFMM, it is important to accomplish the following objectives:

1. Understanding the core elements of the S-matrix method
2. Understanding MATLAB codes that implement mathematical formulas
3. Preparing terminology and common parts of MATLAB codes that will appear in the later chapter explaining FMM, which can be considered to be a full mathematical generalization of 1D multiblock structures

This chapter is organized as follows. In Section 2.1, the S-matrix analysis of a finite-sized single-block structure is studied, through which the definition of the S-matrix method is confirmed. In Section 2.2, an S-matrix analysis of a finite-sized multiblock structure is performed, through which the multiblock interconnection algorithm of the S-matrix method is developed. The interconnection algorithm of left-hand-side and right-hand-side half-infinite blocks and a finite-sized multiblock structure using boundary S-matrix is described. This interconnection algorithm completes the concept and principle of the S-matrix method. In Section 2.3, MATLAB implementation of the mathematical theory is explained. In Section 2.4, practical applications of the S-matrix method to nanophotonics problems—surface plasmon resonance (SPR)—are presented with numerical simulation results.

2.1 Scattering Matrix Analysis of Finite Single-Block Structures

The first problem in this book is the S-matrix analysis of a single-block structure. The S-matrix is a matrix having the reflection and transmission of a single-block structure as its elements, which includes complete information on the optical scattering properties of a single block. The S-matrix is obtained by means of a bidirectional characterization process and is presented in Figure 2.1. As illustrated in Figure 2.1(a), a plane wave moving toward a positive z-direction with a specific incidence angle of θ and a polarization angle of ψ impinges on the surface of a single-block structure at $z = z_-$. $\vec{U}_{TM}$ and $\vec{U}_{TE}$ are the vector components of the incident electric field along the TM and TE polarization directions, respectively. Let us call the process of finding the reflection and transmission coefficients left-to-right directional characterization. In Figure 2.1(b), the right-to-left directional characterization is shown, where the plane wave with the same incidence angle and polarization angle but reverse propagation direction shines on the surface of the single-block structure at $z = z_+$. We can obtain respective reflection and transmission coefficients for the cases of left-to-right and right-to-left directional characterization. Thus, in this single-block analysis, the number of unknown coefficients is four. These reflection and transmission coefficients are referred to as scattering coefficients and their matrix arrangement is the S-matrix. The S-matrix is the most fundamental raw data produced by the electromagnetic analysis of block structures.

The mathematical model of the bidirectional characterization of the single-block structure is derived from Maxwell's equations.

$$\nabla \times \mathbf{E} = jw\mu_0\underline{\underline{\mu}}\mathbf{H} \tag{2.1a}$$

$$\nabla \times \mathbf{H} = -jw\varepsilon_0\underline{\underline{\varepsilon}}\mathbf{E} \tag{2.1b}$$

$$\nabla \cdot \left(\underline{\underline{\varepsilon}}\mathbf{E}\right) = 0 \tag{2.1c}$$

$$\nabla \cdot \left(\underline{\underline{\mu}}\mathbf{H}\right) = 0 \tag{2.1d}$$

where $\mathbf{E} = (\mathbf{E}_x, \mathbf{E}_y, \mathbf{E}_z)$ and $\mathbf{H} = (\mathbf{H}_x, \mathbf{H}_y, \mathbf{H}_z)$ are the electric and magnetic fields, and ω is the angular frequency of incident electromagnetic wave (time dependency representation: $\exp(-j\omega t)$). The material parameters of

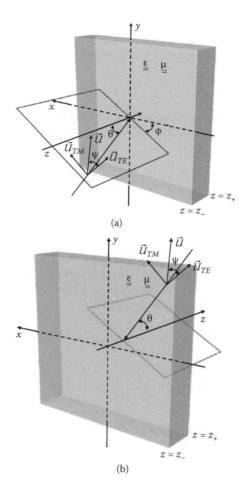

FIGURE 2.1
Bidirectional characterization of a single block with finite thickness of $d = z_+ - z_-$: (a) left-to-right directional characterization and (b) right-to-left directional characterization.

permittivity and permeability are denoted by $\underline{\underline{\varepsilon}}$ and $\underline{\underline{\mu}}$, respectively. ε_0 and μ_0 denote the permittivity and permeability of vacuum. The double sub-bars in the material parameters indicate that they are tensors (actually 3×3 tensors). In the case of uniaxial or biaxial materials, the permittivity and permeability tensors are given as, respectively,

$$\underline{\underline{\varepsilon}} = \begin{pmatrix} \varepsilon_{(x)} & 0 & 0 \\ 0 & \varepsilon_{(y)} & 0 \\ 0 & 0 & \varepsilon_{(z)} \end{pmatrix} \text{ and } \underline{\underline{\mu}} = \begin{pmatrix} \mu_{(x)} & 0 & 0 \\ 0 & \mu_{(y)} & 0 \\ 0 & 0 & \mu_{(z)} \end{pmatrix}$$

The S-matrix analysis of a single-block structure can be divided into three logical steps: (1) eigenmode analysis, (2) S-matrix and coupling coefficient matrix operator analysis of the single-block structure, and (3) field visualization.

2.1.1 Eigenmode Analysis

Consider an optical wave propagating in a single block. Let the wavevector of the optical wave be denoted by $\mathbf{k} = (k_x, k_y, k_z)$, where k_x and k_y are given, respectively, by

$$(k_x, k_y) = \left(\frac{2\pi n}{\lambda} \sin\theta\cos\phi, \frac{2\pi n}{\lambda} \sin\theta\sin\phi \right) \qquad (2.2)$$

where λ, θ, ϕ, and n are free space wavelength, incidence angle, azimuthal angle, and refractive index of the environmental medium, respectively. In general, the z-directional wavenumber, k_z, of the optical wave in the single block can be obtained from the dispersion relation of the optical wave.

We know that the electric and magnetic fields of an eigenmode in a homogeneous medium take the plane wave form as

$$\mathbf{E} = (\mathbf{E}_x, \mathbf{E}_y, \mathbf{E}_z) = (E_x, E_y, E_z)e^{j(k_x x + k_y y + k_z z)} \qquad (2.3a)$$

$$\mathbf{H} = (\mathbf{H}_x, \mathbf{H}_y, \mathbf{H}_z) = j\sqrt{\frac{\varepsilon_0}{\mu_0}}(H_x, H_y, H_z)e^{j(k_x x + k_y y + k_z z)} \qquad (2.3b)$$

where the vector components of (H_x, H_y, H_z) are scaled by $j\sqrt{\varepsilon_0/\mu_0}$ for convenience in calculation. Substituting Equations (2.3a) and (2.3b) into the curl equations (2.1a and 2.1b) results in the algebraic form of Maxwell's equations. In the case of isotropic homogeneous medium with constant permittivity and permeability, Maxwell's equations read as

$$k_0 \mu H_x = jk_z E_y - jk_y E_z \qquad (2.4a)$$

$$k_0 \mu H_y = jk_x E_z - jk_z E_x \qquad (2.4b)$$

$$k_0 \mu H_z = jk_y E_x - jk_x E_y \qquad (2.4c)$$

$$k_0 \varepsilon E_x = jk_z H_y - jk_y H_z \qquad (2.4d)$$

$$k_0 \varepsilon E_y = jk_x H_z - jk_z H_x \qquad (2.4e)$$

$$k_0 \varepsilon E_z = jk_y H_x - jk_x H_y \qquad (2.4f)$$

where k_0 is the wavenumber in free space. The dispersion relation is implicit in an eigenvalue equation formulation of Maxwell's equation. Equations (2.4a) to (2.4f) are arranged by the matrix form

$$
k_0
\begin{pmatrix}
\varepsilon & 0 & 0 & 0 & 0 & 0 \\
0 & \varepsilon & 0 & 0 & 0 & 0 \\
0 & 0 & \varepsilon & 0 & 0 & 0 \\
0 & 0 & 0 & \mu & 0 & 0 \\
0 & 0 & 0 & 0 & \mu & 0 \\
0 & 0 & 0 & 0 & 0 & \mu
\end{pmatrix}
\begin{pmatrix}
E_x \\ E_y \\ E_z \\ H_x \\ H_y \\ H_z
\end{pmatrix}
=
\begin{pmatrix}
0 & 0 & 0 & 0 & jk_z & -jk_y \\
0 & 0 & 0 & -jk_z & 0 & jk_x \\
0 & 0 & 0 & jk_y & -jk_x & 0 \\
0 & jk_z & -jk_y & 0 & 0 & 0 \\
-jk_z & 0 & jk_x & 0 & 0 & 0 \\
jk_y & -jk_x & 0 & 0 & 0 & 0
\end{pmatrix}
\begin{pmatrix}
E_x \\ E_y \\ E_z \\ H_x \\ H_y \\ H_z
\end{pmatrix}
$$

(2.5a)

Here, Equation (2.5a) can be straightforwardly generalized to the form with the anisotropic tensor profiles of permittivity and permeability,

$$
k_0
\begin{pmatrix}
\varepsilon_x & 0 & 0 & 0 & 0 & 0 \\
0 & \varepsilon_y & 0 & 0 & 0 & 0 \\
0 & 0 & \varepsilon_z & 0 & 0 & 0 \\
0 & 0 & 0 & \mu_x & 0 & 0 \\
0 & 0 & 0 & 0 & \mu_y & 0 \\
0 & 0 & 0 & 0 & 0 & \mu_z
\end{pmatrix}
\begin{pmatrix}
E_x \\ E_y \\ E_z \\ H_x \\ H_y \\ H_z
\end{pmatrix}
=
\begin{pmatrix}
0 & 0 & 0 & 0 & jk_z & -jk_y \\
0 & 0 & 0 & -jk_z & 0 & jk_x \\
0 & 0 & 0 & jk_y & -jk_x & 0 \\
0 & jk_z & -jk_y & 0 & 0 & 0 \\
-jk_z & 0 & jk_x & 0 & 0 & 0 \\
jk_y & -jk_x & 0 & 0 & 0 & 0
\end{pmatrix}
\begin{pmatrix}
E_x \\ E_y \\ E_z \\ H_x \\ H_y \\ H_z
\end{pmatrix}
$$

(2.5b)

Furthermore, the matrix can be written by the fully generalized form

$$
k_0
\begin{pmatrix}
\varepsilon_{xx} & \varepsilon_{xy} & \varepsilon_{xz} & 0 & 0 & 0 \\
\varepsilon_{yx} & \varepsilon_{yy} & \varepsilon_{yz} & 0 & 0 & 0 \\
\varepsilon_{zx} & \varepsilon_{zy} & \varepsilon_{zz} & 0 & 0 & 0 \\
0 & 0 & 0 & \mu_{xx} & \mu_{xy} & \mu_{xz} \\
0 & 0 & 0 & \mu_{yx} & \mu_{yy} & \mu_{yz} \\
0 & 0 & 0 & \mu_{zx} & \mu_{zy} & \mu_{zz}
\end{pmatrix}
\begin{pmatrix}
E_x \\ E_y \\ E_z \\ H_x \\ H_y \\ H_z
\end{pmatrix}
=
\begin{pmatrix}
0 & 0 & 0 & 0 & jk_z & -jk_y \\
0 & 0 & 0 & -jk_z & 0 & jk_x \\
0 & 0 & 0 & jk_y & -jk_x & 0 \\
0 & jk_z & -jk_y & 0 & 0 & 0 \\
-jk_z & 0 & jk_x & 0 & 0 & 0 \\
jk_y & -jk_x & 0 & 0 & 0 & 0
\end{pmatrix}
\begin{pmatrix}
E_x \\ E_y \\ E_z \\ H_x \\ H_y \\ H_z
\end{pmatrix}
$$

(2.5c)

From those matrix forms of Maxwell's equations, it is possible to find a specific eigenmode with a specific propagation constant k_z. It is perceived that the eigenvalue equation is parameterized implicitly by four parameters,

ω, k_x, k_y, and k_z, in Equations (2.5a) to (2.5c). This implicit relation between angular frequency ω and wavevector (k_x, k_y, k_z) is called the dispersion relation. This relationship is explicitly expressed as

$$\omega = \omega(k_x, k_y, k_z) \tag{2.6a}$$

$$k_x = k_x(k_y, k_z; \omega) \tag{2.6b}$$

$$k_y = k_y(k_z, k_x; \omega) \tag{2.6c}$$

$$k_z = k_z(k_x, k_y; \omega) \tag{2.6d}$$

The above explicit dispersion relations can be equivalently formulated by matrix eigenvalue equations by rearranging Equations (2.5a) to (2.5c).

First, the dispersion relation of Equation (2.6a) describes that the angular frequency of an electromagnectic eigenmode is determined for a specific set of (k_x, k_y, k_z). For this case, Equation (2.5b) is separated as

$$k_0 \begin{pmatrix} \varepsilon_x & 0 & 0 \\ 0 & \varepsilon_y & 0 \\ 0 & 0 & \varepsilon_z \end{pmatrix} \begin{pmatrix} E_x \\ E_y \\ E_z \end{pmatrix} = \begin{pmatrix} 0 & jk_z & -jk_y \\ -jk_z & 0 & jk_x \\ jk_y & -jk_x & 0 \end{pmatrix} \begin{pmatrix} H_x \\ H_y \\ H_z \end{pmatrix} \tag{2.7a}$$

$$k_0 \begin{pmatrix} \mu_x & 0 & 0 \\ 0 & \mu_y & 0 \\ 0 & 0 & \mu_z \end{pmatrix} \begin{pmatrix} H_x \\ H_y \\ H_z \end{pmatrix} = \begin{pmatrix} 0 & jk_z & -jk_y \\ -jk_z & 0 & jk_x \\ jk_y & -jk_x & 0 \end{pmatrix} \begin{pmatrix} E_x \\ E_y \\ E_z \end{pmatrix} \tag{2.7b}$$

Plugging Equation (2.7b) into Equation (2.7a) leads to the eigenvalue equation for the eigenvalue $(\omega/c)^2$ as

$$\begin{pmatrix} \varepsilon_x^{-1} & 0 & 0 \\ 0 & \varepsilon_y^{-1} & 0 \\ 0 & 0 & \varepsilon_z^{-1} \end{pmatrix} \begin{pmatrix} 0 & jk_z & -jk_y \\ -jk_z & 0 & jk_x \\ jk_y & -jk_x & 0 \end{pmatrix} \begin{pmatrix} \mu_x^{-1} & 0 & 0 \\ 0 & \mu_y^{-1} & 0 \\ 0 & 0 & \mu_z^{-1} \end{pmatrix} \begin{pmatrix} 0 & jk_z & -jk_y \\ -jk_z & 0 & jk_x \\ jk_y & -jk_x & 0 \end{pmatrix} \begin{pmatrix} E_x \\ E_y \\ E_z \end{pmatrix}$$

$$= \left(\frac{\omega}{c}\right)^2 \begin{pmatrix} E_x \\ E_y \\ E_z \end{pmatrix} \tag{2.7c}$$

which is manipulated as

$$
\begin{pmatrix}
\left(\dfrac{\omega}{c}\right)^2 - \left(\dfrac{k_y^2}{\mu_z\varepsilon_x} + \dfrac{k_z^2}{\varepsilon_x\mu_y}\right) & \dfrac{k_x k_y}{\mu_z\varepsilon_x} & \dfrac{k_z k_x}{\varepsilon_x\mu_y} \\[3ex]
\dfrac{k_x k_y}{\mu_z\varepsilon_y} & \left(\dfrac{\omega}{c}\right)^2 - \left(\dfrac{k_z^2}{\mu_x\varepsilon_y} + \dfrac{k_x^2}{\mu_z\varepsilon_y}\right) & \dfrac{k_y k_z}{\mu_x\varepsilon_y} \\[3ex]
\dfrac{k_z k_x}{\mu_y\varepsilon_z} & \dfrac{k_y k_z}{\varepsilon_z\mu_x} & \left(\dfrac{\omega}{c}\right)^2 - \left(\dfrac{k_x^2}{\mu_y\varepsilon_z} + \dfrac{k_y^2}{\varepsilon_z\mu_x}\right)
\end{pmatrix}
\begin{pmatrix} E_x \\ E_y \\ E_z \end{pmatrix} = \begin{pmatrix} 0 \\ 0 \\ 0 \end{pmatrix}
$$

(2.7d)

In the case of homogeneous media, we can find the analytic expression of eigenvalue, i.e., dispersion relation. The condition of the zero determinant of the matrix is equivalent to

$$
\left[\left(\dfrac{\omega}{c}\right)^2\right]^2 - \left\{k_x^2\left(\dfrac{1}{\varepsilon_y\mu_z} + \dfrac{1}{\mu_y\varepsilon_z}\right) + k_y^2\left(\dfrac{1}{\varepsilon_z\mu_x} + \dfrac{1}{\mu_z\varepsilon_x}\right) + k_z^2\left(\dfrac{1}{\varepsilon_x\mu_y} + \dfrac{1}{\mu_x\varepsilon_y}\right)\right\}\left(\dfrac{\omega}{c}\right)^2
$$

$$
+ \left\{\dfrac{k_x^4}{\varepsilon_y\mu_z\mu_y\varepsilon_z} + \dfrac{k_y^4}{\varepsilon_z\mu_x\mu_z\varepsilon_x} + \dfrac{k_z^4}{\varepsilon_x\mu_y\mu_x\varepsilon_y} + \dfrac{k_x^2 k_y^2}{\varepsilon_z\mu_z}\left(\dfrac{1}{\varepsilon_x\mu_y} + \dfrac{1}{\mu_x\varepsilon_y}\right)\right.
$$

$$
\left. + \dfrac{k_y^2 k_z^2}{\varepsilon_x\mu_x}\left(\dfrac{1}{\varepsilon_y\mu_z} + \dfrac{1}{\mu_y\varepsilon_z}\right) + \dfrac{k_z^2 k_x^2}{\varepsilon_y\mu_y}\left(\dfrac{1}{\varepsilon_z\mu_x} + \dfrac{1}{\mu_z\varepsilon_x}\right)\right\} = 0
$$

(2.8)

By solving the quadratic equation with respect to $(\omega/c)^2$, we have the form of dispersion relation

$$
\left(\dfrac{\omega}{c}\right)^2 = \dfrac{1}{2}\left[(X+Y+Z)\pm\sqrt{X'^2+Y'^2+Z'^2-2(X'Y'+Y'Z'+Z'X')}\right]
$$ (2.9a)

where X, Y, Z, X', Y', and Z' are given, respectively, by

$$
X = k_x^2\left(\dfrac{1}{\varepsilon_y\mu_z} + \dfrac{1}{\mu_y\varepsilon_z}\right)
$$ (2.9b)

$$
Y = k_y^2\left(\dfrac{1}{\varepsilon_z\mu_x} + \dfrac{1}{\mu_z\varepsilon_x}\right)
$$ (2.9c)

$$Z = k_z^2 \left(\frac{1}{\varepsilon_x \mu_y} + \frac{1}{\mu_x \varepsilon_y} \right) \tag{2.9d}$$

$$X' = k_x^2 \left(\frac{1}{\varepsilon_y \mu_z} - \frac{1}{\mu_y \varepsilon_z} \right) \tag{2.9e}$$

$$Y' = k_y^2 \left(\frac{1}{\varepsilon_z \mu_x} - \frac{1}{\mu_z \varepsilon_x} \right) \tag{2.9f}$$

$$Z' = k_z^2 \left(\frac{1}{\varepsilon_x \mu_y} - \frac{1}{\mu_x \varepsilon_y} \right) \tag{2.9g}$$

For example, let us consider the nonmagnetic uniaxial dispersion relation, i.e., $\varepsilon_x = \varepsilon_y = \varepsilon_o$, $\varepsilon_z = \varepsilon_e$, and $\mu_x = \mu_y = \mu_z = 1$. Under this condition, we know $Z' = 0$, and thus terms inside the square root can be factorized as

$$\left(\frac{\omega}{c} \right)^2 = \frac{1}{2} \left[(X+Y+Z) \pm \sqrt{X'^2 + Y'^2 - 2X'Y'} \right] = \frac{1}{2} \left[(X+Y+Z) \pm (X'-Y') \right]$$

$$= \frac{k_x^2 + k_y^2}{\varepsilon_e} + \frac{k_z^2}{\varepsilon_0} \quad \text{or} \quad \frac{k_x^2 + k_y^2 + k_z^2}{\varepsilon_o}. \tag{2.10a}$$

Thus the material becomes a uniaxial crystal with the dispersion relations of an ordinary wave and an extraordinary wave:

$$k_0^2 = \frac{k_x^2 + k_y^2 + k_z^2}{\varepsilon_o} , \text{ ordinary wave} \tag{2.10b}$$

$$k_0^2 = \frac{k_x^2 + k_y^2}{\varepsilon_e} + \frac{k_z^2}{\varepsilon_o} , \text{ extraordinary wave} \tag{2.10c}$$

Second, the remaining three dispersion relations of Equations (2.6b) to (2.6d) are represented by the eigenvalue equation with one wavevector component as the eigenvalue. For $k_z = k_z(k_x, k_y; \omega)$ of Equation (2.6d), the eigenvalue equation with one wavevector component as the eigenvalue can be derived. Equation (2.5b) is transformed as

$$j(k_z/k_0)E_y = \mu_x H_x + j(k_y/k_0)E_z \tag{2.11a}$$

$$j(k_z/k_0)E_x = -\mu_y H_y + j(k_x/k_0)E_z \tag{2.11b}$$

$$H_z = \mu_z^{-1}[j(k_y/k_0)E_x - j(k_x/k_0)E_y] \tag{2.11c}$$

$$j(k_z/k_0)H_y = \varepsilon_x E_x + j(k_y/k_0)H_z \tag{2.11d}$$

$$j(k_z/k_0)H_x = -\varepsilon_y E_y + j(k_x/k_0)H_z \tag{2.11e}$$

$$E_z = \varepsilon_z^{-1}[j(k_y/k_0)H_x - j(k_x/k_0)H_y] \tag{2.11f}$$

By substituting Equations (2.11c) and (2.11f) into Equations (2.11a), (2.11b), (2.11d), and (2.11e), we can obtain the eigenvalue equation with the eigenvalue of the z-direction wavevector component k_z:

$$
\begin{pmatrix}
0 & 0 & \bar{k}_y \varepsilon_z^{-1} \bar{k}_x & \mu_x - \bar{k}_y \varepsilon_z^{-1} \bar{k}_y \\
0 & 0 & -\mu_y + \bar{k}_x \varepsilon_z^{-1} \bar{k}_x & -\bar{k}_x \varepsilon_z^{-1} \bar{k}_y \\
\bar{k}_y \mu_z^{-1} \bar{k}_x & \varepsilon_x - \bar{k}_y \mu_z^{-1} \bar{k}_y & 0 & 0 \\
-\varepsilon_y + \bar{k}_x \mu_z^{-1} \bar{k}_x & -\bar{k}_x \mu_z^{-1} \bar{k}_y & 0 & 0
\end{pmatrix}
\begin{pmatrix}
E_y \\
E_x \\
H_y \\
H_x
\end{pmatrix}
= j\bar{k}_z
\begin{pmatrix}
E_y \\
E_x \\
H_y \\
H_x
\end{pmatrix},
\tag{2.12a}
$$

where $\bar{k}_x$, $\bar{k}_y$, and $\bar{k}_z$ are given by $\bar{k}_x = k_x/k_0$, $\bar{k}_y = k_y/k_0$, and $\bar{k}_z = k_z/k_0$, respectively. Since the dimension of the system matrix is four, the total number of obtained eigensolutions is four. Physically, four cases of right-to-left directional TE and TM modes and left-to-right directional TE and TM modes are identified. The eigenvalue equation of Equation (2.12a) is divided into two equations;

$$
\begin{pmatrix}
\bar{k}_y \varepsilon_z^{-1} \bar{k}_x & \mu_x - \bar{k}_y \varepsilon_z^{-1} \bar{k}_y \\
-\mu_y + \bar{k}_x \varepsilon_z^{-1} \bar{k}_x & -\bar{k}_x \varepsilon_z^{-1} \bar{k}_y
\end{pmatrix}
\begin{pmatrix}
H_y \\
H_x
\end{pmatrix}
= j\bar{k}_z
\begin{pmatrix}
E_y \\
E_x
\end{pmatrix}
\tag{2.12b}
$$

$$
\begin{pmatrix}
\bar{k}_y \mu_z^{-1} \bar{k}_x & \varepsilon_x - \bar{k}_y \mu_z^{-1} \bar{k}_y \\
-\varepsilon_y + \bar{k}_x \mu_z^{-1} \bar{k}_x & -\bar{k}_x \mu_z^{-1} \bar{k}_y
\end{pmatrix}
\begin{pmatrix}
E_y \\
E_x
\end{pmatrix}
= j\bar{k}_z
\begin{pmatrix}
H_y \\
H_x
\end{pmatrix}
\tag{2.12c}
$$

Then the half-size eigenvalue equation for electric field components is made by substituting Equation (2.12c) into Equation (2.12b) as

$$
k_0^2
\begin{pmatrix}
\bar{k}_y \varepsilon_z^{-1} \bar{k}_x & \mu_x - \bar{k}_y \varepsilon_z^{-1} \bar{k}_y \\
-\mu_y + \bar{k}_x \varepsilon_z^{-1} \bar{k}_x & -\bar{k}_x \varepsilon_z^{-1} \bar{k}_y
\end{pmatrix}
\begin{pmatrix}
\bar{k}_y \mu_z^{-1} \bar{k}_x & \varepsilon_x - \bar{k}_y \mu_z^{-1} \bar{k}_y \\
-\varepsilon_y + \bar{k}_x \mu_z^{-1} \bar{k}_x & -\bar{k}_x \mu_z^{-1} \bar{k}_y
\end{pmatrix}
\begin{pmatrix}
E_y \\
E_x
\end{pmatrix}
$$

$$
= (jk_z)^2
\begin{pmatrix}
E_y \\
E_x
\end{pmatrix}
\tag{2.13}
$$

In this case for the diagonal anisotropic material, the magnetic field components H_x and H_y are given by

$$\begin{pmatrix} H_y \\ H_x \end{pmatrix} = j\bar{k}_z \begin{pmatrix} \bar{k}_y \varepsilon_z^{-1} \bar{k}_x & \mu_x - \bar{k}_y \varepsilon_z^{-1} \bar{k}_y \\ -\mu_y + \bar{k}_x \varepsilon_z^{-1} \bar{k}_x & -\bar{k}_x \varepsilon_z^{-1} \bar{k}_y \end{pmatrix}^{-1} \begin{pmatrix} E_y \\ E_x \end{pmatrix} \tag{2.14a}$$

The z-directional electric and magnetic field components are taken as

$$E_z = \varepsilon_z^{-1}[j(k_y/k_0)H_x - j(k_x/k_0)H_y] \tag{2.14b}$$

$$H_z = \mu_z^{-1}[j(k_y/k_0)E_x - j(k_x/k_0)E_y] \tag{2.14c}$$

In addition, we can further deal with the case of general anisotropic material, and then Equation (2.5c) is rearranged as the eigenvalue equation with eigenvalue of k_z. Equation (2.5c) is transformed as

$$j(k_z/k_0)E_y = \mu_{xx}H_x + \mu_{xy}H_y + \mu_{xz}H_z + j(k_y/k_0)E_z \tag{2.15a}$$

$$j(k_z/k_0)E_x = -\mu_{yx}H_x - \mu_{yy}H_y - \mu_{yz}H_z + j(k_x/k_0)E_z \tag{2.15b}$$

$$H_z = \mu_{zz}^{-1}[j(k_y/k_0)E_x - j(k_x/k_0)E_y - (\mu_{zx}H_x + \mu_{zy}H_y)] \tag{2.15c}$$

$$j(k_z/k_0)H_y = \varepsilon_{xx}E_x + \varepsilon_{xy}E_y + \varepsilon_{xz}E_z + j(k_y/k_0)H_z \tag{2.15d}$$

$$j(k_z/k_0)H_x = -\varepsilon_{yx}E_x - \varepsilon_{yy}E_y - \varepsilon_{yz}E_z + j(k_x/k_0)H_z \tag{2.15e}$$

$$E_z = \varepsilon_{zz}^{-1}[j(k_y/k_0)H_x - j(k_x/k_0)H_y - (\varepsilon_{zx}E_x + \varepsilon_{zy}E_y)] \tag{2.15f}$$

Plugging Equations (2.15c) and (2.15f) into Equations (2.15a), (2.15b), (2.15d), and (2.15e) leads to the eigenvalue equation with the eigenvalue of the z-direction wavevector component k_z:

$$\begin{pmatrix} -j\mu_{xz}\mu_{zz}^{-1}k_x - jk_y\varepsilon_{zz}^{-1}\varepsilon_{zy} & j\mu_{xz}\mu_{zz}^{-1}k_y - jk_y\varepsilon_{zz}^{-1}\varepsilon_{zx} & \mu_{xy} - \mu_{xz}\mu_{zz}^{-1}\mu_{zy} + k_y\varepsilon_{zz}^{-1}k_x & \mu_{xx} - \mu_{xz}\mu_{zz}^{-1}\mu_{zx} - k_y\varepsilon_{zz}^{-1}k_y \\ j\mu_{yz}\mu_{zz}^{-1}k_x - jk_x\varepsilon_{zz}^{-1}\varepsilon_{zy} & -j\mu_{yz}\mu_{zz}^{-1}k_y - jk_x\varepsilon_{zz}^{-1}\varepsilon_{zx} & -\mu_{yy} + \mu_{yz}\mu_{zz}^{-1}\mu_{zy} + k_x\varepsilon_{zz}^{-1}k_x & -\mu_{yx} + \mu_{yz}\mu_{zz}^{-1}\mu_{zx} - k_x\varepsilon_{zz}^{-1}k_y \\ \varepsilon_{xy} - \varepsilon_{xz}\varepsilon_{zz}^{-1}\varepsilon_{zy} + k_y\mu_{zz}^{-1}k_x & \varepsilon_{xx} - \varepsilon_{xz}\varepsilon_{zz}^{-1}\varepsilon_{zx} - k_y\mu_{zz}^{-1}k_y & -j\varepsilon_{xz}\varepsilon_{zz}^{-1}k_x - jk_y\mu_{zz}^{-1}\mu_{zy} & j\varepsilon_{xz}\varepsilon_{zz}^{-1}k_y - jk_y\mu_{zz}^{-1}\mu_{zx} \\ -\varepsilon_{yy} + \varepsilon_{yz}\varepsilon_{zz}^{-1}\varepsilon_{zy} + k_x\mu_{zz}^{-1}k_x & -\varepsilon_{yx} + \varepsilon_{yz}\varepsilon_{zz}^{-1}\varepsilon_{zx} - k_x\mu_{zz}^{-1}k_y & j\varepsilon_{yz}\varepsilon_{zz}^{-1}k_x - jk_x\mu_{zz}^{-1}\mu_{zy} & -j\varepsilon_{yz}\varepsilon_{zz}^{-1}k_y - jk_x\mu_{zz}^{-1}\mu_{zx} \end{pmatrix}$$

$$\times \begin{pmatrix} E_y \\ E_x \\ H_y \\ H_x \end{pmatrix} = \frac{jk_z}{k_0} \begin{pmatrix} E_y \\ E_x \\ H_y \\ H_x \end{pmatrix} \tag{2.16}$$

A representative example of Equation (2.15) is a liquid crystal structure with an inhomogeneous director profile. In the case for a general anisotropic material, the z-directional electric and magnetic field components are obtained by Equations (2.15c) and (2.15f):

$$E_z = \varepsilon_{zz}^{-1}[j(k_y/k_0)H_x - j(k_x/k_0)H_y - (\varepsilon_{zx}E_x + \varepsilon_{zy}E_y)] \tag{2.17a}$$

$$H_z = \mu_{zz}^{-1}[j(k_y/k_0)E_x - j(k_x/k_0)E_y - (\mu_{zx}H_x + \mu_{zy}H_y)] \tag{2.17b}$$

From these eigenvalue equations, four explicit electromagnetic eigenmodes of optical waves can be found. The eigenmodes can be further classified into two categories, positive (forward) modes and negative (backward) modes. Assuming that eigenvalues hold one of the forms $jk_z^{(g)} = a^{(g)} + jb^{(g)}$, $jk_z^{(g)} = a^{(g)} - jb^{(g)}$, $jk_z^{(g)} = -a^{(g)} + jb^{(g)}$, and $jk_z^{(g)} = -a^{(g)} - jb^{(g)}$, with $a^{(g)} > 0$ and $b^{(g)} > 0$, we refer to the eigenmodes with eigenvalues of $jk_z^{(g)} = a^{(g)} + jb^{(g)}$ or $jk_z^{(g)} = a^{(g)} - jb^{(g)}$ as negative mode and use the notation $k_z^{(g)-}$ with the minus superscript to indicate the negative mode. Eigenmodes with eigenvalues of $jk_z^{(g)} = -a^{(g)} + jb^{(g)}$ and $jk_z^{(g)} = -a^{(g)} - jb^{(g)}$ are referred to as positive mode, and the notation $k_z^{(g)+}$ with the plus superscript is used to indicate the positive mode. In particular, the eigenmodes with pure real eigenvalues of $jk_z^{(g)} = jb^{(g)}$ and $jk_z^{(g)} = -jb^{(g)}$ with $a^{(g)} = 0$ are classified to the positive mode for convenience. The superscript g indicates the mode index. It is necessary that the magnetic field is renormalized by $(H_x, H_y, H_z) \leftarrow j\sqrt{\frac{\varepsilon_0}{\mu_0}}(H_x, H_y, H_z)$. Then the gth internal eigenmodes in the single block are represented as the Bloch wave form:

$$\begin{pmatrix} \mathbf{E}^{(g)} \\ \mathbf{H}^{(g)} \end{pmatrix} = \begin{pmatrix} \mathbf{E}_x^{(g)}, \mathbf{E}_y^{(g)}, \mathbf{E}_z^{(g)} \\ \mathbf{H}_x^{(g)}, \mathbf{H}_y^{(g)}, \mathbf{H}_z^{(g)} \end{pmatrix} = \begin{pmatrix} E_x^{(g)}, E_y^{(g)}, E_z^{(g)} \\ H_x^{(g)}, H_y^{(g)}, H_z^{(g)} \end{pmatrix} e^{j\left(k_x x + k_y y + k_z^{(g)} z\right)} \tag{2.18}$$

These eigenmodes of single block are referred to as Bloch eigenmodes.

2.1.2 S-Matrix and Coupling Coefficient Operator Calculation of a Single Block

In the previous step, the Bloch eigenmodes (four plane waves in the case of 1D structures) in a single block were analyzed in a general fashion. The next task is to find the reflection and transmission fields for an externally incident plane wave. Regarding this issue, we will utilize the S-matrix method as our standard analytical tool. As shown in Figure 2.1, S-matrix analysis of a single block is performed by the bidirectional characterization of scattering property of the target block. The left-to-right directional characterization is to find the reflection and transmission electric fields, $\vec{E}_r$ and $\vec{E}_t$, for an incident field $\vec{E}_i$ that

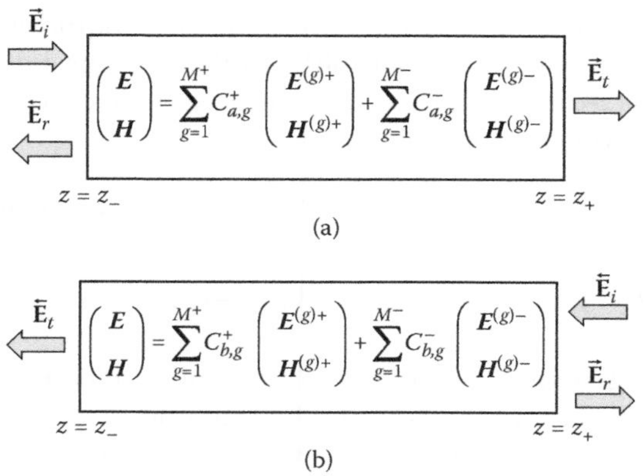

$$\vec{\mathbf{E}}_i$$

$$\vec{\mathbf{E}}_r \qquad \begin{pmatrix} E \\ H \end{pmatrix} = \sum_{g=1}^{M^+} C_{a,g}^+ \begin{pmatrix} E^{(g)+} \\ H^{(g)+} \end{pmatrix} + \sum_{g=1}^{M^-} C_{a,g}^- \begin{pmatrix} E^{(g)-} \\ H^{(g)-} \end{pmatrix} \qquad \vec{\mathbf{E}}_t$$

$$z = z_- \qquad\qquad\qquad\qquad\qquad\qquad z = z_+$$

(a)

$$\vec{\mathbf{E}}_t \qquad \begin{pmatrix} E \\ H \end{pmatrix} = \sum_{g=1}^{M^+} C_{b,g}^+ \begin{pmatrix} E^{(g)+} \\ H^{(g)+} \end{pmatrix} + \sum_{g=1}^{M^-} C_{b,g}^- \begin{pmatrix} E^{(g)-} \\ H^{(g)-} \end{pmatrix} \qquad \vec{\mathbf{E}}_i$$

$$\vec{\mathbf{E}}_r$$

$$z = z_- \qquad\qquad\qquad\qquad\qquad\qquad z = z_+$$

(b)

FIGURE 2.2
Bidirectional characterization: (a) left-to-right directional characterization and (b) right-to-left directional characterization.

impinges at the lefthand interface of $z = z_-$ and propagates toward the right-hand direction (positive z-direction). The fields propagating along the left-to-right direction and the right-to-left direction are symbolized by the superscript arrow symbols $\rightarrow$ and $\leftarrow$, respectively. In the block, a superposition of internal block eigenmodes with specific coupling coefficients, $C_{a,g}^+$ and $C_{a,g}^-$, is excited and transfers electromagnetic energy from the lefthand boundary $z = z_-$ to the righthand boundary at $z = z_+$. The superscript symbols of $+$ and $-$ in $C_{a,g}^+$ and $C_{a,g}^-$ indicate the positive and negative Bloch eigenmodes, respectively, according to the classification of Bloch eigenmodes. In Figure 2.2(a), the field components involved in the left-to-right directional characterization are presented schematically. In Figure 2.2(b), the right-to-left directional characterization of the scattering property is illustrated with the involved field components. The external incident field $\vec{\mathbf{E}}_i$ that impinges at the righthand interface of $z = z_+$ induces the reflection and transmission electromagnetic fields, $\vec{\mathbf{E}}_r$ and $\vec{\mathbf{E}}_t$, and the coupling coefficients of the internal field, $C_{b,g}^+$ and $C_{b,g}^-$.

Information on the field components of bidirectional reflection and transmission coefficients, and coupling coefficients can be used to determine the scattering characteristics of the target block completely. With this general notion in mind, the detailed mathematical framework for analyzing the S-matrix and coupling coefficients of an internal field is developed.

The internal electromagnetic field is represented by the superposition of the Bloch eigenmodes with their own specific coupling coefficients as

$$\begin{pmatrix} \mathbf{E} \\ \mathbf{H} \end{pmatrix} = \sum_{g=1}^{M^+} C_{a,g}^+ \begin{pmatrix} \mathbf{E}^{(g)+} \\ \mathbf{H}^{(g)+} \end{pmatrix} + \sum_{g=1}^{M^-} C_{a,g}^- \begin{pmatrix} \mathbf{E}^{(g)-} \\ \mathbf{H}^{(g)-} \end{pmatrix} \qquad (2.19a)$$

where M^+ and M^- are 2. Here, the representation of the positive Bloch eigenmode and the negative Bloch eigenmode are slightly modified, from Equation (2.18), in the Bloch phase terms as

$$\begin{pmatrix} \mathbf{E}^{(g)+} \\ \mathbf{H}^{(g)+} \end{pmatrix} = \begin{pmatrix} \mathbf{E}_x^{(g)+}, \mathbf{E}_y^{(g)+}, \mathbf{E}_z^{(g)+} \\ \mathbf{H}_x^{(g)+}, \mathbf{H}_y^{(g)+}, \mathbf{H}_z^{(g)+} \end{pmatrix} = \begin{pmatrix} E_x^{(g)+}, E_y^{(g)+}, E_z^{(g)+} \\ H_x^{(g)+}, H_y^{(g)+}, H_z^{(g)+} \end{pmatrix} e^{j\left(k_x x + k_y y + k_z^{(g)+}(z-z_-)\right)} \quad (2.19b)$$

$$\begin{pmatrix} \mathbf{E}^{(g)-} \\ \mathbf{H}^{(g)-} \end{pmatrix} = \begin{pmatrix} \mathbf{E}_x^{(g)-}, \mathbf{E}_y^{(g)-}, \mathbf{E}_z^{(g)-} \\ \mathbf{H}_x^{(g)-}, \mathbf{H}_y^{(g)-}, \mathbf{H}_z^{(g)-} \end{pmatrix} = \begin{pmatrix} E_x^{(g)-}, E_y^{(g)-}, E_z^{(g)-} \\ H_x^{(g)-}, H_y^{(g)-}, H_z^{(g)-} \end{pmatrix} e^{j\left(k_x x + k_y y + k_z^{(g)-}(z-z_+)\right)} \quad (2.19c)$$

The setup using $e^{j(k_x x + k_y y + k_z^{(g)+}(z-z_-))}$ and $e^{j(k_x x + k_y y + k_z^{(g)+}(z-z_+))}$ in the representation of positive modes and negative modes, respectively, is important because this is effective for stabilizing the numerical calculation of field coefficients and several matrix operations that are involved in the numerical analysis.

In the case of left-to-right directional characterization, the representation of the incident optical field, the reflected optical field in the lefthand free space, and the transmitted optical field in the righthand free space are given, respectively, as

$$\begin{pmatrix} \vec{\mathbf{E}}_i \\ \vec{\mathbf{H}}_i \end{pmatrix} = \begin{pmatrix} \vec{\mathbf{E}}_{i,x}, \vec{\mathbf{E}}_{i,y}, \vec{\mathbf{E}}_{i,z} \\ \vec{\mathbf{H}}_{i,x}, \vec{\mathbf{H}}_{i,y}, \vec{\mathbf{H}}_{i,z} \end{pmatrix} = \begin{pmatrix} \vec{E}_{i,x}, \vec{E}_{i,y}, \vec{E}_{i,z} \\ \vec{H}_{i,x}, \vec{H}_{i,y}, \vec{H}_{i,z} \end{pmatrix} e^{j(k_x x + k_y y + k_z (z-z_-))} \quad (2.20a)$$

$$\begin{pmatrix} \vec{\mathbf{E}}_r \\ \vec{\mathbf{H}}_r \end{pmatrix} = \begin{pmatrix} \vec{\mathbf{E}}_{r,x}, \vec{\mathbf{E}}_{r,y}, \vec{\mathbf{E}}_{r,z} \\ \vec{\mathbf{H}}_{r,x}, \vec{\mathbf{H}}_{r,y}, \vec{\mathbf{H}}_{r,z} \end{pmatrix} = \begin{pmatrix} \vec{E}_{r,x}, \vec{E}_{r,y}, \vec{E}_{r,z} \\ \vec{H}_{r,x}, \vec{H}_{r,y}, \vec{H}_{r,z} \end{pmatrix} e^{j(k_x x + k_y y - k_z (z-z_-))} \quad (2.20b)$$

$$\begin{pmatrix} \vec{\mathbf{E}}_t \\ \vec{\mathbf{H}}_t \end{pmatrix} = \begin{pmatrix} \vec{\mathbf{E}}_{t,x}, \vec{\mathbf{E}}_{t,y}, \vec{\mathbf{E}}_{t,z} \\ \vec{\mathbf{H}}_{t,x}, \vec{\mathbf{H}}_{t,y}, \vec{\mathbf{H}}_{t,z} \end{pmatrix} = \begin{pmatrix} \vec{E}_{t,x}, \vec{E}_{t,y}, \vec{E}_{t,z} \\ \vec{H}_{t,x}, \vec{H}_{t,y}, \vec{H}_{t,z} \end{pmatrix} e^{j(k_x x + k_y y + k_z (z-z_+))} \quad (2.20c)$$

The analysis is followed by the principle that the reflection and transmission field in half-infinite free space and the coupling coefficients are obtained so that the boundary conditions of tangential field continuity are satisfied. The tangential components of the incident, reflection, and transmission magnetic fields are solved in Maxwell's equations for the electric field components as follows:

$$\vec{H}_{i,y} = \frac{1}{j\omega\mu_0}\left(\frac{\partial \vec{E}_{i,x}}{\partial z} - \frac{\partial \vec{E}_{i,z}}{\partial x}\right) \Leftrightarrow \vec{H}_{i,y} = \frac{1}{\omega\mu_0}(k_z\vec{E}_{i,x} - k_x\vec{E}_{i,z}) \qquad (2.21a)$$

$$\vec{H}_{i,x} = \frac{1}{j\omega\mu_0}\left(\frac{\partial \vec{E}_{i,z}}{\partial y} - \frac{\partial \vec{E}_{i,y}}{\partial z}\right) \Leftrightarrow \vec{H}_{i,x} = \frac{1}{\omega\mu_0}(k_y\vec{E}_{i,z} - k_z\vec{E}_{i,y}) \qquad (2.21b)$$

$$\vec{H}_{r,y} = \frac{1}{j\omega\mu_0}\left(\frac{\partial \vec{E}_{r,x}}{\partial z} - \frac{\partial \vec{E}_{r,z}}{\partial x}\right) \Leftrightarrow \vec{H}_{r,y} = \frac{1}{\omega\mu_0}(-k_z\vec{E}_{r,x} - k_x\vec{E}_{r,z}) \qquad (2.21c)$$

$$\vec{H}_{r,x} = \frac{1}{j\omega\mu_0}\left(\frac{\partial \vec{E}_{r,z}}{\partial y} - \frac{\partial \vec{E}_{r,y}}{\partial z}\right) \Leftrightarrow \vec{H}_{r,x} = \frac{1}{\omega\mu_0}(k_y\vec{E}_{r,z} + k_z\vec{E}_{r,y}) \qquad (2.21d)$$

$$\vec{H}_{t,y} = \frac{1}{j\omega\mu_0}\left(\frac{\partial \vec{E}_{t,x}}{\partial z} - \frac{\partial \vec{E}_{t,z}}{\partial x}\right) \Leftrightarrow \vec{H}_{t,y} = \frac{1}{\omega\mu_0}(k_z\vec{E}_{t,x} - k_x\vec{E}_{t,z}) \qquad (2.21e)$$

$$\vec{H}_{t,x} = \frac{1}{j\omega\mu_0}\left(\frac{\partial \vec{E}_{t,z}}{\partial y} - \frac{\partial \vec{E}_{t,y}}{\partial z}\right) \Leftrightarrow \vec{H}_{t,x} = \frac{1}{\omega\mu_0}(k_y\vec{E}_{t,z} - k_z\vec{E}_{t,y}) \qquad (2.21f)$$

The tangential H-field coefficients, $\vec{H}_{i,y}$, $\vec{H}_{i,x}$, are represented by the three E-field coefficients, $\vec{E}_{i,x}$, $\vec{E}_{i,y}$, $\vec{E}_{i,z}$. The pairs of $\vec{H}_{r,y}$ and $\vec{H}_{r,x}$, and $\vec{H}_{t,y}$ and $\vec{H}_{t,x}$, are represented by the corresponding three electric field coefficients. Considering the following plane wave condition in free space, we have

$$k_x\vec{E}_{i,x} + k_y\vec{E}_{i,y} + k_z\vec{E}_{i,z} = 0 \Leftrightarrow \vec{E}_{i,z} = -\frac{k_x\vec{E}_{i,x} + k_y\vec{E}_{i,y}}{k_z} \qquad (2.22a)$$

$$k_x\vec{E}_{r,x} + k_y\vec{E}_{r,y} - k_z\vec{E}_{r,z} = 0 \Leftrightarrow \vec{E}_{r,z} = \frac{k_x\vec{E}_{r,x} + k_y\vec{E}_{r,y}}{k_z} \qquad (2.22b)$$

$$k_x\vec{E}_{t,x} + k_y\vec{E}_{t,y} + k_z\vec{E}_{t,z} = 0 \Leftrightarrow \vec{E}_{t,z} = -\frac{k_x\vec{E}_{t,x} + k_y\vec{E}_{t,y}}{k_z} \qquad (2.22c)$$

Hence the z-directional components, $\vec{E}_{i,z}$, $\vec{E}_{r,z}$, and $\vec{E}_{t,z}$, can be eliminated in Equations (2.21a) to (2.21f) by substituting these terms by Equations (2.22a) to (2.22c). According to the tangential (transversal) field continuation condition, the boundary condition at the left boundary of $z = z_-$ can be expressed by the matrix equation

$$
\begin{pmatrix}
I & 0 & I & 0 \\
0 & I & 0 & I \\
\dfrac{1}{\omega\mu_0}\dfrac{k_x k_y}{k_z} & \dfrac{1}{\omega\mu_0}\dfrac{\left(k_z^2 + k_x^2\right)}{k_z} & -\dfrac{1}{\omega\mu_0}\dfrac{k_x k_y}{k_z} & -\dfrac{1}{\omega\mu_0}\dfrac{\left(k_z^2 + k_x^2\right)}{k_z} \\
-\dfrac{1}{\omega\mu_0}\dfrac{\left(k_y^2 + k_z^2\right)}{k_z} & -\dfrac{1}{\omega\mu_0}\dfrac{k_y k_x}{k_z} & \dfrac{1}{\omega\mu_0}\dfrac{\left(k_y^2 + k_z^2\right)}{k_z} & \dfrac{1}{\omega\mu_0}\dfrac{k_y k_x}{k_z}
\end{pmatrix}
\begin{pmatrix}
\vec{E}_{i,y} \\
\vec{E}_{i,x} \\
\vec{E}_{r,y} \\
\vec{E}_{r,x}
\end{pmatrix}
$$

$$
=
\begin{pmatrix}
E_y^{(1)+} & E_y^{(2)+} & E_y^{(1)-} e^{jk_z^{(1)-}(z_- - z_+)} & E_y^{(2)-} e^{jk_z^{(2)-}(z_- - z_+)} \\
E_x^{(1)+} & E_x^{(2)+} & E_x^{(1)-} e^{jk_z^{(1)-}(z_- - z_+)} & E_x^{(2)-} e^{jk_z^{(2)-}(z_- - z_+)} \\
H_y^{(1)+} & H_y^{(2)+} & H_y^{(1)-} e^{jk_z^{(1)-}(z_- - z_+)} & H_y^{(2)-} e^{jk_z^{(2)-}(z_- - z_+)} \\
H_x^{(1)+} & H_x^{(2)+} & H_x^{(1)-} e^{jk_z^{(1)-}(z_- - z_+)} & H_x^{(2)-} e^{jk_z^{(2)-}(z_- - z_+)}
\end{pmatrix}
\begin{pmatrix}
C_{a,1}^+ \\
C_{a,2}^+ \\
C_{a,1}^- \\
C_{a,2}^-
\end{pmatrix}.
\tag{2.23a}
$$

Also, the boundary condition at the right boundary of $z = z_+$ is given by the matrix equation

$$
\begin{pmatrix}
E_y^{(1)+} e^{jk_z^{(1)+}(z_+ - z_-)} & E_y^{(2)+} e^{jk_z^{(2)+}(z_+ - z_-)} & E_y^{(1)-} & E_y^{(2)-} \\
E_x^{(1)+} e^{jk_z^{(1)+}(z_+ - z_-)} & E_x^{(2)+} e^{jk_z^{(2)+}(z_+ - z_-)} & E_x^{(1)-} & E_x^{(2)-} \\
H_y^{(1)+} e^{jk_z^{(1)+}(z_+ - z_-)} & H_y^{(2)+} e^{jk_z^{(2)+}(z_+ - z_-)} & H_y^{(1)-} & H_y^{(2)-} \\
H_x^{(1)+} e^{jk_z^{(1)+}(z_+ - z_-)} & H_x^{(2)+} e^{jk_z^{(2)+}(z_+ - z_-)} & H_x^{(1)-} & H_x^{(2)-}
\end{pmatrix}
\begin{pmatrix}
C_{a,1}^+ \\
C_{a,2}^+ \\
C_{a,1}^- \\
C_{a,2}^-
\end{pmatrix}
$$

$$
=
\begin{pmatrix}
I & 0 & I & 0 \\
0 & I & 0 & I \\
\dfrac{1}{\omega\mu_0}\dfrac{k_x k_y}{k_z} & \dfrac{1}{\omega\mu_0}\dfrac{\left(k_z^2 + k_x^2\right)}{k_z} & -\dfrac{1}{\omega\mu_0}\dfrac{k_x k_y}{k_z} & -\dfrac{1}{\omega\mu_0}\dfrac{\left(k_z^2 + k_x^2\right)}{k_z} \\
-\dfrac{1}{\omega\mu_0}\dfrac{\left(k_y^2 + k_z^2\right)}{k_z} & -\dfrac{1}{\omega\mu_0}\dfrac{k_y k_x}{k_z} & \dfrac{1}{\omega\mu_0}\dfrac{\left(k_y^2 + k_z^2\right)}{k_z} & \dfrac{1}{\omega\mu_0}\dfrac{k_y k_x}{k_z}
\end{pmatrix}
\begin{pmatrix}
\vec{E}_{t,y} \\
\vec{E}_{t,x} \\
0 \\
0
\end{pmatrix}.
\tag{2.23b}
$$

The boundary condition matching equations (2.23a and 2.23b) are written in the form of more compact matrix operator equations as

$$
\begin{pmatrix}
\underline{\underline{W}}_h & \underline{\underline{W}}_h \\
\underline{\underline{V}}_h & -\underline{\underline{V}}_h
\end{pmatrix}
\begin{pmatrix}
\vec{E}_i \\
\vec{E}_r
\end{pmatrix}
=
\begin{pmatrix}
\underline{\underline{W}}^+(0) & \underline{\underline{W}}^-(z_- - z_+) \\
\underline{\underline{V}}^+(0) & \underline{\underline{V}}^-(z_- - z_+)
\end{pmatrix}
\begin{pmatrix}
\underline{C}_a^+ \\
\underline{C}_a^-
\end{pmatrix}
\tag{2.24a}
$$

$$\begin{pmatrix} \underline{\underline{W}}^+(z_+ - z_-) & \underline{\underline{W}}^-(0) \\ \underline{\underline{V}}^+(z_+ - z_-) & \underline{\underline{V}}^-(0) \end{pmatrix} \begin{pmatrix} \underline{C}_a^+ \\ \underline{C}_a^- \end{pmatrix} = \begin{pmatrix} \underline{\underline{W}}_h & \underline{\underline{W}}_h \\ \underline{\underline{V}}_h & -\underline{\underline{V}}_h \end{pmatrix} \begin{pmatrix} \vec{E}_t \\ 0 \end{pmatrix} \tag{2.24b}$$

where $\underline{\underline{W}}_h$ and $\underline{\underline{V}}_h$ are 4×4 matrices given, respectively, by

$$\underline{\underline{W}}_h = \begin{pmatrix} 1 & 0 \\ 0 & 1 \end{pmatrix} \tag{2.24c}$$

$$\underline{\underline{V}}_h = \begin{pmatrix} \dfrac{1}{\omega\mu_0}\dfrac{k_x k_y}{k_z} & \dfrac{1}{\omega\mu_0}\dfrac{\left(k_z^2 + k_x^2\right)}{k_z} \\[3mm] -\dfrac{1}{\omega\mu_0}\dfrac{\left(k_y^2 + k_z^2\right)}{k_z} & -\dfrac{1}{\omega\mu_0}\dfrac{k_y k_x}{k_z} \end{pmatrix} \tag{2.24d}$$

$\underline{\underline{W}}^+(z)$ and $\underline{\underline{V}}^+(z)$ are 4×2 matrices indicating the part of the positive modes in Equations (2.23a) and (2.23b), given, respectively, by

$$\underline{\underline{W}}^+(z) = \begin{pmatrix} E_y^{(1)+} e^{jk_z^{(1)+}z} & E_y^{(2)+} e^{jk_z^{(2)+}z} \\ E_x^{(1)+} e^{jk_z^{(1)+}z} & E_x^{(2)+} e^{jk_z^{(2)+}z} \end{pmatrix} \tag{2.24e}$$

$$\underline{\underline{V}}^+(z) = \begin{pmatrix} H_y^{(1)+} e^{jk_z^{(1)+}z} & H_y^{(2)+} e^{jk_z^{(2)+}z} \\ H_x^{(1)+} e^{jk_z^{(1)+}z} & H_x^{(2)+} e^{jk_z^{(2)+}z} \end{pmatrix} \tag{2.24f}$$

$\underline{\underline{W}}^-(z)$ and $\underline{\underline{V}}^-(z)$ are 4×2 matrices indicating the part of the negative modes given, respectively, by

$$\underline{\underline{W}}^-(z) = \begin{pmatrix} E_y^{(1)-} e^{jk_z^{(1)-}z} & E_y^{(2)-} e^{jk_z^{(2)-}z} \\ E_x^{(1)-} e^{jk_z^{(1)-}z} & E_x^{(2)-} e^{jk_z^{(2)-}z} \end{pmatrix} \tag{2.24g}$$

$$\underline{\underline{V}}^-(z) = \begin{pmatrix} H_y^{(1)-} e^{jk_z^{(1)-}z} & H_y^{(2)-} e^{jk_z^{(2)-}z} \\ H_x^{(1)-} e^{jk_z^{(1)-}z} & H_x^{(2)-} e^{jk_z^{(2)-}z} \end{pmatrix} \tag{2.24h}$$

For specific input $\vec{E}_i$, the reflection, transmission, and internal electromagnetic field distribution can be obtained by solving Equations (2.24a) and (2.24b).

The S-matrix method is founded on the operator representation of the above process, which is a more general approach for dealing with electromagnetic scattering problems. The principle of S-matrix analysis can be easily understood with the analysis of a 1D single block. Define the input field operator as

$$\underline{\vec{U}} = \begin{pmatrix} 1 & 0 \\ 0 & 1 \end{pmatrix}$$ (2.25a)

Then, the input field can be represented by the linear combination of the eigen-column vector of the input field operator as

$$\underline{\vec{E}_i} = \begin{pmatrix} \vec{E}_{i,y} \\ \vec{E}_{i,x} \end{pmatrix} = \vec{E}_{i,y} \begin{pmatrix} 1 \\ 0 \end{pmatrix} + \vec{E}_{i,x} \begin{pmatrix} 0 \\ 1 \end{pmatrix}$$ (2.25b)

For the input field with $(\vec{E}_{i,y}, \vec{E}_{i,x}) = (1,0)$, the reflection and transmission coefficients are supposed to be $(\vec{E}_{r,y}, \vec{E}_{r,x}) = (\vec{e}_{r,y}^{(1)}, \vec{e}_{r,x}^{(1)})$ and $(\vec{E}_{t,y}, \vec{E}_{t,x}) = (\vec{e}_{t,y}^{(1)}, \vec{e}_{t,x}^{(1)})$, respectively. And for the other orthogonal input field with $(\vec{E}_{i,y}, \vec{E}_{i,x}) = (0, 1)$, the reflection and transmission coefficients are $(\vec{E}_{r,y}, \vec{E}_{r,x}) = (\vec{e}_{r,y}^{(2)}, \vec{e}_{r,x}^{(2)})$ and $(\vec{E}_{t,y}, \vec{E}_{t,x}) = (\vec{e}_{t,y}^{(2)}, \vec{e}_{t,x}^{(2)})$, respectively.

Then by arranging the reflection coefficients and transmission coefficients as the coloum vectors, we can define the reflection coefficient matrix operator and transmission coefficient matrix operator, respectively, as

$$\underline{\vec{R}} = \begin{pmatrix} \vec{e}_{r,y}^{(1)} & \vec{e}_{r,y}^{(2)} \\ \vec{e}_{r,x}^{(1)} & \vec{e}_{r,x}^{(2)} \end{pmatrix}$$ (2.25c)

$$\underline{\vec{T}} = \begin{pmatrix} \vec{e}_{t,y}^{(1)} & \vec{e}_{t,y}^{(2)} \\ \vec{e}_{t,x}^{(1)} & \vec{e}_{t,x}^{(2)} \end{pmatrix}$$ (2.25d)

These operators contain the information of the external scattering responses to orthogonal unit inputs. Considering the linearity of Equations (2.24a) and (2.24b), the reflection coefficient is related to the input field coefficient $\vec{E}_i$ by the reflection coefficient operator $\underline{\vec{R}}$ as

$$\underline{\vec{E}_r} = \vec{E}_{i,y} \begin{pmatrix} \vec{e}_{r,y}^{(1)} \\ \vec{e}_{r,x}^{(1)} \end{pmatrix} + \vec{E}_{i,x} \begin{pmatrix} \vec{e}_{r,y}^{(2)} \\ \vec{e}_{r,x}^{(2)} \end{pmatrix} = \begin{pmatrix} \vec{e}_{r,y}^{(1)} & \vec{e}_{r,y}^{(2)} \\ \vec{e}_{r,x}^{(1)} & \vec{e}_{r,x}^{(2)} \end{pmatrix} \begin{pmatrix} \vec{E}_{i,y} \\ \vec{E}_{i,x} \end{pmatrix} = \underline{\vec{R}} \begin{pmatrix} \vec{E}_{i,y} \\ \vec{E}_{i,x} \end{pmatrix}$$ (2.25e)

$$\underline{\vec{E}_t} = \vec{E}_{i,y} \begin{pmatrix} \vec{e}_{t,y}^{(1)} \\ \vec{e}_{t,x}^{(1)} \end{pmatrix} + \vec{E}_{i,x} \begin{pmatrix} \vec{e}_{t,y}^{(2)} \\ \vec{e}_{t,x}^{(2)} \end{pmatrix} = \begin{pmatrix} \vec{e}_{t,y}^{(1)} & \vec{e}_{t,y}^{(2)} \\ \vec{e}_{t,x}^{(1)} & \vec{e}_{t,x}^{(2)} \end{pmatrix} \begin{pmatrix} \vec{E}_{i,y} \\ \vec{E}_{i,x} \end{pmatrix} = \underline{\vec{T}} \begin{pmatrix} \vec{E}_{i,y} \\ \vec{E}_{i,x} \end{pmatrix}$$ (2.25f)

Furthermore, with these linear operations, we can write the operator mathematics form as

$$
\begin{pmatrix} \underline{\underline{W}}_h & \underline{\underline{W}}_h \\ \underline{\underline{V}}_h & -\underline{\underline{V}}_h \end{pmatrix} \begin{pmatrix} \vec{\tilde{U}} \\ \vec{\tilde{R}} \end{pmatrix} = \begin{pmatrix} \underline{\underline{W}}^+(0) & \underline{\underline{W}}^-(z_- - z_+) \\ \underline{\underline{V}}^+(0) & \underline{\underline{V}}^-(z_- - z_+) \end{pmatrix} \begin{pmatrix} \underline{\underline{C}}_a^+ \\ \underline{\underline{C}}_a^- \end{pmatrix}
\tag{2.26a}
$$

$$
\begin{pmatrix} \underline{\underline{W}}^+(z_+ - z_-) & \underline{\underline{W}}^-(0) \\ \underline{\underline{V}}^+(z_+ - z_-) & \underline{\underline{V}}^-(0) \end{pmatrix} \begin{pmatrix} \underline{\underline{C}}_a^+ \\ \underline{\underline{C}}_a^- \end{pmatrix} = \begin{pmatrix} \underline{\underline{W}}_h & \underline{\underline{W}}_h \\ \underline{\underline{V}}_h & -\underline{\underline{V}}_h \end{pmatrix} \begin{pmatrix} \vec{\tilde{T}} \\ \vec{0} \end{pmatrix}
\tag{2.26b}
$$

In the analysis, by finding unit responses for unit inputs instead of calculating reflection and transmission fields for a specific input field, we have obtained the operator form of Maxwell's equation by the algebraic equations of Equations (2.26a) and (2.26b). $\vec{\tilde{U}}$ is the input operator, actually, an 4×4 identity matrix. $\vec{\tilde{R}}$ and $\vec{\tilde{T}}$ are the reflection coefficient matrix operator and the transmission coefficient matrix operator, respectively. The coupling coefficient matrix operators $\underline{\underline{C}}_a^+$ and $\underline{\underline{C}}_a^-$ are obtained as, from Equations (2.26a) and (2.26b),

$$
\begin{pmatrix} \underline{\underline{C}}_a^+ \\ \underline{\underline{C}}_a^- \end{pmatrix} = \begin{pmatrix} \underline{\underline{W}}_h^{-1}\underline{\underline{W}}^+(0) + \underline{\underline{V}}_h^{-1}\underline{\underline{V}}^+(0) & \underline{\underline{W}}_h^{-1}\underline{\underline{W}}^-(z_- - z_+) + \underline{\underline{V}}_h^{-1}\underline{\underline{V}}^-(z_- - z_+) \\ \underline{\underline{W}}_h^{-1}\underline{\underline{W}}^+(z_+ - z_-) - \underline{\underline{V}}_h^{-1}\underline{\underline{V}}^+(z_+ - z_-) & \underline{\underline{W}}_h^{-1}\underline{\underline{W}}^-(0) - \underline{\underline{V}}_h^{-1}\underline{\underline{V}}^-(0) \end{pmatrix}^{-1}
$$
$$
\times \begin{pmatrix} 2\vec{\underline{\underline{U}}} \\ \underline{\underline{0}} \end{pmatrix}
\tag{2.27a}
$$

The reflection and transmission operators $\vec{\tilde{R}}$ and $\vec{\tilde{T}}$ are obtained, respectively, by

$$
\vec{\tilde{R}} = \underline{\underline{W}}_h^{-1}\left[\underline{\underline{W}}^+(0)\underline{\underline{C}}_a^+ + \underline{\underline{W}}^-(z_- - z_+)\underline{\underline{C}}_a^- - \underline{\underline{W}}_h\vec{\underline{\underline{U}}} \right]
\tag{2.27b}
$$

$$
\vec{\tilde{T}} = \underline{\underline{W}}_h^{-1}\left[\underline{\underline{W}}^+(z_+ - z_-)\underline{\underline{C}}_a^+ + \underline{\underline{W}}^-(0)\underline{\underline{C}}_a^- \right]
\tag{2.27c}
$$

The obtained matrix operators, $\vec{\tilde{R}}$, $\vec{\tilde{T}}$, $\underline{\underline{C}}_a^+$, and $\underline{\underline{C}}_a^-$, provide complete information on the left-to-right directional characteristics of the single block. By the same manner as the reflection and transmission matrix operators, the coupling coefficients of internal Bloch eigenmodes excited by an external plane wave $(\vec{E}_{i,y}, \vec{E}_{i,x})$ can be described by the operator form. The coupling coefficients for left-to-right directional input are given by

$$
\begin{pmatrix} C_{a,1}^+ \\ C_{a,2}^+ \end{pmatrix} = \underline{\underline{C}}_a^+ \begin{pmatrix} \vec{E}_{i,y} \\ \vec{E}_{i,x} \end{pmatrix}
\tag{2.27d}
$$

$$\begin{pmatrix} C_{b,1}^- \\ C_{b,2}^- \end{pmatrix} = \underline{\underline{C}}_a^- \begin{pmatrix} \vec{E}_{i,y} \\ \vec{E}_{i,x} \end{pmatrix} \tag{2.27e}$$

Hence we can define the coupling coefficient matrix operator $\underline{\underline{C}}_a$ by

$$\underline{\underline{C}}_a = \begin{pmatrix} \underline{\underline{C}}_a^+ \\ \underline{\underline{C}}_a^- \end{pmatrix} \tag{2.27f}$$

This operator form description of the left-to-right directional characterization is illustrated schematically in Figure 2.3(a).

By the same manner, the right-to-left characterization is performed. In the case of the right-to-left characterization, the excitation field $\bar{\mathbf{E}}_i$, the reflection field $\bar{\mathbf{E}}_r$, and the transmission field $\bar{\mathbf{E}}_t$ are given, respectively, by

$$\begin{pmatrix} \bar{\mathbf{E}}_i \\ \bar{\mathbf{H}}_i \end{pmatrix} = \begin{pmatrix} \bar{\mathbf{E}}_{i,x}, \bar{\mathbf{E}}_{i,y}, \bar{\mathbf{E}}_{i,z} \\ \bar{\mathbf{H}}_{i,x}, \bar{\mathbf{H}}_{i,y}, \bar{\mathbf{H}}_{i,z} \end{pmatrix} = \begin{pmatrix} \bar{E}_{i,x}, \bar{E}_{i,y}, \bar{E}_{i,z} \\ \bar{H}_{i,x}, \bar{H}_{i,y}, \bar{H}_{i,z} \end{pmatrix} e^{j(k_x x + k_y y - k_z(z - z_+))} \tag{2.28a}$$

$$\begin{pmatrix} \bar{\mathbf{E}}_r \\ \bar{\mathbf{H}}_r \end{pmatrix} = \begin{pmatrix} \bar{\mathbf{E}}_{r,x}, \bar{\mathbf{E}}_{r,y}, \bar{\mathbf{E}}_{r,z} \\ \bar{\mathbf{H}}_{r,x}, \bar{\mathbf{H}}_{r,y}, \bar{\mathbf{H}}_{r,z} \end{pmatrix} = \begin{pmatrix} \bar{E}_{r,x}, \bar{E}_{r,y}, \bar{E}_{r,z} \\ \bar{H}_{r,x}, \bar{H}_{r,y}, \bar{H}_{r,z} \end{pmatrix} e^{j(k_x x + k_y y + k_z(z - z_+))} \tag{2.28b}$$

$$\begin{pmatrix} \bar{\mathbf{E}}_t \\ \bar{\mathbf{H}}_t \end{pmatrix} = \begin{pmatrix} \bar{\mathbf{E}}_{t,x}, \bar{\mathbf{E}}_{t,y}, \bar{\mathbf{E}}_{t,z} \\ \bar{\mathbf{H}}_{t,x}, \bar{\mathbf{H}}_{t,y}, \bar{\mathbf{H}}_{t,z} \end{pmatrix} = \begin{pmatrix} \bar{E}_{t,x}, \bar{E}_{t,y}, \bar{E}_{t,z} \\ \bar{H}_{t,x}, \bar{H}_{t,y}, \bar{H}_{t,z} \end{pmatrix} e^{j(k_x x + k_y y - k_z(z - z_-))} \tag{2.28c}$$

The tangential components of the incident, reflection, and transmission magnetic fields are solved for the electric field components as follows:

$$\bar{\mathbf{H}}_{i,y} = \frac{1}{j\omega\mu_0}\left(\frac{\partial \bar{\mathbf{E}}_{i,x}}{\partial z} - \frac{\partial \bar{\mathbf{E}}_{i,z}}{\partial x}\right) \Leftrightarrow \bar{H}_{i,y} = \frac{1}{\omega\mu_0}(-k_z\bar{E}_{i,x} - k_x\bar{E}_{i,z}), \tag{2.29a}$$

$$\bar{\mathbf{H}}_{i,x} = \frac{1}{j\omega\mu_0}\left(\frac{\partial \bar{\mathbf{E}}_{i,z}}{\partial y} - \frac{\partial \bar{\mathbf{E}}_{i,y}}{\partial z}\right) \Leftrightarrow \bar{H}_{i,x} = \frac{1}{\omega\mu_0}(k_y\bar{E}_{i,z} + k_z\bar{E}_{i,y}), \tag{2.29b}$$

$$\bar{\mathbf{H}}_{r,y} = \frac{1}{j\omega\mu_0}\left(\frac{\partial \bar{\mathbf{E}}_{r,x}}{\partial z} - \frac{\partial \bar{\mathbf{E}}_{r,z}}{\partial x}\right) \Leftrightarrow \bar{H}_{r,y} = \frac{1}{\omega\mu_0}(k_z\bar{E}_{r,x} - k_x\bar{E}_{r,z}), \tag{2.29c}$$

$$\tilde{\mathbf{H}}_{r,x} = \frac{1}{j\omega\mu_0}\left(\frac{\partial \tilde{\mathbf{E}}_{r,z}}{\partial y} - \frac{\partial \tilde{\mathbf{E}}_{r,y}}{\partial z}\right) \Leftrightarrow \tilde{H}_{r,x} = \frac{1}{\omega\mu_0}(k_y\tilde{E}_{r,z} - k_z\tilde{E}_{r,y}), \qquad (2.29\text{d})$$

$$\tilde{\mathbf{H}}_{t,y} = \frac{1}{j\omega\mu_0}\left(\frac{\partial \tilde{\mathbf{E}}_{t,x}}{\partial z} - \frac{\partial \tilde{\mathbf{E}}_{t,z}}{\partial x}\right) \Leftrightarrow \tilde{H}_{t,y} = \frac{1}{\omega\mu_0}(-k_z\tilde{E}_{t,x} - k_x\tilde{E}_{t,z}), \qquad (2.29\text{e})$$

$$\tilde{\mathbf{H}}_{t,x} = \frac{1}{j\omega\mu_0}\left(\frac{\partial \tilde{\mathbf{E}}_{t,z}}{\partial y} - \frac{\partial \tilde{\mathbf{E}}_{t,y}}{\partial z}\right) \Leftrightarrow \tilde{H}_{t,x} = \frac{1}{\omega\mu_0}(k_y\tilde{E}_{t,z} + k_z\tilde{E}_{t,y}). \qquad (2.29\text{f})$$

The plane wave condition in free space gives

$$k_x\tilde{E}_{i,x} + k_y\tilde{E}_{i,y} - k_z\tilde{E}_{i,z} = 0 \Leftrightarrow \tilde{E}_{i,z} = \frac{k_x\tilde{E}_{i,x} + k_y\tilde{E}_{i,y}}{k_z} \qquad (2.30\text{a})$$

$$k_x\tilde{E}_{r,x} + k_y\tilde{E}_{r,y} + k_z\tilde{E}_{r,z} = 0 \Leftrightarrow \tilde{E}_{r,z} = -\frac{k_x\tilde{E}_{r,x} + k_y\tilde{E}_{r,y}}{k_z} \qquad (2.30\text{b})$$

$$k_x\tilde{E}_{t,x} + k_y\tilde{E}_{t,y} - k_z\tilde{E}_{t,z} = 0 \Leftrightarrow \tilde{E}_{t,z} = -\frac{k_x\tilde{E}_{t,x} + k_y\tilde{E}_{t,y}}{k_z} \qquad (2.30\text{c})$$

According to the tangential (transverse) field continuation condition, the boundary condition at the left boundary of $z = z_-$ can be expressed by the matrix equation

$$\begin{pmatrix} I & 0 & I & 0 \\ 0 & I & 0 & I \\ \dfrac{1}{\omega\mu_0}\dfrac{k_xk_y}{k_z} & \dfrac{1}{\omega\mu_0}\dfrac{\left(k_z^2 + k_x^2\right)}{k_z} & -\dfrac{1}{\omega\mu_0}\dfrac{k_xk_y}{k_z} & -\dfrac{1}{\omega\mu_0}\dfrac{\left(k_z^2 + k_x^2\right)}{k_z} \\ -\dfrac{1}{\omega\mu_0}\dfrac{\left(k_y^2 + k_z^2\right)}{k_z} & -\dfrac{1}{\omega\mu_0}\dfrac{k_yk_x}{k_z} & \dfrac{1}{\omega\mu_0}\dfrac{\left(k_y^2 + k_z^2\right)}{k_z} & \dfrac{1}{\omega\mu_0}\dfrac{k_yk_x}{k_z} \end{pmatrix} \begin{pmatrix} 0 \\ 0 \\ \tilde{E}_{t,y} \\ \tilde{E}_{t,x} \end{pmatrix}$$

$$= \begin{pmatrix} E_y^{(1)+} & E_y^{(2)+} & E_y^{(1)-}e^{jk_z^{(1)-}(z_--z_+)} & E_y^{(2)-}e^{jk_z^{(2)-}(z_--z_+)} \\ E_x^{(1)+} & E_x^{(2)+} & E_x^{(1)-}e^{jk_z^{(1)-}(z_--z_+)} & E_x^{(2)-}e^{jk_z^{(2)-}(z_--z_+)} \\ H_y^{(1)+} & H_y^{(2)+} & H_y^{(1)-}e^{jk_z^{(1)-}(z_--z_+)} & H_y^{(2)-}e^{jk_z^{(2)-}(z_--z_+)} \\ H_x^{(1)+} & H_x^{(2)+} & H_x^{(1)-}e^{jk_z^{(1)-}(z_--z_+)} & H_x^{(2)-}e^{jk_z^{(2)-}(z_--z_+)} \end{pmatrix} \begin{pmatrix} c_{b,1}^+ \\ c_{b,2}^+ \\ c_{b,1}^- \\ c_{b,2}^- \end{pmatrix}. \qquad (2.31\text{a})$$

The boundary condition at the right boundary of $z = z_+$ is expressed by the matrix equation

$$
\begin{pmatrix}
E_y^{(1)+}e^{jk_z^{(1)+}(z_+-z_-)} & E_y^{(2)+}e^{jk_z^{(2)+}(z_+-z_-)} & E_y^{(1)-} & E_y^{(2)-} \\
E_x^{(1)+}e^{jk_z^{(1)+}(z_+-z_-)} & E_x^{(2)+}e^{jk_z^{(2)+}(z_+-z_-)} & E_x^{(1)-} & E_x^{(2)-} \\
H_y^{(1)+}e^{jk_z^{(1)+}(z_+-z_-)} & H_y^{(2)+}e^{jk_z^{(2)+}(z_+-z_-)} & H_y^{(1)-} & H_y^{(2)-} \\
H_x^{(1)+}e^{jk_z^{(1)+}(z_+-z_-)} & H_x^{(2)+}e^{jk_z^{(2)+}(z_+-z_-)} & H_x^{(1)-} & H_x^{(2)-}
\end{pmatrix}
\begin{pmatrix}
c_{b,1}^+ \\
c_{b,2}^+ \\
c_{b,1}^- \\
c_{b,2}^-
\end{pmatrix}
$$

$$
=
\begin{pmatrix}
I & 0 & I & 0 \\
0 & I & 0 & I \\
\dfrac{1}{\omega\mu_0}\dfrac{k_x k_y}{k_z} & \dfrac{1}{\omega\mu_0}\dfrac{\left(k_z^2+k_x^2\right)}{k_z} & -\dfrac{1}{\omega\mu_0}\dfrac{k_x k_y}{k_z} & -\dfrac{1}{\omega\mu_0}\dfrac{\left(k_z^2+k_x^2\right)}{k_z} \\
-\dfrac{1}{\omega\mu_0}\dfrac{\left(k_y^2+k_z^2\right)}{k_z} & -\dfrac{1}{\omega\mu_0}\dfrac{k_y k_x}{k_z} & \dfrac{1}{\omega\mu_0}\dfrac{\left(k_y^2+k_z^2\right)}{k_z} & \dfrac{1}{\omega\mu_0}\dfrac{k_y k_x}{k_z}
\end{pmatrix}
\begin{pmatrix}
\vec{E}_{t,y} \\
\vec{E}_{t,x} \\
\vec{E}_{i,y} \\
\vec{E}_{i,x}
\end{pmatrix}.
$$

$$(2.31b)$$

These equations are rewritten by the operator mathematics form

$$
\begin{pmatrix}
\underline{\underline{W}}_h & \underline{\underline{W}}_h \\
\underline{\underline{V}}_h & -\underline{\underline{V}}_h
\end{pmatrix}
\begin{pmatrix}
\underline{\underline{0}} \\
\underline{\underline{\bar{T}}}
\end{pmatrix}
=
\begin{pmatrix}
\underline{\underline{W}}^+(0) & \underline{\underline{W}}^-(z_- - z_+) \\
\underline{\underline{V}}^+(0) & \underline{\underline{V}}^-(z_- - z_+)
\end{pmatrix}
\begin{pmatrix}
\underline{\underline{C}}_b^+ \\
\underline{\underline{C}}_b^-
\end{pmatrix}
\qquad (2.32a)
$$

$$
\begin{pmatrix}
\underline{\underline{W}}^+(z_+ - z_-) & \underline{\underline{W}}^-(0) \\
\underline{\underline{V}}^+(z_+ - z_-) & \underline{\underline{V}}^-(0)
\end{pmatrix}
\begin{pmatrix}
\underline{\underline{C}}_b^+ \\
\underline{\underline{C}}_b^-
\end{pmatrix}
=
\begin{pmatrix}
\underline{\underline{W}}_h & \underline{\underline{W}}_h \\
\underline{\underline{V}}_h & -\underline{\underline{V}}_h
\end{pmatrix}
\begin{pmatrix}
\underline{\underline{\vec{R}}} \\
\underline{\underline{\bar{U}}}
\end{pmatrix}
\qquad (2.32b)
$$

where $\underline{\underline{\bar{U}}}$ is an 4×4 identity matrix. $\underline{\underline{\vec{R}}}$ and $\underline{\underline{\bar{T}}}$ are the reflection coefficient matrix operator and transmission coefficient matrix operator, respectively. The coupling coefficient matrix operators $\underline{\underline{C}}_b^+$ and $\underline{\underline{C}}_b^-$ are obtained as, from Equations (2.32a) and (2.32b),

$$
\begin{pmatrix}
\underline{\underline{C}}_b^+ \\
\underline{\underline{C}}_b^-
\end{pmatrix}
=
\begin{pmatrix}
\underline{\underline{W}}_h^{-1}\underline{\underline{W}}^+(0) + \underline{\underline{V}}_h^{-1}\underline{\underline{V}}^+(0) & \underline{\underline{W}}_h^{-1}\underline{\underline{W}}^-(z_- - z_+) + \underline{\underline{V}}_h^{-1}\underline{\underline{V}}^-(z_- - z_+) \\
\underline{\underline{W}}_h^{-1}\underline{\underline{W}}^+(z_+ - z_-) - \underline{\underline{V}}_h^{-1}\underline{\underline{V}}^+(z_+ - z_-) & \underline{\underline{W}}_h^{-1}\underline{\underline{W}}^-(0) - \underline{\underline{V}}_h^{-1}\underline{\underline{V}}^-(0)
\end{pmatrix}^{-1}
$$

$$
\times
\begin{pmatrix}
\underline{\underline{0}} \\
2\underline{\underline{\bar{U}}}
\end{pmatrix}
\qquad (2.33a)
$$

The reflection and transmission operators $\vec{\underline{\underline{R}}}$ and $\vec{\underline{\underline{T}}}$ are obtained, respectively, by

$$\vec{\underline{\underline{R}}} = \underline{\underline{W}}_h^{-1}\left[\underline{\underline{W}}^+(z_+ - z_-)\underline{\underline{C}}_b^+ + \underline{\underline{W}}^-(0)\underline{\underline{C}}_b^- - \underline{\underline{W}}_h\underline{\underline{U}} \right] \qquad (2.33b)$$

$$\vec{\underline{\underline{T}}} = \underline{\underline{W}}_h^{-1}\left[\underline{\underline{W}}^+(0)\underline{\underline{C}}_b^+ + \underline{\underline{W}}^-(z_- - z_+)\underline{\underline{C}}_b^- \right] \qquad (2.33c)$$

The obtained matrix operators, $\vec{\underline{\underline{R}}}$, $\vec{\underline{\underline{T}}}$, $\underline{\underline{C}}_b^+$, and $\underline{\underline{C}}_b^-$, provide complete information on the right-to-left directional characteristics of the single block. The coupling coefficient matrix operator $\underline{\underline{C}}_b$ is defined by

$$\underline{\underline{C}}_b = \begin{pmatrix} \underline{\underline{C}}_b^+ \\ \underline{\underline{C}}_b^- \end{pmatrix} \qquad (2.33d)$$

This operator form description of the left-to-right directional characterization is illustrated schematically in Figure 2.3(b). The S-matrix **S** is defined by

$$\mathbf{S} = \begin{pmatrix} \vec{\underline{\underline{T}}} & \vec{\underline{\underline{R}}} \\ \vec{\underline{\underline{R}}} & \vec{\underline{\underline{T}}} \end{pmatrix} \qquad (2.33e)$$

2.1.3 Field Visualization

An important step in an electromagnetic analysis is to visualize the analyzed electromagnetic field distribution. The field visualization is modeled separately for left-to-right directional field and right-to-left directional field excitations.

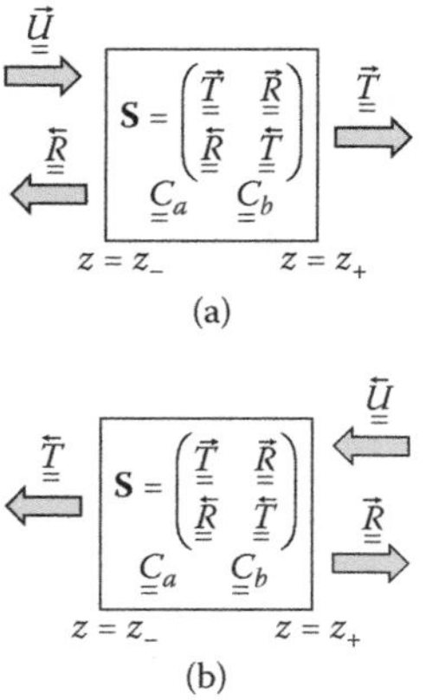

(a)

(b)

FIGURE 2.3
Operator form description of bidirectional characterization of single-block (a) left-to-right directional characterization and (b) right-to-left directional characterization.

2.1.3.1 Left-to-Right Field Visualization

When the incident field $\bar{\mathbf{E}}_i$ is in the left half-infinite free space, the reflection field coefficient inside the left half-infinite free space and the transmission field coefficient inside the right half-infinite free space are given, respectively, as

$$\begin{pmatrix} \bar{E}_{r,y} \\ \bar{E}_{r,x} \end{pmatrix} = \underline{\bar{R}} \begin{pmatrix} \bar{E}_{i,y} \\ \bar{E}_{i,x} \end{pmatrix}$$

and

$$\begin{pmatrix} \vec{E}_{t,y} \\ \vec{E}_{t,x} \end{pmatrix} = \underline{\vec{T}} \begin{pmatrix} \vec{E}_{i,y} \\ \vec{E}_{i,x} \end{pmatrix}.$$

The coupling coefficients of the internal optical field distribution are represented by

$$\underline{C}_a^+ = \underline{\underline{C}}_a^+ \begin{pmatrix} \vec{E}_{i,y} \\ \vec{E}_{i,x} \end{pmatrix}$$

and

$$\underline{C}_a^- = \underline{\underline{C}}_a^- \begin{pmatrix} \vec{E}_{i,y} \\ \vec{E}_{i,x} \end{pmatrix}.$$

The total field distributions $\mathbf{E}$ and $\mathbf{H}$ are obtained in the respective spatial regions (Figure 2.4):

1. For $z < z_-$

$$\begin{pmatrix} \mathbf{E} \\ \mathbf{H} \end{pmatrix} = \begin{pmatrix} \vec{E}_i \\ \vec{H}_i \end{pmatrix} + \begin{pmatrix} \bar{E}_r \\ \bar{H}_r \end{pmatrix} = \begin{pmatrix} \vec{E}_{i,x} & \vec{E}_{i,y} & \vec{E}_{i,z} \\ \vec{H}_{i,x} & \vec{H}_{i,y} & \vec{H}_{i,z} \end{pmatrix} e^{j(k_x x + k_y y + k_z(z - z_-))}$$

$$+ \begin{pmatrix} \bar{E}_{r,x} & \bar{E}_{r,y} & \bar{E}_{r,z} \\ \bar{H}_{r,x} & \bar{H}_{r,y} & \bar{H}_{r,z} \end{pmatrix} e^{j(k_x x + k_y y - k_z(z - z_-))}. \tag{2.34a}$$

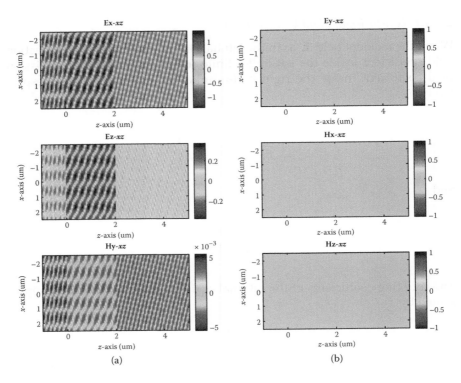

FIGURE 2.4
Electric and magnetic field distributions of a TM mode plane wave obliquely incident on a finite-sized dielectric slab with the thickness of 2 μm along the left-to-right direction. (MATLAB code: *FMM_1D_SMM.m.*)

2. For $z_- \leq z < z_+$

$$
\begin{pmatrix} \mathbf{E} \\ \mathbf{H} \end{pmatrix} = \sum_{g=1}^{M^+} C_{a,g}^+ \begin{pmatrix} \mathbf{E}^{(g)+} \\ \mathbf{H}^{(g)+} \end{pmatrix} + \sum_{g=1}^{M^-} C_{a,g}^- \begin{pmatrix} \mathbf{E}^{(g)-} \\ \mathbf{H}^{(g)-} \end{pmatrix}
$$

$$
= \sum_{g=1}^{M^+} C_{a,g}^+ \begin{pmatrix} E_x^{(g)+} & E_y^{(g)+} & E_z^{(g)+} \\ H_x^{(g)+} & H_y^{(g)+} & H_z^{(g)+} \end{pmatrix} e^{j\left(k_x x + k_y y + k_z^{(g)+}(z-z_-)\right)} \qquad (2.34b)
$$

$$
+ \sum_{g=1}^{M^-} C_{a,g}^- \begin{pmatrix} E_x^{(g)-} & E_y^{(g)-} & E_z^{(g)-} \\ H_x^{(g)-} & H_y^{(g)-} & H_z^{(g)-} \end{pmatrix} e^{j\left(k_x x + k_y y + k_z^{(g)-}(z-z_+)\right)},
$$

3. For $z_+ \leq z$

$$\begin{pmatrix} \mathbf{E} \\ \mathbf{H} \end{pmatrix} = \begin{pmatrix} \vec{\mathbf{E}}_t \\ \vec{\mathbf{H}}_t \end{pmatrix} = \begin{pmatrix} \vec{E}_{t,x} & \vec{E}_{t,y} & \vec{E}_{t,z} \\ \vec{H}_{t,x} & \vec{H}_{t,y} & \vec{H}_{t,z} \end{pmatrix} e^{j(k_x x + k_y y + k_z(z-z_+))} \qquad (2.34c)$$

2.1.3.2 Right-to-Left Field Visualization

When the incident field, $\bar{\mathbf{E}}_i$, is in the right half-infinite free space, the reflection field coefficient inside the left half-infinite free space and the transmission field coefficient inside the right half-infinite free space are given, respectively, as

$$\begin{pmatrix} \vec{E}_{r,y} \\ \vec{E}_{r,x} \end{pmatrix} = \vec{\underline{R}} \begin{pmatrix} \bar{E}_{i,y} \\ \bar{E}_{i,x} \end{pmatrix}$$

and

$$\begin{pmatrix} \bar{E}_{t,y} \\ \bar{E}_{t,x} \end{pmatrix} = \bar{\underline{T}} \begin{pmatrix} \bar{E}_{i,y} \\ \bar{E}_{i,x} \end{pmatrix}.$$

And the coupling coefficients of the internal optical field distribution are represented by

$$\underline{C}_b^+ = \underline{\underline{C}}_b^+ \begin{pmatrix} \bar{E}_{i,y} \\ \bar{E}_{i,x} \end{pmatrix}$$

and

$$\underline{C}_b^- = \underline{\underline{C}}_b^- \begin{pmatrix} \bar{E}_{i,y} \\ \bar{E}_{i,x} \end{pmatrix}.$$

The total field distributions $\mathbf{E}$ and $\mathbf{H}$ are obtained in the respective spatial regions (Figure 2.5):

1. For $z = z_-$

$$\begin{pmatrix} \mathbf{E} \\ \mathbf{H} \end{pmatrix} = \begin{pmatrix} \bar{\mathbf{E}}_t \\ \bar{\mathbf{H}}_t \end{pmatrix} = \begin{pmatrix} \bar{E}_{t,x} & \bar{E}_{t,y} & \bar{E}_{t,z} \\ \bar{H}_{t,x} & \bar{H}_{t,y} & \bar{H}_{t,z} \end{pmatrix} e^{j(k_x x + k_y y - k_z(z-z_-))} \qquad (2.35a)$$

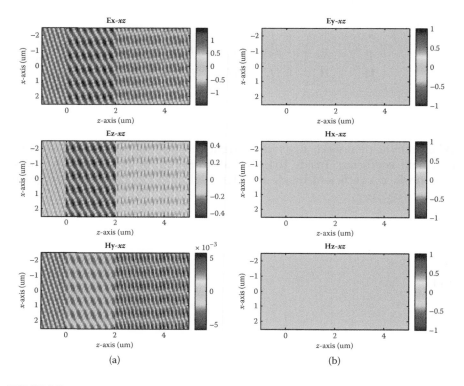

FIGURE 2.5
Electric and magnetic field distributions of a TM mode plane wave obliquely incident on a finite-sized dielectric slab with the thickness of 2 μm along the right-to-left direction. (MATLAB code: *FMM_1D_SMM.m*.)

2. For $z_- \leq z < z_+$

$$
\begin{pmatrix} \mathbf{E} \\ \mathbf{H} \end{pmatrix} = \sum_{g=1}^{M^+} C_{b,g}^+ \begin{pmatrix} \mathbf{E}^{(g)+} \\ \mathbf{H}^{(g)+} \end{pmatrix} + \sum_{g=1}^{M^-} C_{b,g}^- \begin{pmatrix} \mathbf{E}^{(g)-} \\ \mathbf{H}^{(g)-} \end{pmatrix}
$$

$$
= \sum_{g=1}^{M^+} C_{b,g}^+ \begin{pmatrix} E_x^{(g)+} & E_y^{(g)+} & E_z^{(g)+} \\ H_x^{(g)+} & H_y^{(g)+} & H_z^{(g)+} \end{pmatrix} e^{j\left(k_x x + k_y y + k_z^{(g)+}(z-z_-)\right)}
$$

$$
+ \sum_{g=1}^{M^-} C_{b,g}^- \begin{pmatrix} E_x^{(g)-} & E_y^{(g)-} & E_z^{(g)-} \\ H_x^{(g)-} & H_y^{(g)-} & H_z^{(g)-} \end{pmatrix} e^{j\left(k_x x + k_y y + k_z^{(g)-}(z-z_+)\right)} \qquad (2.35b)
$$

3. For $z_+ \leq z$

$$
\begin{pmatrix} \mathbf{E} \\ \mathbf{H} \end{pmatrix} = \begin{pmatrix} \bar{\mathbf{E}}_i \\ \bar{\mathbf{H}}_i \end{pmatrix} + \begin{pmatrix} \bar{\mathbf{E}}_r \\ \bar{\mathbf{H}}_r \end{pmatrix} = \begin{pmatrix} \vec{E}_{i,x} & \vec{E}_{i,y} & \vec{E}_{i,z} \\ \vec{H}_{i,x} & \vec{H}_{i,y} & \vec{H}_{i,z} \end{pmatrix} e^{j(k_x x + k_y y - k_z(z - z_+))}
$$

$$
+ \begin{pmatrix} \vec{E}_{r,x} & \vec{E}_{r,y} & \vec{E}_{r,z} \\ \vec{H}_{r,x} & \vec{H}_{r,y} & \vec{H}_{r,z} \end{pmatrix} e^{j(k_x x + k_y y + k_z(z - z_+))} \qquad\qquad (2.35c)
$$

2.2 Scattering Matrix Analysis of Collinear Multiblock Structures

In this section, the S-matrix analysis of a multiblock structure that is a collinear stack comprised of several single blocks shown in Figure 2.6 is treated in depth. The material parameters for permittivity and permeability (as well as the thickness) of the nth single block in a multiblock structure can be arbitrarily given and denoted by $\underline{\underline{\varepsilon}}^{(n)}$ and $\underline{\underline{\mu}}^{(n)}$, respectively. The complete characterization of the multiblock structure is also performed by bidirectional characterization, as illustrated in Figure 2.6. The objective of the bidirectional characterization of a multiblock structure is to find (1) the S-matrix for the multiblock structure and (2) Bloch eigenmodes with coupling coefficient matrix operators for each block. The S-matrix and coupling coefficient matrix operators provide complete information regarding the multiblock structure. The elemental S-matrices, coupling coefficient matrix operators, and Bloch eigenmodes of single blocks comprising multiblocks are the fundamental raw data in electromagnetic analyses of multiblock structures.

In this section, two types of S-matrix forms are conceptually distinguished for convenience. One is the block S-matrix for the characterization of a finite-sized block. The other is the boundary S-matrix for the characterization of the boundary. Detailed information regarding these terms will be manifested in the following paragraphs.

2.2.1 Two-Block Interconnection

The key algorithm of the analysis of a multiblock structure is the interconnection of two single blocks. It is assumed that two single blocks with elemental S-matrices and coupling coefficient matrix operators for two single blocks can be completely prepared by the analysis described in the previous section. The bidirectional characterization of a cascaded two-block structure can then be algorithmically performed without directly solving the

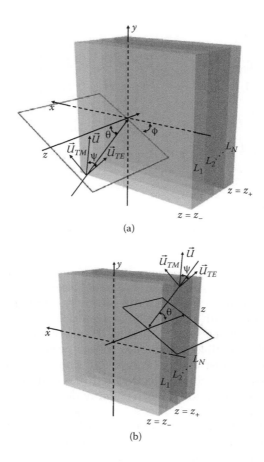

FIGURE 2.6
Bidirectional characterization of a multiblock: (a) left-to-right directional characterization and (b) right-to-left directional characterization.

total structure. The two-block interconnection algorithm based on the elemental S-matrix analysis is illustrated in Figure 2.7. The zero-distance interconnection is an important concept in the S-matrix analysis of a cascaded two-block structure. It indicates that the two-block structure is composed of two independent single blocks with the distance between them being infinitesimally equal to zero, but in the electromagnetic sense, infinite multiple electromagnetic reflections occur between them. The concept of zero-distance interconnection is presented in Figure 2.7, where the two-block interconnection takes the limiting process that the distance between two blocks denoted by d approaches zero in the sense of limitation. In Figure 2.7(a), the left-to-right directional characterization of the interconnected two single blocks is shown. Considering physical infinite multiple reflections and transmissions in the interface, we can construct a rigorous algorithm to represent the total

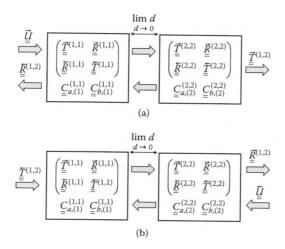

FIGURE 2.7
Bidirectional characterization of cascaded two-block structure: (a) left-to-right directional characterization and (b) right-to-left directional characterization.

reflection and transmission fields by the sum of multiply generated partial reflection and transmission fields for both directional characterizations.

This process can be described by a simple geometric series. Consider two single blocks with finite size and respective S-matrices, $\mathbf{S}^{(1,1)}$ and $\mathbf{S}^{(2,2)}$. The S-matrix of the interconnected structure, $\mathbf{S}^{(1,2)}$, and its coupling coefficient matrix operator is obtained by the following multiple reflection and transmission processes. The first and second components of the superscript indicate the left first block index and right final block index of the multiblock structure, respectively.

The reflection and transmission matrix operators, $\underline{\underline{\tilde{R}}}^{(1,2)}$, $\underline{\underline{\tilde{T}}}^{(1,2)}$, $\underline{\underline{\vec{R}}}^{(1,2)}$, and $\underline{\underline{\vec{T}}}^{(1,2)}$ of $\mathbf{S}^{(1,2)}$ of the multiblocks, are obtained as

$$\underline{\underline{\tilde{R}}}^{(1,2)} = \underline{\underline{\tilde{R}}}^{(1,1)} + \underline{\underline{\tilde{T}}}^{(1,1)} \sum_{k=0}^{\infty} \left(\underline{\underline{\tilde{R}}}^{(2,2)} \underline{\underline{\vec{R}}}^{(1,1)} \right)^{k} \underline{\underline{\tilde{R}}}^{(2,2)} \underline{\underline{\vec{T}}}^{(1,1)} = \underline{\underline{\tilde{R}}}^{(1,1)}$$

$$+ \underline{\underline{\tilde{T}}}^{(1,1)} \left(\underline{\underline{I}} - \underline{\underline{\tilde{R}}}^{(2,2)} \underline{\underline{\vec{R}}}^{(1,1)} \right)^{-1} \underline{\underline{\tilde{R}}}^{(2,2)} \underline{\underline{\vec{T}}}^{(1,1)} \tag{2.36a}$$

$$\underline{\underline{\vec{T}}}^{(1,2)} = \underline{\underline{\vec{T}}}^{(2,2)} \sum_{k=0}^{\infty} \left(\underline{\underline{\vec{R}}}^{(1,1)} \underline{\underline{\tilde{R}}}^{(2,2)} \right)^{k} \underline{\underline{\vec{T}}}^{(1,1)} = \underline{\underline{\vec{T}}}^{(2,2)} \left(\underline{\underline{I}} - \underline{\underline{\vec{R}}}^{(1,1)} \underline{\underline{\tilde{R}}}^{(2,2)} \right)^{-1} \underline{\underline{\vec{T}}}^{(1,1)} \tag{2.36b}$$

$$\underline{\underline{\vec{R}}}^{(1,2)} = \underline{\underline{\vec{R}}}^{(2,2)} + \underline{\underline{\vec{T}}}^{(2,2)} \sum_{k=0}^{\infty} \left(\underline{\underline{\vec{R}}}^{(1,1)} \underline{\underline{\tilde{R}}}^{(2,2)} \right)^{k} \underline{\underline{\vec{R}}}^{(1,1)} \underline{\underline{\tilde{T}}}^{(2,2)}$$

$$= \underline{\underline{\vec{R}}}^{(2,2)} + \underline{\underline{\vec{T}}}^{(2,2)} \left(\underline{\underline{I}} - \underline{\underline{\vec{R}}}^{(1,1)} \underline{\underline{\tilde{R}}}^{(2,2)} \right)^{-1} \underline{\underline{\vec{R}}}^{(1,1)} \underline{\underline{\tilde{T}}}^{(2,2)}, \tag{2.36c}$$

$$\vec{T}^{(1,2)} = \vec{T}^{(1,1)} \sum_{k=0}^{\infty} \left(\vec{R}^{(2,2)} \vec{R}^{(1,1)} \right)^k \vec{T}^{(2,2)} = \vec{T}^{(1,1)} \left(\underline{I} - \vec{R}^{(2,2)} \vec{R}^{(1,1)} \right)^{-1} \vec{T}^{(2,2)} \quad (2.36d)$$

In Equations (2.36a) to (2.36d), the infinite summation terms represent infinite multiple reflections between two blocks, which stably converge to the equations in the second equalities. The zero-distance condition makes no phase retardation factors in the multiple reflection and transmission processes. The stable convergence of the geometric sequence of multiple reflections is confirmed naturally from a physical sense because there cannot be an electromagnetic blowup in this natural process. Consequently, the S-matrix of the interconnected block is given by

$$\mathbf{S}^{(1,2)} = \begin{pmatrix} \vec{T}^{(1,1)} & \vec{R}^{(1,1)} \\ \vec{R}^{(1,1)} & \vec{T}^{(1,1)} \end{pmatrix} \qquad (2.36e)$$

The above-obtained relations are referred to as the Redheffer star product relation of the S-matrix, and the S-matrix of block structure is referred to as a block S-matrix. This relationship is symbolized by the Redheffer star product of the block S-matrices as

$$\mathbf{S}^{(1,2)} = \mathbf{S}^{(1,1)} * \mathbf{S}^{(2,2)} \qquad (2.37)$$

It should be noted that the Redheffer star product has the physical origin on the infinite sum of multiple reflections and transmissions between adjacent multiblocks.

The internal coupling coefficient matrix operators of the combined multiblock, $\mathbf{C}_{a,(1,2)}^{(1,2)}$ and $\mathbf{C}_{b,(1,2)}^{(1,2)}$, are updated when two blocks are interconnected through the infinite multiple reflections. $\mathbf{C}_{a,(1,2)}^{(1,2)}$ and $\mathbf{C}_{b,(1,2)}^{(1,2)}$ represent the set of the coupling coefficient matrix operators of the first and second blocks, which are defined by

$$\mathbf{C}_{a,(1,2)}^{(1,2)} = \left\{ \underline{C}_{a,(1)}^{(1,2)}, \underline{C}_{a,(2)}^{(1,2)} \right\} \qquad (2.38a)$$

$$\mathbf{C}_{b,(1,2)}^{(1,2)} = \left\{ \underline{C}_{b,(1)}^{(1,2)}, \underline{C}_{b,(2)}^{(1,2)} \right\} \qquad (2.38b)$$

where $\{ \underline{C}_{a,(1)}^{(1,2)}, \underline{C}_{b,(1)}^{(1,2)} \}$ and $\{ \underline{C}_{a,(2)}^{(1,2)}, \underline{C}_{b,(2)}^{(1,2)} \}$ are the respective coupling coefficient matrix operators corresponding to the first and second blocks. The superscript and subscript indicate the range of total block and the local block, respectively. The algorithm of the internal coupling coefficients is derived from the notion of infinite multiple reflections as

$$\underline{C}_{a,(1)}^{(1,2)} = \underline{C}_{a,(1)}^{(1,1)} + \underline{C}_{b,(1)}^{(1,1)} \left(\underline{I} - \vec{R}^{(2,2)} \vec{R}^{(1,1)} \right)^{-1} \vec{R}^{(2,2)} \vec{T}^{(1,1)} \qquad (2.39a)$$

$$\underline{\underline{C}}_{b,(1)}^{(1,2)} = \underline{\underline{C}}_{b,(1)}^{(1,1)} \left(\underline{\underline{I}} - \vec{\underline{\underline{R}}}^{(2,2)} \bar{\underline{\underline{R}}}^{(1,1)} \right)^{-1} \bar{\underline{\underline{T}}}^{(2,2)} \qquad (2.39b)$$

$$\underline{\underline{C}}_{a,(2)}^{(1,2)} = \underline{\underline{C}}_{a,(2)}^{(2,2)} \left(\underline{\underline{I}} - \vec{\underline{\underline{R}}}^{(1,1)} \bar{\underline{\underline{R}}}^{(2,2)} \right)^{-1} \vec{\underline{\underline{T}}}^{(1,1)} \qquad (2.39c)$$

$$\underline{\underline{C}}_{b,(2)}^{(1,2)} = \underline{\underline{C}}_{b,(2)}^{(2,2)} + \underline{\underline{C}}_{a,(2)}^{(2,2)} \left(\underline{\underline{I}} - \vec{\underline{\underline{R}}}^{(1,1)} \bar{\underline{\underline{R}}}^{(2,2)} \right)^{-1} \vec{\underline{\underline{R}}}^{(1,1)} \bar{\underline{\underline{T}}}^{(2,2)} \qquad (2.39d)$$

The relationship is symbolized by the star product of the coupling coefficient matrix operator

$$\left(\mathbf{C}_{a,(1,2)}^{(1,2)}, \mathbf{C}_{b,(1,2)}^{(1,2)} \right) = \left(\underline{\underline{C}}_{a,(1)}^{(1,1)}, \underline{\underline{C}}_{b,(1)}^{(1,1)} \right) * \left(\underline{\underline{C}}_{a,(2)}^{(2,2)}, \underline{\underline{C}}_{b,(2)}^{(2,2)} \right) \qquad (2.39e)$$

2.2.2 N-Block Interconnection with Parallelism

The interconnection algorithm of two single blocks has been derived. By the mathematical induction, the S-matrix interconnection can be straightforwardly extended to N-block interconnection. The generalized Redheffer star product of the S-matrix and coupling coefficient matrix operator of the N-block structure can be defined in the recursive sense. The key benefit in the generalization is parallelism. Let us refer to the S-matrix of the finite-sized multiblock structure as block S-matrix.

Before the formulation of the S-matrix method is unfolded, the review of the bidirectional characterization of a partial multiblock is necessary. In Figure 2.8, the bidirectional characterization of a partial multiblock $M^{(n,n+m)}$ composed of consecutive m blocks, $L_n \sim L_{n+m}$, is illustrated, where the superscript $(n, n+m)$ indicates the first and last indices of blocks composing the partial multiblock $M^{(n,n+m)}$. It is assumed that the partial multiblock $M^{(n,n+m)}$ is surrounded by a homogeneous and isotropic medium, i.e., free space, and the thickness of the homogeneous medium is set to zero for the domain decomposition. The bidirectional characterization of the partial multiblock $M^{(n,n+m)}$ is related to obtaining the reflected, transmitted, and internal field distributions excited by two independent incident fields propagating in a left-to-right direction and a right-to-left direction, which are illustrated in Figure 2.8(a) and (b), respectively. The meanings of the other conventions and symbols seen in Figure 2.8 are explained in the following paragraphs.

Let us first consider the left-to-right directional characterization of the partial multiblock $M^{(n,n+m)}$. $l_{n,n+m}$ is the thickness of the partial multiblock $M^{(n,n+m)}$ given by $l_{n,n+m} = d_n + d_{n+1} + \cdots d_{n+m}$, where d_k is the thickness of the kth block. Let the electric and magnetic field representations in the $(n + k)$th block L_{n+k} of the partial multiblock $M^{(n,n+m)}$ be denoted by $\mathbf{E}_{(n+k)}^{(n,n+m)}$ and $\mathbf{H}_{(n+k)}^{(n,n+m)}$, respectively. $l_{1,n+k}$ is given by $l_{1,n+k} = d_1 + d_2 + \cdots d_{n+k}$. The coupling coefficients

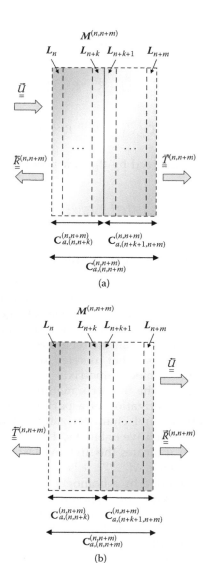

FIGURE 2.8
Bidirectional characterization of a multiblock for obtaining the block S-matrices: (a) left-to-right directional characterization and (b) right-to-left directional characterization.

of the internal eigenmodes excited by the incident plane wave in the $(n + k)$th block L_{n+k} are distinguished and denoted by $\underline{C}^{(n,n+m)}_{(n+k)}$, respectively.

Using the same procedure as was used for the two-block interconnection, the Redheffer star product relation of partial multiblocks can be intuitively derived based on the physical process of infinite multiple reflections between adjacent partial multiblocks as indicated in Figure 2.9. The infinite multiple reflections between two multiblock systems, $M^{(n,n+m)}$ and $M^{(n+m+1,n+m+l)}$, are mathematically described by the geometric series of multiple reflections as

$$\overleftarrow{\underline{\underline{R}}}^{(n,n+m+l)} = \overleftarrow{\underline{\underline{R}}}^{(n,n+m)} + \overleftarrow{\underline{\underline{T}}}^{(n,n+m)} \left(\underline{\underline{I}} - \overleftarrow{\underline{\underline{R}}}^{(n+m+1,n+m+l)} \overrightarrow{\underline{\underline{R}}}^{(n,n+m)} \right)^{-1}$$

$$\times \overleftarrow{\underline{\underline{R}}}^{(n+m+1,n+m+l)} \overrightarrow{\underline{\underline{T}}}^{(n,n+m)} \tag{2.40a}$$

$$\overrightarrow{\underline{\underline{T}}}^{(n,n+m+l)} = \overrightarrow{\underline{\underline{T}}}^{(n+m+1,n+m+l)} \left(\underline{\underline{I}} - \overrightarrow{\underline{\underline{R}}}^{(n,n+m)} \overleftarrow{\underline{\underline{R}}}^{(n+m+1,n+m+l)} \right)^{-1} \overrightarrow{\underline{\underline{T}}}^{(n,n+m)} \tag{2.40b}$$

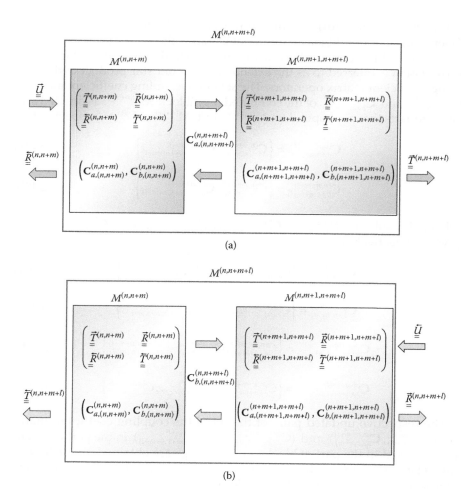

FIGURE 2.9
Derivation of the Redheffer star product relation through (a) the left-to-right directional characterization and (b) the right-to-left directional characterization of the combined multiblock.

$$\underline{\underline{\vec{R}}}^{(n,n+m+l)} = \underline{\underline{\vec{R}}}^{(n+m+1,n+m+l)} + \underline{\underline{\vec{T}}}^{(n+m+1,n+m+l)}\left(\underline{\underline{I}} - \underline{\underline{\vec{R}}}^{(n,n+m)}\underline{\underline{\tilde{R}}}^{(n+m+1,n+m+l)}\right)^{-1}$$

$$\times \underline{\underline{\vec{R}}}^{(n,n+m)}\underline{\underline{\vec{T}}}^{(n+m+1,n+m+l)} \tag{2.40c}$$

$$\underline{\underline{\vec{T}}}^{(n,n+m+l)} = \underline{\underline{\vec{T}}}^{(n,n+m)}\left(\underline{\underline{I}} - \underline{\underline{\tilde{R}}}^{(n+m+1,n+m+l)}\underline{\underline{\vec{R}}}^{(n,n+m)}\right)^{-1}\underline{\underline{\vec{T}}}^{(n+m+1,n+m+l)} \tag{2.40d}$$

This relationship is symbolized by the star product of the block S-matrices as

$$\mathbf{S}^{(n,n+m+l)} = \mathbf{S}^{(n,n+m)} * \mathbf{S}^{(n+m+1,n+m+l)} \tag{2.41}$$

We also see that, during multiple reflections, the internal coupling coefficient matrices of all blocks are updated, since the whole process is linear and the coupling coefficient is linearly proportional to the external field. The formulation is referred to as the extended Redheffer star product of the coupling coefficient matrix operators. The set of the coupling coefficient matrices $\mathbf{C}_{a,(n,n+m+l)}^{(n,n+m+l)}$ and $\mathbf{C}_{b,(n,n+m+l)}^{(n,n+m+l)}$ of the combined multiblock $M^{(n,n+m+l)}$ is divided into two subparts that correspond to $M^{(n,n+m)}$ and $M^{(n+m+1,n+m+l)}$ as, respectively,

$$\mathbf{C}_{a,(n,n+m+l)}^{(n,n+m+l)} = \left\{\mathbf{C}_{a,(n,n+m)}^{(n,n+m+l)}, \mathbf{C}_{a,(n+m+1,n+m+l)}^{(n,n+m+l)}\right\} \tag{2.42a}$$

$$\mathbf{C}_{b,(n,n+m+l)}^{(n,n+m+l)} = \left\{\mathbf{C}_{b,(n,n+m)}^{(n,n+m+l)}, \mathbf{C}_{b,(n+m+1,n+m+l)}^{(n,n+m+l)}\right\} \tag{2.42b}$$

where the subsets $\mathbf{C}_{a,(n,n+m)}^{(n,n+m+l)}$, $\mathbf{C}_{b,(n,n+m)}^{(n,n+m+l)}$, $\mathbf{C}_{a,(n+m+1,n+m+l)}^{(n,n+m+l)}$, and $\mathbf{C}_{b,(n+m+1,n+m+l)}^{(n,n+m+l)}$ are given, respectively, by

$$\mathbf{C}_{a,(n,n+m)}^{(n,n+m+l)} = \left\{\underline{\underline{C}}_{a,(n)}^{(n,n+m+l)}, \underline{\underline{C}}_{a,(n+1)}^{(n,n+m+l)}, \dots, \underline{\underline{C}}_{a,(n+m)}^{(n,n+m+l)}\right\} \tag{2.42c}$$

$$\mathbf{C}_{b,(n,n+m)}^{(n,n+m+l)} = \left\{\underline{\underline{C}}_{b,(n)}^{(n,n+m+l)}, \underline{\underline{C}}_{b,(n+1)}^{(n,n+m+l)}, \dots, \underline{\underline{C}}_{b,(n+m)}^{(n,n+m+l)}\right\} \tag{2.42d}$$

$$\mathbf{C}_{a,(n+m+1,n+m+l)}^{(n,n+m+l)} = \left\{\underline{\underline{C}}_{a,(n+m+1)}^{(n,n+m+l)}, \underline{\underline{C}}_{a,(n+m+2)}^{(n,n+m+l)}, \dots, \underline{\underline{C}}_{a,(n+m+l)}^{(n,n+m+l)}\right\} \tag{2.42e}$$

$$\mathbf{C}_{b,(n+m+1,n+m+l)}^{(n,n+m+l)} = \left\{\underline{\underline{C}}_{b,(n+m+1)}^{(n,n+m+l)}, \underline{\underline{C}}_{b,(n+m+2)}^{(n,n+m+l)}, \dots, \underline{\underline{C}}_{b,(n+m+l)}^{(n,n+m+l)}\right\} \tag{2.42f}$$

The respective updated pairs of the set of the coupling coefficient matrices $(\mathbf{C}_{a,(n,n+m)}^{(n,n+m+l)}, \mathbf{C}_{b,(n,n+m)}^{(n,n+m+l)})$ and $(\mathbf{C}_{a,(n+m+1,n+m+l)}^{(n,n+m+l)}, \mathbf{C}_{b,(n+m+1,n+m+l)}^{(n,n+m+l)})$ in the part of blocks $L_n - L_{n+m}$ and in the part of blocks $L_{n+m+1} - L_{n+m+l}$ are obtained by the following relations. For k in the range of $n \le k \le n+m$, the coupling coefficient matrices, $\underline{\underline{C}}_{a,(k)}^{(n,n+m+l)}$ and $\underline{\underline{C}}_{b,(k)}^{(n,n+m+l)}$, are derived as

$$\underline{\underline{C}}_{a,(k)}^{(n,n+m+l)} = \underline{\underline{C}}_{a,(k)}^{(n,n+m)} + \underline{\underline{C}}_{b,(k)}^{(n,n+m)}\left(\underline{\underline{I}} - \underline{\underline{\tilde{R}}}^{(n+m+1,n+m+l)}\underline{\underline{\vec{R}}}^{(n,n+m)}\right)^{-1}\underline{\underline{\tilde{R}}}^{(n+m+1,n+m+l)}\underline{\underline{\vec{T}}}^{(n,n+m)} \tag{2.43a}$$

$$\underline{\underline{C}}_{b,(k)}^{(n,n+m+l)} = \underline{\underline{C}}_{b,(k)}^{(n,n+m)}\left(\underline{\underline{I}} - \vec{\underline{\underline{R}}}^{(n+m+1,n+m+l)}\tilde{\underline{\underline{R}}}^{(n,n+m)}\right)^{-1}\vec{\underline{\underline{T}}}^{(n+m+1,n+m+l)} \qquad (2.43b)$$

For k in the range of $n+m+1 \le k \le n+m+l$, the coupling coefficient matrices, $\underline{\underline{C}}_{a,(k)}^{(n,n+m+l)}$ and $\underline{\underline{C}}_{b,(k)}^{(n,n+m+l)}$, are derived as

$$\underline{\underline{C}}_{a,(k)}^{(n,n+m+l)} = \underline{\underline{C}}_{a,(k)}^{(n+m+1,n+m+l)}\left(\underline{\underline{I}} - \vec{\underline{\underline{R}}}^{(n,n+m)}\tilde{\underline{\underline{R}}}^{(n+m+1,n+m+l)}\right)^{-1}\vec{\underline{\underline{T}}}^{(n,n+m)} \qquad (2.43c)$$

$$\underline{\underline{C}}_{b,(k)}^{(n,n+m+l)} = \underline{\underline{C}}_{b,(k)}^{(n+m+1,n+m+l)} + \underline{\underline{C}}_{a,(k)}^{(n+m+1,n+m+l)}\left(\underline{\underline{I}} - \vec{\underline{\underline{R}}}^{(n,n+m)}\tilde{\underline{\underline{R}}}^{(n+m+1,n+m+l)}\right)^{-1}$$

$$\times \vec{\underline{\underline{R}}}^{(n,n+m)}\tilde{\underline{\underline{T}}}^{(n+m+1,n+m+l)} \qquad (2.43d)$$

These relationships can be symbolized by the extended star product as

$$\left(\mathbf{C}_{a,(n,n+m+l)}^{(n,n+m+l)}, \mathbf{C}_{b,(n,n+m+l)}^{(n,n+m+l)}\right) = \left(\mathbf{C}_{a,(n,n+m)}^{(n,n+m)}, \mathbf{C}_{b,(n,n+m)}^{(n,n+m)}\right) * \left(\mathbf{C}_{a,(n+m+1,n+m+l)}^{(n+m+1,n+m+l)}, \mathbf{C}_{b,(n+m+1,n+m+l)}^{(n+m+1,n+m+l)}\right)$$

$$(2.44)$$

In addition, considering the physical process of infinite multiple reflections and transmissions between adjacent multiblocks, it can be seen that associative rules definitely exist in the derived star product relation of the coupling coefficients as well as in the star product relation of the S-matrix components. These associative rules enable the parallel computation of the internal coupling coefficients.

Consequently, with the aid of the associative rule, the block S-matrix and the coupling coefficient matrix of the multiblock $M^{(1,N)}$ shown in Figure 2.10(a) can be obtained by

$$\mathbf{S}^{(1,N)} = \mathbf{S}^{(1,1)} * \mathbf{S}^{(2,2)} * \cdots * \mathbf{S}^{(N-1,N-1)} * \mathbf{S}^{(N,N)} \qquad (2.45a)$$

$$\left(\mathbf{C}_{a,(1,N)}^{(1,N)}, \mathbf{C}_{b,(1,N)}^{(1,N)}\right) = \left(\underline{\underline{C}}_{a,(1)}^{(1,1)}, \underline{\underline{C}}_{b,(1)}^{(1,1)}\right) * \left(\underline{\underline{C}}_{a,(2)}^{(2,2)}, \underline{\underline{C}}_{b,(2)}^{(2,2)}\right) * \cdots * \left(\underline{\underline{C}}_{a,(N-1)}^{(N-1,N-1)}, \underline{\underline{C}}_{b,(N-1)}^{(N-1,N-1)}\right)$$

$$* \left(\underline{\underline{C}}_{a,(N)}^{(N,N)}, \underline{\underline{C}}_{b,(N)}^{(N,N)}\right) \qquad (2.45b)$$

The complete associative relation of the block S-matrix established by the proposed additional recursion relations of the coupling coefficient operators is the cornerstone for constructing the generalized S-matrix method describing general nanophotonic networks, which will be addressed in Chapter 6.

2.2.3 Half-Infinite Block Interconnection

In the previous section, we analyzed and performed analysis of the so-called block S-matrix of a finite-sized multiblock body. The finite-sized multiblock structure is surrounded by two half-infinite spaces, which are

Multi-block

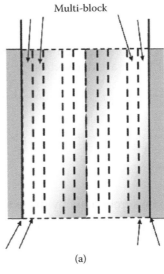

(a)

$$\mathbf{S}^{(1,N)} = \begin{pmatrix} \underline{\underline{T}}^{(1,N)} & \underline{\underline{R}}^{(1,N)} \\ \underline{\underline{R}}^{(1,N)} & \underline{\underline{T}}^{(1,N)} \end{pmatrix}$$

$$\mathbf{C}_{a,(1,N)}^{(1,N)} = \left\{ \underline{\underline{C}}_{a,(1)}^{(1,N)}, \ \underline{\underline{C}}_{a,(2)}^{(1,N)}, \cdots, \underline{\underline{C}}_{a,(N)}^{(1,N)} \right\}$$

$$\mathbf{C}_{b,(1,N)}^{(1,N)} = \left\{ \underline{\underline{C}}_{b,(1)}^{(1,N)}, \ \underline{\underline{C}}_{b,(2)}^{(1,N)}, \cdots, \underline{\underline{C}}_{b,(N)}^{(1,N)} \right\}$$

(b)

FIGURE 2.10
(a) Multiblock $M^{(1,N)}$ and (b) the multiblock S-matrix and coupling coefficient matrix operator.

referred to as the left half-infinite block and right half-infinite block. The left half-infinite space and right half-infinite space are considered to be final building blocks of the general multiblock structure. In this section, the boundary S-matrix and half-infinite block interconnection are addressed, which is the final element of completing the S-matrix method of a general 1D multiblock structure.

In Figure 2.11, the general multibock structure and its construction process are presented conceptually. The n h block and n th boundary are denoted by L_n and B_n, respectively. The total structure is composed of lefthand half-infinite region L_0, multiblock body $M^{(1,N)}$, and righthand half-infinite region L_{N+1}.

Let us define the boundary S-matrix as the reflection and transmission field coefficient operators at the boundaries B_0 and B_N of the multiblock

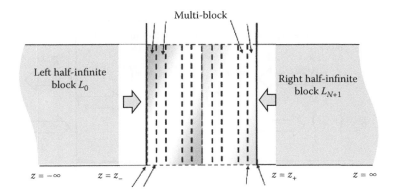

FIGURE 2.11
Process of half-infinite block interconnection.

structure shown in Figure 2.12(a). The bidirectional characterization of the boundaries B_0 and B_N for obtaining the boundary S-matrices is illustrated in Figure 2.12(a) and (b), respectively. As mentioned previously, the input and output blocks, L_0 and L_{N+1}, are actually half-infinite blocks.

The boundary S-matrices must be manifested to connect these half-infinite blocks to the finite body of the multiblock $M^{(1,N)}$. If two boundary S-matrices are prepared, we can use the star product to connect these half-infinite blocks to the finite body of multiblock. The boundary S-matrix of the left half-infinite block is given by

$$\mathbf{S}^{(0,0)} = \begin{pmatrix} \vec{\underline{T}}^{(0,0)} & \vec{\underline{R}}^{(0,0)} \\ \overleftarrow{\underline{R}}^{(0,0)} & \overleftarrow{\underline{T}}^{(0,0)} \end{pmatrix} \tag{2.46}$$

The boundary conditions at $z = z_c$ for left-to-right and right-to-left directional characterizations are described, respectively, as

$$\begin{pmatrix} \underline{W}^{(0)+}(z_c) & \underline{W}^{(0)-}(z_c) \\ \underline{V}^{(0)+}(z_c) & \underline{V}^{(0)-}(z_c) \end{pmatrix} \begin{pmatrix} \vec{\underline{U}} \\ \vec{\underline{R}}^{(0,0)} \end{pmatrix} = \begin{pmatrix} \underline{W}_h & \underline{W}_h \\ \underline{V}_h & -\underline{V}_h \end{pmatrix} \begin{pmatrix} \vec{\underline{T}}^{(0,0)} \\ \underline{0} \end{pmatrix} \tag{2.47a}$$

$$\begin{pmatrix} \underline{W}^{(0)+}(z_c) & \underline{W}^{(0)-}(z_c) \\ \underline{V}^{(0)+}(z_c) & \underline{V}^{(0)-}(z_c) \end{pmatrix} \begin{pmatrix} \underline{0} \\ \overleftarrow{\underline{T}}^{(0,0)} \end{pmatrix} = \begin{pmatrix} \underline{W}_h & \underline{W}_h \\ \underline{V}_h & -\underline{V}_h \end{pmatrix} \begin{pmatrix} \overleftarrow{\underline{R}}^{(0,0)} \\ \overleftarrow{\underline{U}} \end{pmatrix} \tag{2.47b}$$

The functions $\underline{W}_h$, $\underline{V}_h$, $\underline{W}^{(0)+}(z_c)$, $\underline{W}^{(0)-}(z_c)$, $\underline{V}^{(0)+}(z_c)$, and $\underline{V}^{(0)-}(z_c)$ are defined in Equations (2.24c) to (2.24h). By solving Equations (2.47a) and

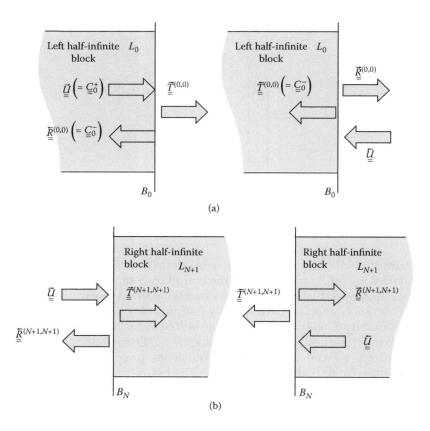

FIGURE 2.12
Bidirectional characterization of (a) the boundary B_0 and (b) the boundary B_N.

(2.47b), we have the S-matrix elements, $\vec{\underline{\underline{T}}}^{(0,0)}$, $\vec{\underline{\underline{R}}}^{(0,0)}$, $\overleftarrow{\underline{\underline{T}}}^{(0,0)}$, and $\overleftarrow{\underline{\underline{R}}}^{(0,0)}$, respectively, as

$$\vec{\underline{\underline{R}}}^{(0,0)} = -\left[\left(\underline{\underline{W}}_h\right)^{-1}\underline{\underline{W}}^{(0)-}(z_c) - \left(\underline{\underline{V}}_h\right)^{-1}\underline{\underline{V}}^{(0)-}(z_c)\right]^{-1}\left[\left(\underline{\underline{W}}_h\right)^{-1}\underline{\underline{W}}^{(0)+}(z_c) - \left(\underline{\underline{V}}_h\right)^{-1}\underline{\underline{V}}^{(0)+}(z_c)\right]$$

(2.47c)

$$\vec{\underline{\underline{T}}}^{(0,0)} = \left[\left(\underline{\underline{W}}^{(0)-}(z_c)\right)^{-1}\underline{\underline{W}}_h - \left(\underline{\underline{V}}^{(0)-}(z_c)\right)^{-1}\underline{\underline{V}}_h\right]^{-1}$$

$$\times \left[\left(\underline{\underline{W}}^{(0)-}(z_c)\right)^{-1}\underline{\underline{W}}^{(0)+}(z_c) - \left(\underline{\underline{V}}^{(0)-}(z_c)\right)^{-1}\underline{\underline{V}}^{(0)+}(z_c)\right]$$

(2.47d)

$$\overleftarrow{\underline{\underline{T}}}^{(0,0)} = 2\left[\left(\underline{\underline{W}}_h\right)^{-1}\underline{\underline{W}}^{(0)-}(z_c) - \left(\underline{\underline{V}}_h\right)^{-1}\underline{\underline{V}}^{(0)-}(z_c)\right]^{-1}$$

(2.47e)

$$\vec{\underline{R}}^{(0,0)} = -\left[\left(\underline{W}^{(0)-}(z_c)\right)^{-1}\underline{W}_h - \left(\underline{V}^{(0)-}(z_c)\right)^{-1}\underline{V}_h\right]^{-1}\left[\left(\underline{W}^{(0)-}(z_c)\right)^{-1}\underline{W}_h + \left(\underline{V}^{(0)-}(z_c)\right)^{-1}\underline{V}_h\right]$$

(2.47f)

The boundary S-matrix of the right half-infinite block is given by

$$\mathbf{S}^{(N+1,N+1)} = \begin{pmatrix} \vec{\underline{T}}^{(N+1,N+1)} & \vec{\underline{R}}^{(N+1,N+1)} \\ \underleftarrow{\underline{R}}^{(N+1,N+1)} & \underleftarrow{\underline{T}}^{(N+1,N+1)} \end{pmatrix}$$

(2.48)

The boundary condition at $z = z_c$ for the left-to-right and right-to-left directional characterizations is described, respectively, as

$$\begin{pmatrix} \underline{W}_h & \underline{W}_h \\ \underline{V}_h & -\underline{V}_h \end{pmatrix}\begin{pmatrix} \vec{\underline{U}} \\ \underleftarrow{\underline{R}}^{(N+1,N+1)} \end{pmatrix} = \begin{pmatrix} \underline{W}^{(N+1)+}(z_c) & \underline{W}^{(N+1)-}(z_c) \\ \underline{V}^{(N+1)+}(z_c) & \underline{V}^{(N+1)-}(z_c) \end{pmatrix}\begin{pmatrix} \vec{\underline{T}}^{(N+1,N+1)} \\ \underline{0} \end{pmatrix}$$

(2.49a)

$$\begin{pmatrix} \underline{W}_h & \underline{W}_h \\ \underline{V}_h & -\underline{V}_h \end{pmatrix}\begin{pmatrix} \underline{0} \\ \underleftarrow{\underline{T}}^{(N+1,N+1)} \end{pmatrix} = \begin{pmatrix} \underline{W}^{(N+1)+}(z_c) & \underline{W}^{(N+1)-}(z_c) \\ \underline{V}^{(N+1)+}(z_c) & \underline{V}^{(N+1)-}(z_c) \end{pmatrix}\begin{pmatrix} \vec{\underline{R}}^{(N+1,N+1)} \\ \underleftarrow{\underline{U}} \end{pmatrix}$$

(2.49b)

The S-matrix elements, $\vec{\underline{T}}^{(N+1,N+1)}$, $\vec{\underline{R}}^{(N+1,N+1)}$, $\underleftarrow{\underline{T}}^{(N+1,N+1)}$, and $\underleftarrow{\underline{R}}^{(N+1,N+1)}$, are given, respectively, as

$$\vec{\underline{T}}^{(N+1,N+1)} = 2\left[\left(\underline{W}_h\right)^{-1}\underline{W}^{(N+1)+}(z_c) + \left(\underline{V}_h\right)^{-1}\underline{V}^{(N+1)+}(z_c)\right]^{-1}$$

(2.49c)

$$\vec{\underline{R}}^{(N+1,N+1)} = -\left[\left(\underline{W}^{(N+1)+}(z_c)\right)^{-1}\underline{W}_h + \left(\underline{V}^{(N+1)+}(z_c)\right)^{-1}\underline{V}_h\right]^{-1}$$
$$\times\left[\left(\underline{W}^{(N+1)+}(z_c)\right)^{-1}\underline{W}_h - \left(\underline{V}^{(N+1)+}(z_c)\right)^{-1}\underline{V}_h\right]$$

(2.49d)

$$\underleftarrow{\underline{T}}^{(N+1,N+1)} = \left[\left(\underline{W}^{(N+1)+}(z_c)\right)^{-1}\underline{W}_h + \left(\underline{V}^{(N+1)+}(z_c)\right)^{-1}\underline{V}_h\right]^{-1}$$
$$\times\left[\left(\underline{W}^{(N+1)+}(z_c)\right)^{-1}\underline{W}^{(N+1)-}(z_c) - \left(\underline{V}^{(N+1)+}(z_c)\right)^{-1}\underline{V}^{(N+1)-}(z_c)\right]$$

(2.49e)

$$\underleftarrow{\underline{R}}^{(N+1,N+1)} = -\left[\left(\underline{W}_h\right)^{-1}\underline{W}^{(N+1)+}(z_c) + \left(\underline{V}_h\right)^{-1}\underline{V}^{(N+1)+}(z_c)\right]^{-1}$$
$$\times\left[\left(\underline{W}_h\right)^{-1}\underline{W}^{(N+1)-}(z_c) + \left(\underline{V}_h\right)^{-1}\underline{V}^{(N+1)-}(z_c)\right]$$

(2.49f)

With the above results, it is now possible to construct the total S-matrix of the whole multiblock $M^{(0,N+1)}$. At the first step, the S-matrix $\mathbf{S}^{(0,N)}$ of the multiblock $M^{(0,N)}$ is derived by the Redheffer star product of $\mathbf{S}^{(0,0)}$ of Equation (2.46) and $\mathbf{S}^{(1,N)}$ of Equation (2.45) as

$$\mathbf{S}^{(0,N)} = \mathbf{S}^{(0,0)} * \mathbf{S}^{(1,N)} \tag{2.50a}$$

The coupling coefficient matrices $\underline{C}_{a,(k)}^{(0,N)}$ and $\underline{C}_{b,(k)}^{(0,N)}$ ($1 \le k \le N$) of the blocks $L_1 \sim L_N$ in $M^{(0,N)}$ are obtained by, from Equations (2.43c) and (2.43d),

$$\left(\underline{C}_{a,(k)}^{(0,N)}, \underline{C}_{b,(k)}^{(0,N)} \right)$$

$$= \left(\underline{C}_{a,(k)}^{(1,N)} \left(\underline{I} - \vec{\underline{R}}^{(0,0)} \tilde{\underline{R}}^{(1,N)} \right)^{-1} \vec{\underline{T}}^{(0,0)}, \underline{C}_{b,(k)}^{(1,N)} + \underline{C}_{a,(k)}^{(1,N)} \left(\underline{I} - \vec{\underline{R}}^{(0,0)} \tilde{\underline{R}}^{(1,N)} \right)^{-1} \vec{\underline{R}}^{(0,0)} \tilde{\underline{T}}^{(1,N)} \right)$$

$$\tag{2.50b}$$

Next, the total S-matrix $\mathbf{S}^{(0,N+1)}$ of the multiblock $M^{(0,N+1)}$ is obtained by the Redheffer star product of $\mathbf{S}^{(0,N)}$ of Equation (2.50a) and $\mathbf{S}^{(N+1,N+1)}$ of Equation (2.48) as

$$\mathbf{S}^{(0,N+1)} = \mathbf{S}^{(0,N)} * \mathbf{S}^{(N+1,N+1)} \tag{2.51a}$$

The final coupling coefficient matrix $\underline{C}_{a,(k)}^{(0,N+1)}$ and $\underline{C}_{b,(k)}^{(0,N+1)}$ ($1 \le k \le N$) of the blocks $L_1 - L_N$ in $M^{(0,N+1)}$ is obtained from Equations (2.43a) and (2.43b):

$$\left(\underline{C}_{a,(k)}^{(0,N+1)}, \underline{C}_{b,(k)}^{(0,N+1)} \right) = \left(\underline{C}_{a,(k)}^{(0,N)} + \underline{C}_{b,(k)}^{(0,N)} \left(\underline{I} - \tilde{\underline{R}}^{(N+1,N+1)} \vec{\underline{R}}^{(0,N)} \right)^{-1} \tilde{\underline{R}}^{(N+1,N+1)} \vec{\underline{T}}^{(0,N+1)}, \right.$$

$$\left. \underline{C}_{b,(k)}^{(0,N)} \left(\underline{I} - \tilde{\underline{R}}^{(N+1,N+1)} \vec{\underline{R}}^{(0,N)} \right)^{-1} \tilde{\underline{T}}^{(N+1,N+1)} \right). \tag{2.51b}$$

The S-matrix $\mathbf{S}^{(0,N+1)}$ and the coupling coefficient matrices, $\underline{C}_{a,(k)}^{(0,N+1)}$ and $\underline{C}_{b,(k)}^{(0,N+1)}$, ($1 \le k \le N$) of the blocks $L_1 - L_N$ in $M^{(0,N+1)}$, provide the complete characterization of the multiblock $M^{(0,N+1)}$.

Figure 2.13 shows a general multiblock structure with $N + 2$ blocks and $N + 1$ boundaries between adjacent blocks and the symbolic expression of total S-matrix and total coupling coefficient matrix operators that have been produced consequently.

2.2.4 Field Visualization

The total electromagnetic field distribution is calculated separately for left-to-right directional field and right-to-left directional field excitations.

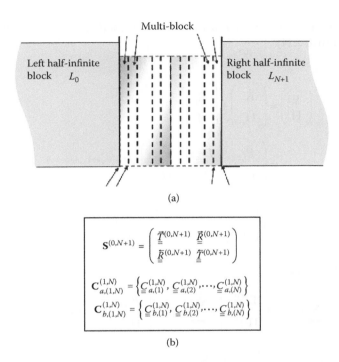

FIGURE 2.13

(a) Multiblock structure. (b) S-matrix and coupling.

2.2.4.1 Left-to-Right Field Visualization

When the incident field, $\vec{E}_i$, in the left half-infinite block impinges on the multiblock structure $M^{(1, N)}$ at the boundary B_0, the reflection field coefficient inside the left half-infinite block and the transmission field coefficient inside the right half-infinite block are given, respectively, as

$$\begin{pmatrix} \vec{E}_{r,y} \\ \vec{E}_{r,x} \end{pmatrix} = \underline{\underline{\vec{R}}}^{(0,N+1)} \begin{pmatrix} \vec{E}_{i,y} \\ \vec{E}_{i,x} \end{pmatrix} \tag{2.52a}$$

$$\begin{pmatrix} \vec{E}_{t,y} \\ \vec{E}_{t,x} \end{pmatrix} = \underline{\underline{\vec{T}}}^{(0,N+1)} \begin{pmatrix} \vec{E}_{i,y} \\ \vec{E}_{i,x} \end{pmatrix} \tag{2.52b}$$

And the coupling coefficients of the internal optical field distribution are represented by

$$\left\{ \underline{C}^{(1,N)}_{a,(1)}, \underline{C}^{(1,N)}_{a,(2)}, \dots, \underline{C}^{(1,N)}_{a,(N)} \right\} = \left\{ \underline{C}^{(1,N)}_{a,(1)} \begin{pmatrix} \vec{E}_{i,y} \\ \vec{E}_{i,x} \end{pmatrix}, \underline{C}^{(1,N)}_{a,(2)} \begin{pmatrix} \vec{E}_{i,y} \\ \vec{E}_{i,x} \end{pmatrix}, \dots, \underline{C}^{(1,N)}_{a,(N)} \begin{pmatrix} \vec{E}_{i,y} \\ \vec{E}_{i,x} \end{pmatrix} \right\} \tag{2.52c}$$

The total field distribution **E** and **H** are obtained in the respective spatial regions (Figure 2.14).

1. For $z < z_-$

$$
\begin{pmatrix} \mathbf{E} \\ \mathbf{H} \end{pmatrix} = \begin{pmatrix} \vec{\mathbf{E}}_i \\ \vec{\mathbf{H}}_i \end{pmatrix} + \begin{pmatrix} \overleftarrow{\mathbf{E}}_r \\ \overleftarrow{\mathbf{H}}_r \end{pmatrix}
$$

$$
= \begin{pmatrix} \vec{E}_{i,x} & \vec{E}_{i,y} & \vec{E}_{i,z} \\ \vec{H}_{i,x} & \vec{H}_{i,y} & \vec{H}_{i,z} \end{pmatrix} e^{j(k_x x + k_y y + k_{z,(0)}(z-z_-))}
$$

$$
+ \begin{pmatrix} \overleftarrow{E}_{r,x} & \overleftarrow{E}_{r,y} & \overleftarrow{E}_{r,z} \\ \overleftarrow{H}_{r,x} & \overleftarrow{H}_{r,y} & \overleftarrow{H}_{r,z} \end{pmatrix} e^{j(k_x x + k_y y - k_{z,(0)}(z-z_-))} \tag{2.53a}
$$

2. For $z_- \leq z < z_+$

$$
\begin{pmatrix} \mathbf{E} \\ \mathbf{H} \end{pmatrix} = \begin{cases} \displaystyle\sum_{g=1}^{M^+} C_{a,(1),g}^+ \begin{pmatrix} \mathbf{E}_{(1)}^{(g)+} \\ \mathbf{H}_{(1)}^{(g)+} \end{pmatrix} + \sum_{g=1}^{M^-} C_{a,(1),g}^- \begin{pmatrix} \mathbf{E}_{(1)}^{(g)-} \\ \mathbf{H}_{(1)}^{(g)-} \end{pmatrix} & \text{for } 0 \leq z \leq l_{1,1} \\[3em] \displaystyle\sum_{g=1}^{M^+} C_{a,(2),g}^+ \begin{pmatrix} \mathbf{E}_{(2)}^{(g)+} \\ \mathbf{H}_{(2)}^{(g)+} \end{pmatrix} + \sum_{g=1}^{M^-} C_{a,(2),g}^- \begin{pmatrix} \mathbf{E}_{(2)}^{(g)-} \\ \mathbf{H}_{(2)}^{(g)-} \end{pmatrix} & \text{for } l_{1,1} \leq z \leq l_{1,2} \\[2em] \qquad\qquad\qquad\qquad \vdots \\[1em] \displaystyle\sum_{g=1}^{M^+} C_{a,(N),g}^+ \begin{pmatrix} \mathbf{E}_{(N)}^{(g)+} \\ \mathbf{H}_{(N)}^{(g)+} \end{pmatrix} + \sum_{g=1}^{M^-} C_{a,(N),g}^- \begin{pmatrix} \mathbf{E}_{(N)}^{(g)-} \\ \mathbf{H}_{(N)}^{(g)-} \end{pmatrix} & \text{for } l_{1,N-1} \leq z \leq l_{1,N} \end{cases} \tag{2.53b}
$$

3. For $z_+ \leq z$

$$
\begin{pmatrix} \mathbf{E} \\ \mathbf{H} \end{pmatrix} = \begin{pmatrix} \vec{\mathbf{E}}_t \\ \vec{\mathbf{H}}_t \end{pmatrix} = \begin{pmatrix} \vec{E}_{t,x} & \vec{E}_{t,y} & \vec{E}_{t,z} \\ \vec{H}_{t,x} & \vec{H}_{t,y} & \vec{H}_{t,z} \end{pmatrix} e^{j(k_x x + k_y y + k_{z,(N+1)}(z-z_+))} \tag{2.53c}
$$

2.2.4.2 Right-to-Left Field Visualization

When the incident field, $\overleftarrow{\mathbf{E}}_i$, in the left half-infinite block impinges on the multiblock structure $M^{(1,N)}$ at the boundary B_N, the reflection field coefficient inside the left half-infinite block and the transmission field coefficient inside the right half-infinite block are given, respectively, as

$$
\begin{pmatrix} \overleftarrow{E}_{r,y} \\ \overleftarrow{E}_{r,x} \end{pmatrix} = \vec{\underline{\underline{R}}}^{(0,N+1)} \begin{pmatrix} \overleftarrow{E}_{i,y} \\ \overleftarrow{E}_{i,x} \end{pmatrix} \tag{2.54a}
$$

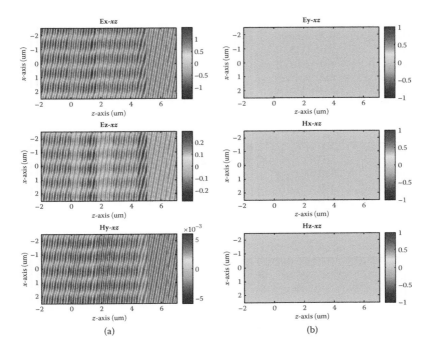

FIGURE 2.14
Electric and magnetic field distributions of a TM mode plane wave obliquely incident on a finite-sized dielectric multiblock slab with the thickness of 2 μm along the left-to-right direction. (MATLAB code: *FMM_1D_SMM.m*.)

$$\begin{pmatrix} \tilde{E}_{t,y} \\ \tilde{E}_{t,x} \end{pmatrix} = \tilde{\tilde{T}}^{(0,N+1)} \begin{pmatrix} \tilde{E}_{i,y} \\ \tilde{E}_{i,x} \end{pmatrix} \tag{2.54b}$$

And the coupling coefficients of the internal optical field distribution are represented by

$$\left\{ \underline{C}_{b,(1)}^{(1,N)}, \underline{C}_{b,(2)}^{(1,N)}, \cdots, \underline{C}_{b,(N)}^{(1,N)} \right\} = \left\{ \underline{\underline{C}}_{b,(1)}^{(1,N)} \begin{pmatrix} \tilde{E}_{i,y} \\ \tilde{E}_{i,x} \end{pmatrix}, \underline{\underline{C}}_{b,(2)}^{(1,N)} \begin{pmatrix} \tilde{E}_{i,y} \\ \tilde{E}_{i,x} \end{pmatrix}, \cdots, \underline{\underline{C}}_{b,(N)}^{(1,N)} \begin{pmatrix} \tilde{E}_{i,y} \\ \tilde{E}_{i,x} \end{pmatrix} \right\} \tag{2.54c}$$

The total field distribution **E** and **H** are obtained in the respective spatial regions (Figure 2.15).

1. For $z < z_-$

$$\begin{pmatrix} \mathbf{E} \\ \mathbf{H} \end{pmatrix} = \begin{pmatrix} \tilde{\mathbf{E}}_t \\ \tilde{\mathbf{H}}_t \end{pmatrix} = \begin{pmatrix} \tilde{E}_{t,x} & \tilde{E}_{t,y} & \tilde{E}_{t,z} \\ \tilde{H}_{t,x} & \tilde{H}_{t,y} & \tilde{H}_{t,z} \end{pmatrix} e^{j(k_x x + k_y y - k_{z,(0)}(z-z_-))} \tag{2.55a}$$

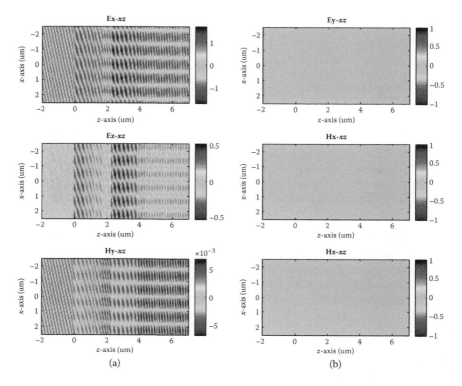

FIGURE 2.15
Electric and magnetic field distributions of a TM mode plane wave obliquely incident on a finite-sized dielectric multiblock slab with the thickness of 2 µm along the right-to-left direction. (MATLAB code: *FMM_1D_SMM.m*.)

2. For $z_- \leq z < z_+$

$$
\begin{pmatrix} \mathbf{E} \\ \mathbf{H} \end{pmatrix} = \begin{cases} \displaystyle\sum_{g=1}^{M^+} C^+_{b,(1),g} \begin{pmatrix} \mathbf{E}^{(g)+}_{(1)} \\ \mathbf{H}^{(g)+}_{(1)} \end{pmatrix} + \sum_{g=1}^{M^-} C^-_{b,(1),g} \begin{pmatrix} \mathbf{E}^{(g)-}_{(1)} \\ \mathbf{H}^{(g)-}_{(1)} \end{pmatrix} & \text{for } 0 \leq z \leq l_{1,1} \\[3em] \displaystyle\sum_{g=1}^{M^+} C^+_{b,(2),g} \begin{pmatrix} \mathbf{E}^{(g)+}_{(2)} \\ \mathbf{H}^{(g)+}_{(2)} \end{pmatrix} + \sum_{g=1}^{M^-} C^-_{b,(2),g} \begin{pmatrix} \mathbf{E}^{(g)-}_{(2)} \\ \mathbf{H}^{(g)-}_{(2)} \end{pmatrix} & \text{for } l_{1,1} \leq z \leq l_{1,2} \\[2em] \qquad\qquad\qquad\qquad\qquad \vdots \\[2em] \displaystyle\sum_{g=1}^{M^+} C^+_{b,(N),g} \begin{pmatrix} \mathbf{E}^{(g)+}_{(N)} \\ \mathbf{H}^{(g)+}_{(N)} \end{pmatrix} + \sum_{g=1}^{M^-} C^-_{b,(N),g} \begin{pmatrix} \mathbf{E}^{(g)-}_{(N)} \\ \mathbf{H}^{(g)-}_{(N)} \end{pmatrix} & \text{for } l_{1,N-1} \leq z \leq l_{1,N} \end{cases}
$$

$$(2.55b)$$

3. For $z_+ \leq z$

$$
\begin{pmatrix} \mathbf{E} \\ \mathbf{H} \end{pmatrix} = \begin{pmatrix} \vec{\bar{E}}_i \\ \vec{\bar{H}}_i \end{pmatrix} + \begin{pmatrix} \vec{\bar{E}}_r \\ \vec{\bar{H}}_r \end{pmatrix}
$$

$$
= \begin{pmatrix} \bar{E}_{i,x} & \bar{E}_{i,y} & \bar{E}_{i,z} \\ \bar{H}_{i,x} & \bar{H}_{i,y} & \bar{H}_{i,z} \end{pmatrix} e^{j(k_x x + k_y y - k_{z,(N+1)}(z - z_+))}
$$

$$
+ \begin{pmatrix} \vec{E}_{r,x} & \vec{E}_{r,y} & \vec{E}_{r,z} \\ \vec{H}_{r,x} & \vec{H}_{r,y} & \vec{H}_{r,z} \end{pmatrix} e^{j(k_x x + k_y y + k_{z,(N+1)}(z - z_+))} \tag{2.55c}
$$

2.2.4.3 Energy Conservation

The electromagnetic energy is conserved in the case of a lossless medium. The magnitude of the Poynting vector of the incidence field is equal to the sum of the magnitudes of the Poynting vectors of reflection and transmission fields.

1. Left-to-right directional characterization: The incident electromagnetic power at the boundary B_0 normalized by 1 is equal to the sum of the reflection power normal to the boundary B_0 and the transmission power normal to the boundary B_N.

$$
\left(\left| \bar{E}_{r,x} \right|^2 + \left| \bar{E}_{r,y} \right|^2 + \left| \bar{E}_{r,z} \right|^2 \right) \mathrm{Re} \left(k_z^{(0)} / k_z^{(0)} \right)
$$

$$
+ \left(\left| \bar{E}_{t,x} \right|^2 + \left| \bar{E}_{t,y} \right|^2 + \left| \bar{E}_{t,z} \right|^2 \right) \mathrm{Re} \left(k_z^{(N+1)} / k_z^{(0)} \right) = 1 \tag{2.56a}
$$

2. Right-to-left directional characterization: The incident electromagnetic power at the boundary B_N normalized by 1 is equal to the sum of the reflection power normal to the boundary B_N and the transmission power normal to the boundary B_0.

$$
\left(\left| \bar{E}_{r,x} \right|^2 + \left| \bar{E}_{r,y} \right|^2 + \left| \bar{E}_{r,z} \right|^2 \right) \mathrm{Re} \left(k_z^{(0)} / k_z^{(N+1)} \right)
$$

$$
+ \left(\left| \bar{E}_{t,x} \right|^2 + \left| \bar{E}_{t,y} \right|^2 + \left| \bar{E}_{t,z} \right|^2 \right) \mathrm{Re} \left(k_z^{(N+1)} / k_z^{(N+1)} \right) = 1 \tag{2.56b}
$$

2.3 MATLAB® Implementation

MATLAB implementation of the above-described mathematical mode is provided with MATLAB code *FMM_1D_SMM.m*. The whole code is presented as follows:

MATLAB Code 2.1: *FMM_1D_SMM.m*

```
1    % FMM standard code written by H. Kim
2    % semi-infinite homogeneous space I - multi-block -
     semi-infinite homogeneous space II
3
4    clear all;
5    close all;
6    clc;
7
8    addpath( [pwd '\PRCWA_COM'] );
9    addpath( [pwd '\FIELD_VISUAL'] );
10   addpath( [pwd '\STRUCTURE'] );
11
12
13   %% STEP 1 : wavevector setting and structure modeling
14
15   % length unit
16   global nm;    % nano
17   global um;    % micro
18   global mm;    % mili
19
20   % basic parameters
21   global k0;                          % wavenumber
22   global c0; global w0;        % speed of light, angular
                                        frequency
23   global eps0; global mu0;     % vacuum permittivity &
                                        permeability
24
25   % refractive index, permittivity, permeability in zero-
     thickness buffer layer
26   global n0; global epr0; global mur0;
27   % refractive index, permittivity, permeability in
     homogeneous space I
28   global ni; global epri; global muri;
29   % refractive index, permittivity, permeability in
     homogeneous space II
30   global nf; global eprf; global murf
31
```

```
32    % x-directional supercell period, y-directional supercell
         period
33    global Tx; global Ty;
34    global nx; global ny;
35    % # of x-direction Fourier harmonics, # of y-directional
         Fourier harmonics
36    global NBx; global NBy;
37    global num_hx; global num_hy;
38    global kx_vc; global ky_vc; global kz_vc;
39
40    % input output free space
41    global kix; global kiy; global kiz;
42    global kfz;
43    global kx_ref; global ky_ref; global kz_ref;
44    global kx_tra; global ky_tra; global kz_tra;
45
46    nm=1e-9;
47    lambda=532*nm;
48
49    % direct_=1 left-to-right characterization,
50    % direct_=2 right-to-left characterization
51    direct_ =2; % 1 = left-to-right , 2 = right-to-left
52
53    PRCWA_basic;                      % 3D structure
54    PRCWA_Gen_K;                      % zero-thickness buffer
55
56    % The example structure is a multi-block structure with
         randomly generated
57    % refractive indices
58
59    PRCWA_Gen_Random_Multilayer; %
60
61    %% STEP 2 Block S-matrix computation of single block
         structures
62
63    L=1;
64
65    Ta=zeros(2*L,2*L,Nlay); % left to right
66    Ra=zeros(2*L,2*L,Nlay); % left to right
67    Tb=zeros(2*L,2*L,Nlay); % right to left
68    Rb=zeros(2*L,2*L,Nlay); % right to left
69    Ca=zeros(4*L,2*L,Nlay); % left to right
70    Cb=zeros(4*L,2*L,Nlay); % right to left
71    tCa=zeros(4*L,2*L,Nlay); % left to right
72    tCb=zeros(4*L,2*L,Nlay); % right to left
73
```

```
74   Diagonal_SMM;                    % diagonal anisotropy
75   %Off_diagonal_tensor_SMM;  % off-diagonal anisotropy
76
77   %% STEP3 S-matrix method
78   % The obtained S-matrix and coupling coefficient matrix
       operator of single
79   % blocks are combined to generate the S-matrix and
       coupling coefficient
80   % operator of interconnected multi-block structures by
       the Redheffer star
81   % product...
82
83   I=eye(2*L,2*L);
84   T_temp1a=Ta(:,:,1);
85   R_temp1a=Ra(:,:,1);
86   T_temp1b=Tb(:,:,1);
87   R_temp1b=Rb(:,:,1);
88
89   %% Important
90    tCa=Ca;
91    tCb=Cb;
92    %%
93
94   for laynt=2:Nlay
95
96       %laynt
97
98       T_temp2a=Ta(:,:,laynt);
99       R_temp2a=Ra(:,:,laynt);
100      T_temp2b=Tb(:,:,laynt);
101      R_temp2b=Rb(:,:,laynt);
102
103      RRa=(R_temp1a+T_temp1b*inv(I-R_temp2a*R_temp1b)*R_
             temp2a*T_temp1a);
104      TTa=T_temp2a*inv(I-R_temp1b*R_temp2a)*T_temp1a;
105      RRb=(R_temp2b+T_temp2a*inv(I-R_temp1b*R_temp2a)*R_
             temp1b*T_temp2b);
106      TTb=T_temp1b*inv(I-R_temp2a*R_temp1b)*T_temp2b;
107
108    for k=1:laynt-1
109
110    tCa(:,:,k)=Ca(:,:,k)+Cb(:,:,k)*inv(I-R_temp2a*R_
       temp1b)*R_temp2a*T_temp1a;
111    tCb(:,:,k)=Cb(:,:,k)*inv(I-R_temp2a*R_temp1b)*T_temp2b;
112
113    end; % for k
```

```
114
115     tCa(:,:,laynt)=Ca(:,:,laynt)*inv(I-R_temp1b*R_
        temp2a)*T_temp1a;
116     tCb(:,:,laynt)=Cb(:,:,laynt)+Ca(:,:,laynt)*inv(I-R_
        temp1b*R_temp2a)*R_temp1b*T_temp2b;
117
118     T_temp1a=TTa;
119     R_temp1a=RRa;
120     T_temp1b=TTb;
121     R_temp1b=RRb;
122
123     Ca=tCa;
124     Cb=tCb;
125    end; % laynt
126
127    TTa=T_temp1a; % left-to-right transmission operator
128    RRa=R_temp1a; % left-to-right reflection operator
129    TTb=T_temp1b; % right-to-left transmission operator
130    RRb=R_temp1b; % right-to-left reflection operator
131
132    %% STEP4 Half-infinite block interconnection & Field
           visualization
133
134    switch direct_
135
136      case 1    % left-to-right
137           % polarization angle : TM=0, TE=pi/2
138           psi=0;
139           tm_Ux=cos(psi)*cos(theta)*cos(phi)-
             sin(psi)*sin(phi);
140           tm_Uy=cos(psi)*cos(theta)*sin(phi)+sin(psi)
             *cos(phi);
141           tm_Uz=-cos(psi)*sin(theta);
142
143           psi=pi/2;
144           te_Ux=cos(psi)*cos(theta)*cos(phi)-
             sin(psi)*sin(phi);
145           te_Uy=cos(psi)*cos(theta)*sin(phi)+sin(psi)
             *cos(phi);
146           te_Uz=-cos(psi)*sin(theta);
147
148           % Ux=cos(psi)*cos(theta)*cos(phi)-
             sin(psi)*sin(phi);
149           % Uy=cos(psi)*cos(theta)*sin(phi)+sin(psi)
             *cos(phi);
150           % Uz=-cos(psi)*sin(theta);
```

```
151
152     case 2        % right-to-left
153                        % polarization angle : TM=0, TE=pi/2
154          psi=0;
155          tm_Ux=cos(psi)*cos(theta)*cos(phi)-
                sin(psi)*sin(phi);
156          tm_Uy=cos(psi)*cos(theta)*sin(phi)+sin(psi)
                *cos(phi);
157          tm_Uz=cos(psi)*sin(theta);
158
159          psi=pi/2;
160          te_Ux=cos(psi)*cos(theta)*cos(phi)-
                sin(psi)*sin(phi);
161          te_Uy=cos(psi)*cos(theta)*sin(phi)+sin(psi)
                *cos(phi);
162          te_Uz=cos(psi)*sin(theta);
163
164          % Ux=cos(psi)*cos(theta)*cos(phi)-
                sin(psi)*sin(phi);
165          % Uy=cos(psi)*cos(theta)*sin(phi)+sin(psi)
                *cos(phi);
166          % Uz=cos(psi)*sin(theta);
167
168     end;
169
170     Bdr_Smat_case1; % 1. homogeneous space - grating -
                                              homogeneous
                                              space
```

The important constituents of the MATLAB functions are listed in the following table. In particular, the readers have to understand the following five MATLAB functions:

1. *FMM_single_block_tensor.m*
2. *FMM_single_block.m*
3. *Bdr_SMat_infr_outfr.m*
4. *Bdr_SMat_wg.m*
5. *Bdr_SMat_wg_tensor.m*

These MATLAB functions are the analysis functions of block and boundary S-matrices of target structures.

File Name	Description
FMM_1D_SMM.m	Main routine of S-matrix method with field visualization
PRCWA_COM/Directory	*Common Library*
FMM_single_block.m	S-matrix calculation of single block with diagonal anisotropic material
sWp_gen.m	Transversal boundary electric field distribution at z=z+
sWm_gen.m	Transversal boundary electric field distribution at z=z−
sVp_gen.m	Transversal boundary magnetic field distribution at z=z+
sVm_gen.m	Transversal boundary magnetic field distribution at z=z−
FMM_single_block_tensor.m	S-matrix calculation of single block with general off-diagonal tensor anisotropic material
Wp_gen.m	Transversal boundary electric field distribution at z=z+
Wm_gen.m	Transversal boundary electric field distribution at z=z−
Vp_gen.m	Transversal boundary magnetic field distribution at z=z+
Vm_gen.m	Transversal boundary magnetic field distribution at z=z−
Bdr_SMat_infr_outfr.m	Boundary S-matrix calculation of left and right half-infinite isotropic homogeneous spaces
Bdr_SMat_wg.m	Boundary S-matrix calculation of left and right half-infinite inhomogeneous spaces with diagonal anisotropic material
Bdr_SMat_wg_tensor.m	Boundary S-matrix calculation of left and right half-infinite inhomogeneous spaces with general off-diagonal tensor anisotropic material
PRCWA_basic.m	Basic setting of Fourier harmonic orders, supercell period, buffer block material
PRCWA_Gen_K.m	Setting wavevector grid of Fourier harmonics
fun_*.m	Permittivity functions of Ag, Au, Si, PMMA for wavelength
Odd_*.m	FFT shift functions of odd number sampling signal for Fourier transform
FIELD_VISUAL/Directory	*Field Visualization*
Field_visualization_3D_xz_*.m	Cross-section profile of six electromagnetic field components at x-z plane
Field_visualization_3D_yz_*.m	Cross-section profile of six electromagnetic field components at y-z plane
Field_visualization_3D_xy.m	Cross-section profile of six electromagnetic field components at x-y plane
STRUCTURE/Directory	*Example Structures*
PRCWA_Gen_Random_ Multilayer.m	Setting the Toeplitz matrices of permittivity and permeability tensors
Grating_gen_Random_ Multilayer.m	Setting permittivity and permeability tensors

The code structure is divided into the following four steps:

Step 1: Wavevector setting and structure modeling (lines 13~59 in *FMM_1D_SMM.m*). In *PRCWA_basic.m*, the permittivity and permeability of the left and right half-infinite spaces are determined. Equation (2.2) is coded in *PRCWA_Gen_K.m*. The material structure described in the left-hand side of Equations (2.5a) to (2.5c) is implemented in *PRCWA_Gen_Random_Multilayer.m* and *Grating_gen_Random_Multilayer.m*.

Step 2: Block S-matrix computation of single-block structures (lines 61~77 in *FMM_1D_SMM.m*). The block S-matrix computation of a single-block structure with diagonal anisotropic material is performed by *FMM_single_block.m* in *Diagonal_SMM.m*, while the block S-matrix of a single-block structure with general off-diagonal anisotropic material is analyzed by *FMM_single_block_tensor.m* in *Off-diagonal_tensor_SMM.m*. In these MATLAB codes, *Exx*, *Eyy*, *Ezz*, *Axx*, *Ayy*, and *Azz* correspond to ε_{xx}, ε_{yy}, ε_{zz}, α_{xx}, α_{yy}, and α_{zz} respectively. *Gxx*, *Gyy*, *Gzz*, *Bxx*, *Byy*, and *Bzz* correspond to μ_{xx}, μ_{yy}, μ_{zz}, β_{xx}, β_{yy}, and β_{zz}, respectively. Equations (2.12a) to (2.12c) are coded in lines 100~112 in *FMM_single_block.m*. Equation (2.13) is written in line 111.

MATLAB Code 2.2: *FMM_single_block.m*

```
100    % System Matrix
101
102    SA=zeros(2*L, 2*L);
103    SB=zeros(2*L,2*L);
104
105    SA=[ (Ky)*inv(Ezz)*(Kx)              BG_x-(Ky)*inv(Ezz)*(Ky)
106         (Kx)*inv(Ezz)*(Kx)-GB_y        -(Kx)*inv(Ezz)*(Ky)];
107
108    SB=[ (Ky)*inv(Gzz)*(Kx)              AE_x-(Ky)*inv(Gzz)*(Ky)
109         (Kx)*inv(Gzz)*(Kx)-EA_x        -(Kx)*inv(Gzz)*(Ky)];
110
111    St=k0^2*SA*SB;
112    clear SB;
```

On the other hand, the Maxwell eigenvalue equation of Equation (2.16) for general anisotropic media is analyzed by *FMM_single_block_tensor.m* in *Off_diagonal_tensor_SMM.m*. The MATLAB code of Equation (2.16) is presented in lines 140~168 in *FMM_single_block_tensor.m*.

MATLAB Code 2.3: *FMM_single_block_tensor.m*

```
140   % System Matrix
141
142   St11= -j*Gxz*inv(Gzz)*Kx - j*Ky*inv(Ezz)*Ezy;
143   St12= j*Gxz*inv(Gzz)*Ky - j*Ky*inv(Ezz)*Ezx;
144   St13= Gxy - Gxz*inv(Gzz)*Gzy + Ky*inv(Ezz)*Kx;
145   St14= BG_x - Gxz*inv(Gzz)*Gzx - Ky*inv(Ezz)*Ky;
146
147   St21= j*Gyz*inv(Gzz)*Kx-j*Kx*inv(Ezz)*Ezy;
148   St22= -j*Gyz*inv(Gzz)*Ky-j*Kx*inv(Ezz)*Ezx;
149   St23= -GB_y+Gyz*inv(Gzz)*Gzy + Kx*inv(Ezz)*Kx;
150   St24= -Gyx+Gyz*inv(Gzz)*Gzx - Kx*inv(Ezz)*Ky;
151
152   St31= Exy - Exz*inv(Ezz)*Ezy+Ky*inv(Gzz)*Kx;
153   St32= AE_x - Exz*inv(Ezz)*Ezx-Ky*inv(Gzz)*Ky;
154   St33= -j*Exz*inv(Ezz)*Kx-j*Ky*inv(Gzz)*Gzy;
155   St34= j*Exz*inv(Ezz)*Ky-j*Ky*inv(Gzz)*Gzx;
156
157   St41= -EA_y+Eyz*inv(Ezz)*Ezy+Kx*inv(Gzz)*Kx;
158   St42= -Eyx+Eyz*inv(Ezz)*Ezx-Kx*inv(Gzz)*Ky;
159   St43= j*Eyz*inv(Ezz)*Kx-j*Kx*inv(Gzz)*Gzy;
160   St44= -j*Eyz*inv(Ezz)*Ky-j*Kx*inv(Gzz)*Gzx;
161
162   St=[ St11 St12 St13 St14
163        St21 St22 St23 St24
164        St31 St32 St33 St34
165        St41 St42 St43 St44];
166
167   St=k0*St;
```

Equations (2.27a) to (2.27c) related to the left-to-right characterization and Equations (2.33a) to (2.33c) related to the right-to-left characterization are coded in lines 243–284 of *FMM_single_block.m*.

MATLAB Code 2.4: *FMM_single_block.m*

```
243   % left-to-right
244   U=eye(2*L,2*L);
245
246   S11=inv(Wh)*Wp_zm+inv(Vh)*Vp_zm;
247   S12=inv(Wh)*Wm_zm+inv(Vh)*Vm_zm;
248   S21=inv(Wh)*Wp_zp-inv(Vh)*Vp_zp;
249   S22=inv(Wh)*Wm_zp-inv(Vh)*Vm_zp;
250
251   S=[S11 S12; S21 S22];
252
253   clear S11;
```

```
254    clear S12;
255    clear S21;
256    clear S22;
257    D=[2*U;zeros(2*L,2*L)];
258
259    CCa=inv(S)*D;
260    Cap=CCa(1:2*L,:);
261    Cam=CCa(2*L+1:4*L,:);
262    Ra=inv(Wh)*(Wp_zm*Cap+Wm_zm*Cam-Wh*U);
263    Ta=inv(Wh)*(Wp_zp*Cap+Wm_zp*Cam);
264
265    % right-to-left
266
267    S11=inv(Wh)*Wp_zm+inv(Vh)*Vp_zm;
268    S12=inv(Wh)*Wm_zm+inv(Vh)*Vm_zm;
269    S21=inv(Wh)*Wp_zp-inv(Vh)*Vp_zp;
270    S22=inv(Wh)*Wm_zp-inv(Vh)*Vm_zp;
271
272    S=[S11 S12 ; S21 S22];
273    clear S11;
274    clear S12;
275    clear S21;
276    clear S22;
277
278    D=[zeros(2*L,2*L);2*U];
279    CCb=inv(S)*D;
280    Cbp=CCb(1:2*L,:);
281    Cbm=CCb(2*L+1:4*L,:);
282
283    Rb=inv(Wh)*(Wp_zp*Cbp+Wm_zp*Cbm-Wh*U);
284    Tb=inv(Wh)*(Wp_zm*Cbp+Wm_zm*Cbm);
```

At lines 191~192 of *FMM_single_block.m*, z_- and z_+ are set by *zm* and *zp*, respectively. In lines 205~206, the necessary variables to make the system matrix of Equations (2.27) and (2.33) are prepared as follows.

$\underline{\underline{W}}^+(0)$	Wp_zm	sWp_gen(pW,pevalue,pcnt,L,zm-zm)
$\underline{\underline{W}}^-(z_- - z_+)$	Wm_zm	sWm_gen(mW,mevalue,mcnt,L,zm-zp)
$\underline{\underline{V}}^+(0)$	Vp_zm	sVp_gen(pV,pevalue,pcnt,L,zm-zm)
$\underline{\underline{V}}^-(z_- - z_+)$	Vm_zm	sVm_gen(mV,mevalue,mcnt,L,zm-zp)
$\underline{\underline{W}}^+(z_+ - z_-)$	Wp_zp	sWp_gen(pW,pevalue,pcnt,L,zm-zm)

$\underline{\underline{V}}^{+}(z_{+}-z_{-})$	Vp_zp	sWm_gen(mW,mevalue,mcnt,L,zm-zp)
$\underline{\underline{W}}^{-}(0)$	Wm_zp	sVp_gen(pV,pevalue,pcnt,L,zm-zm)
$\underline{\underline{V}}^{-}(0)$	Vm_zp	sVm_gen(mV,mevalue,mcnt,L,zm-zp)

$\underline{\underline{W}}_{h}$ and $\underline{\underline{V}}_{h}$ are implemented in lines 238~239. It is noteworthy that the final sentence is KII divided by (w0*mu0) at line 239. Equation (2.27a) is solved and $\underline{\underline{C}}_{a}$ is obtained at line 259 by *CCa*. Equations (2.27b) and (2.27c) are solved and $\underline{\underline{\vec{R}}}$ and $\underline{\underline{\vec{T}}}$ are obtained at lines 262 and 263, respectively, by *Ra* and *Ta*. In the MATLAB code of *FMM_single_block.m*, Equations (2.32a)~(2.33d) are solved and $\underline{\underline{\vec{R}}}$, $\underline{\underline{\vec{T}}}$, and $\underline{\underline{C}}_{b}$ are obtained in lines 267~284. In the case of *FMM_single_block_tensor.m*, the exact same codes are used in lines 285~316. In *FMM_single_block.m* the calculation of Fourier coefficients of pseudo-Fourier series is performed at lines 197~236. In *FMM_single_block_tensor.m* the same calculation is done at lines 283~308.

MATLAB Code 2.5: *FMM_single_block.m*

```
165  % Fourier coefficients (pfEx,pfEy,pfEz,pfHx,pfHy,pfHz) ,
        (mfEx,mfEy,mfEz,mfHx,mfHy,mfHz)
166
167     pfEy=pW(1:L,:);
168     pfEx=pW(L+1:2*L,:);
169     pfHy=pV(1:L,:);
170     pfHx=pV(L+1:2*L,:);
171     pfEz=inv(Ezz)*(j*Ky*pfHx-j*Kx*pfHy);
172     pfHz=inv(Gzz)*(j*Ky*pfEx-j*Kx*pfEy);
173
174     pfHy=j*(eps0/mu0)^0.5*pfHy;
175     pfHx=j*(eps0/mu0)^0.5*pfHx;
176     pfHz=j*(eps0/mu0)^0.5*pfHz;
177
178
179     mfEy=mW(1:L,:);
180     mfEx=mW(L+1:2*L,:);
181     mfHy=mV(1:L,:);
182     mfHx=mV(L+1:2*L,:);
183     mfEz=inv(Ezz)*(j*Ky*mfHx-j*Kx*mfHy);
184     mfHz=inv(Gzz)*(j*Ky*mfEx-j*Kx*mfEy);
185
186     mfHy=j*(eps0/mu0)^0.5*mfHy;
187     mfHx=j*(eps0/mu0)^0.5*mfHx;
188     mfHz=j*(eps0/mu0)^0.5*mfHz;
```

MATLAB Code 2.6: *FMM_single_block_tensor.m*

```
200    % Fourier coefficients (pfEx,pfEy,pfEz,pfHx,pfHy,pfHz)
       , (mfEx,mfEy,mfEz,mfHx,mfHy,mfHz)
201
202       pfEy=pW(1:L,:);
203       pfEx=pW(L+1:2*L,:);
204       pfHy=pW(2*L+1:3*L,:);
205       pfHx=pW(3*L+1:4*L,:);
206       pfEz=inv(Ezz)*(j*Ky*pfHx-j*Kx*pfHy-Ezx*pfEx-
          Ezy*pfEy);
207       pfHz=inv(Gzz)*(j*Ky*pfEx-j*Kx*pfEy-Gzx*pfHx-
          Gzy*pfHy);
208
209       pfHy=j*(eps0/mu0)^0.5*pfHy;
210       pfHx=j*(eps0/mu0)^0.5*pfHx;
211       pfHz=j*(eps0/mu0)^0.5*pfHz;
212
213       mfEy=mW(1:L,:);
214       mfEx=mW(L+1:2*L,:);
215       mfHy=mW(2*L+1:3*L,:);
216       mfHx=mW(3*L+1:4*L,:);
217       mfEz=inv(Ezz)*(j*Ky*mfHx-j*Kx*mfHy-Ezx*mfEx-
          Ezy*mfEy);
218       mfHz=inv(Gzz)*(j*Ky*mfEx-j*Kx*mfEy-Gzx*mfHx-
          Gzy*mfHy);
219
220       mfHy=j*(eps0/mu0)^0.5*mfHy;
221       mfHx=j*(eps0/mu0)^0.5*mfHx;
222       mfHz=j*(eps0/mu0)^0.5*mfHz;
```

The outputs of *FMM_single_block_tensor.m* and *FMM_single_block.m* are given by:

S-Matrix and Coupling Coefficient Operator		Positive Eigenmodes		Negative Eigenmodes	
$\underline{\underline{\vec{T}}}$	Ta	$E_{x,m,n}^{(g)+}$	pfEx	$E_{x,m,n}^{(g)-}$	mfEx
$\underline{\underline{\vec{R}}}$	Ra	$E_{y,m,n}^{(g)+}$	pfEy	$E_{y,m,n}^{(g)-}$	mfEy
$\underline{\underline{\overleftarrow{T}}}$	Tb	$E_{z,m,n}^{(g)+}$	pfEz	$E_{z,m,n}^{(g)-}$	mfEz
$\underline{\underline{\overleftarrow{R}}}$	Rb	$H_{x,m,n}^{(g)+}$	pfHx	$H_{x,m,n}^{(g)-}$	mfHx
$\underline{\underline{C}}_a$	CCa	$H_{y,m,n}^{(g)+}$	pfHy	$H_{y,m,n}^{(g)-}$	mfHy
$\underline{\underline{C}}_b$	CCb	$H_{z,m,n}^{(g)+}$	pfHz	$H_{z,m,n}^{(g)-}$	mfHz
		$k_z^{(g)+}$	pevalue	$k_z^{(g)-}$	mevalue

The boundary S-matrix analysis of free space is performed in *Bdr_SMat_infr_outfr.m*.

Step 3: S-matrix method (lines 79~136 in *FMM_1D_SMM.m*). The recursive algorithm of the S-matrix method of Equations (2.40a) to (2.40d) is coded in lines 106, 107, 109, and 110 of *FMM_1D_SMM.m*. Equations (2.43a) and (2.43b) correspond to lines 115 and 116 in *FMM_1D_SMM.m*. Equations (2.43c) and (2.43d) are coded in lines 120 and 121, respectively.

Step 4: Half-infinite block interconnection and field visualization. The direction of the directional characterization is selectable by the parameter *direct_* (line 52 of *FMM_1D_SMM.m*). For the cases of the left-to-right and right-to-left directional characterizations, the parameter *direct_* is set to *direct_=1* and *direct_=2*, respectively. The interconnection of the left half-infinite block, right half-infinite block, and finite-sized block described by Equations (2.50a), (2.50b), (2.51a), and (2.51b) is implemented in *Bdr_Smat_case1.m* (see line 176 of *FMM_1D_SMM.m* and lines 5~60 of *Bdr_Smat_case1.m*).

MATLAB Code 2.7: *Bdr_Smat_case1.m*

```
1    %% input + body
2        I=eye(2*L,2*L);
3        T_temp1a=Lfree_Tf;      %outTb
4        R_temp1a=Lfree_Rb;      %outRf
5        T_temp1b=Lfree_Tb;      %outTf
6        R_temp1b=Lfree_Rf;      %outRb
7
8    %%% Important
9        tCa=Ca;
10       tCb=Cb;
11       T_temp2a=TTa;
12       R_temp2a=RRa;
13       T_temp2b=TTb;
14       R_temp2b=RRb;
15
16   RRa=(R_temp1a+T_temp1b*inv(I-R_temp2a*R_temp1b)*R_
     temp2a*T_temp1a);
17   TTa=T_temp2a*inv(I-R_temp1b*R_temp2a)*T_temp1a;
18   %
19   RRb=(R_temp2b+T_temp2a*inv(I-R_temp1b*R_temp2a)*R_
     temp1b*T_temp2b);
20   TTb=T_temp1b*inv(I-R_temp2a*R_temp1b)*T_temp2b;
21
22   for k=1:Nlay
23   tCa(:,:,k)=Ca(:,:,k)*inv(I-R_temp1b*R_temp2a)*T_temp1a;
```

```
24    tCb(:,:,k)=Cb(:,:,k)+Ca(:,:,k)*inv(I-R_temp1b*R_
      temp2a)*R_temp1b*T_temp2b;
25    end;
26    Ca=tCa;
27    Cb=tCb;
28
29    %% body + output
30        T_temp1a=TTa;
31        R_temp1a=RRa;
32        T_temp1b=TTb;
33        R_temp1b=RRb;
34    %
35        T_temp2a=Rfree_Tf;
36        R_temp2a=Rfree_Rb;
37        T_temp2b=Rfree_Tb;
38        R_temp2b=Rfree_Rf;
39
40    RRa=(R_temp1a+T_temp1b*inv(I-R_temp2a*R_temp1b)*R_
      temp2a*T_temp1a);
41    TTa=T_temp2a*inv(I-R_temp1b*R_temp2a)*T_temp1a;
42    %
43    RRb=(R_temp2b+T_temp2a*inv(I-R_temp1b*R_temp2a)*R_
      temp1b*T_temp2b);
44    TTb=T_temp1b*inv(I-R_temp2a*R_temp1b)*T_temp2b;
45
46
47    for k=1:Nlay
48    %
49    tCa(:,:,k)=Ca(:,:,k)+Cb(:,:,k)*inv(I-R_temp2a*R_
      temp1b)*R_temp2a*T_temp1a;
50    tCb(:,:,k)=Cb(:,:,k)*inv(I-R_temp2a*R_temp1b)*T_temp2b;
51    %
52    end; % for k
53    Ca=tCa;
54    Cb=tCb;
```

The left-to-right field visualization described in Equations (2.52a)~(2.53c) is implemented in *Field_visualization_3D_xz_case1_Lfree_Rfree_leftright.m*. The right-to-left field visualization described in Equations (2.54a)~(2.55c) is implemented in *Field_visualization_3D_xz_case1_Lfree_Rfree_rightleft.m*. In both visualization script files, the determination of polarization of whether TE or TM mode is selected and the calculation of the reflection and transmission efficiencies are performed.

3

Fourier Modal Method

In the previous chapter, the principle and mathematical framework of the S-matrix method was established by an analysis of 1D multiblock structures. The term *1D* means that the structures have permittivity and permeability profiles that vary along the z-direction. The principle and mathematical framework of the Fourier modal method (FMM) is described in this chapter. We present FMM with a logical extension of the previous 1D structure analysis, keeping in mind that FMM is actually a mathematical generalization of the previous analysis on 1D structures to 2D/3D structures. In Section 3.1, the S-matrix analysis of a finite single-block structure with FMM is performed, through which the basis of FMM is constructed. The Fourier modal representation of optical eigenmodes is the key for all later discussions regarding FMM. The more general formulation of a Fourier representation for an optical field is discussed in Chapter 5. The elementary block S-matrix of a single-block structure is analyzed within the framework of FMM. FMM of a single block with only a transversal permittivity profile variation is the cornerstone for the following body of multiblock FMM and the local FMM (LFMM) dealt with in this book. In Section 3.2, an S-matrix analysis of a multiblock structure with FMM is performed. The theoretical unification of the S-matrix method elucidated in Section 2.2 and FMM founded in Section 3.1 is developed. The algorithm for interconnecting with the block S-matrix and boundary S-matrix of multiblock structures is established. In Section 3.3, MATLAB implementation of the mathematical theory is discussed. In Section 3.4, a practical application of FMM to extraordinary optical transmission (EOT) phenomena is presented with numerical simulation results and a physical interpretation.

3.1 Fourier Modal Analysis of Single-Block Structures

The S-matrix analysis of a single-block structure with periodic transversal permittivity and permeability profiles is the most important building block of the Fourier modal method. Analyses of other complex structures as well as multiblock structures is based on the basic building block of the theory, i.e., the S-matrix analysis of a single-block structure. The Fourier modal analysis and the S-matrix analysis of a single-block structure are a prerequisite for studying various advanced issues of FMM.

The S-matrix analysis of a single-block structure is the first problem in this chapter. Here the term *single block* is used to denote a finite-sized slab structure with only transversal periodic variations in permittivity and permeability profiles. Let us first consider the case of nontensorial permittivity and permeability profiles. Since there is no variation on the z-variable, the permittivity and permeability profiles, $\varepsilon(x, y)$ and $\mu(x, y)$, can be approximately represented by the truncated Fourier series forms, respectively, as

$$\varepsilon(x,y) = \sum_{m=-2M}^{2M} \sum_{n=-2N}^{2N} \varepsilon_{m,n} e^{j(mG_x x + nG_y y)} \tag{3.1a}$$

$$\mu(x,y) = \sum_{m=-2M}^{2M} \sum_{n=-2N}^{2N} \mu_{m,n} e^{j(mG_x x + nG_y y)} \tag{3.1b}$$

where G_x and G_y are the x-direction and the y-direction primitive reciprocal vectors given by $G_x = 2\pi/T_x$ and $G_y = 2\pi/T_y$, respectively. T_x and T_y are the x-directional and y-directional periods of the permittivity and permeability profiles, respectively. In the structure modeling of FMM, the Fourier coefficients $\varepsilon_{m,n}$ and $\mu_{m,n}$ should be prepared. The single block can be classified into (1) x-directional 1D binary grating structure $\varepsilon(x) = \varepsilon(x + nT_x)$, (2) y-directional 1D binary grating structure $\varepsilon(y) = \varepsilon(y + nT_y)$, and (3) xy-directional 2D binary grating structure $\varepsilon(x, y) = \varepsilon(x + mT_x, y + nT_y)$. The exemplary structures of the respective cases are illustrated in Figure 3.1.

Using the procedure used for a homogeneous single-block structure in the previous chapter, the complete characterization of scattering properties of single blocks can be accomplished by the bidirectional characterization as presented in Figure 3.2. It should be noted that the single block has the permittivity and permeability profiles of the general tensor form of $\underline{\varepsilon}(x,y)$ and $\underline{\mu}(x,y)$ in Figure 3.2.

Maxwell's equations for a single block with transversal permittivity and permeability variations are given by

$$\nabla \times \mathbf{E} = j\omega\mu_0\mu(x,y)\mathbf{H} \tag{3.2a}$$

$$\nabla \times \mathbf{H} = -j\omega\varepsilon_0\varepsilon(x,y)\mathbf{E} \tag{3.2b}$$

$$\nabla \cdot [\varepsilon(x,y)\mathbf{E}] = 0 \tag{3.2c}$$

$$\nabla \cdot [\mu(x,y)\mathbf{H}] = 0 \tag{3.2d}$$

In Equations (3.2a) to (3.2d), the material of a single block is assumed to be isotropic for convenience, but the formulation will be extended to general

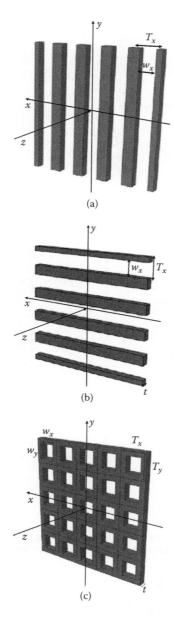

FIGURE 3.1
(a) Periodic 1D binary grating with $\varepsilon(x) = \varepsilon(x + nT_x)$, (b) periodic 1D binary grating with $\varepsilon(y) = (y + nT_y)$, and (c) periodic 2D rectangular grating with $\varepsilon(x,y) = \varepsilon(x + mT_x, y + nT_y)$.

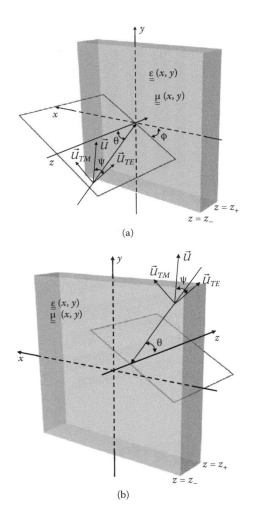

(a)

(b)

FIGURE 3.2
Bidirectional characterization of a single block with finite thickness of $d = z_+ - z_-$. (a) Left-to-right directional characterization and (b) right-to-left directional characterization.

anisotropic media later. Analogous to the analysis of the homogeneous single block in the previous chapter, the S-matrix analysis of a single-block structure is comprised of three steps: (1) eigenmode analysis, (2) S-matrix and coupling coefficient matrix operator analysis, and (3) field visualization.

3.1.1 Eigenmode Analysis

If an optical plane wave with the wavevector of $(k_{x,0}, k_{y,0}, k_{z,0})$ illuminates a diffraction grating with x-directional and y-directional periods, T_x and T_y, the optical field is reflected along fan-shaped discrete diffraction channels (diffraction orders) and is also transmitted to fan-shaped discrete diffraction channels. If the surrounding space is free space and the discrete set of wave-vector of the $(s, t)^{\text{th}}$ diffraction channel is denoted by $(k_{x,st}, k_{y,st}, k_{z,st})$, then the wavevector $(k_{x,st}, k_{y,st}, k_{z,st})$ is taken as

$$(k_{x,st}, k_{y,st}, k_{z,st}) = \left(k_{x,0} + \frac{2\pi s}{T_x}, k_{y,0} + \frac{2\pi t}{T_y}, \pm \sqrt{ \left(\frac{2\pi}{\lambda} \right)^2 - \left(k_{x,0} + \frac{2\pi s}{T_x} \right)^2 - \left(k_{y,0} + \frac{2\pi t}{T_y} \right)^2 } \right)$$

(3.3)

where λ is the operating wavelength. From the Bloch theorem, the Bloch eigenmode is represented by the pseudoperiodic vector function,

$$\mathbf{E} = (\mathbf{E}_x, \mathbf{E}_y, \mathbf{E}_z) = e^{j(k_{x,0}x + k_{y,0}y + k_z z)}(E_x(x,y), E_y(x,y), E_z(x,y))$$

(3.4a)

$$\mathbf{H} = (\mathbf{H}_x, \mathbf{H}_y, \mathbf{H}_z) = e^{j(k_{x,0}x + k_{y,0}y + k_z z)}(H_x(x,y), H_y(x,y), H_z(x,y))$$

(3.4b)

where the six envelope components of $E_x(x, y)$, $E_y(x, y)$, $E_z(x, y)$, $H_x(x, y)$, $H_y(x, y)$, and $H_z(x, y)$ are periodic functions with the x-directional and y-directional periods of T_x and T_y, respectively. Since, for single-block structures, the z-dependent term is not included in the structure representations, the six envelope field functions can be approximated by means of a 2D Fourier series with finite harmonic orders, i.e., a symmetrically truncated Fourier series. The truncation order of the field can usually be chosen greater than or equal to the truncation order of the Fourier series representation of the structure. However, for convenience, the truncation orders of x-directional and y-directional harmonics are set to M and N, respectively. The convergence of the field profile is a criterion of determining the values of M and N. Consequently, the Bloch eigenmodes in the single block, $\mathbf{E}$ and $\mathbf{H}$, within periodic media can be represented by the pseudo-Fourier series:

$$\mathbf{E} = (\mathbf{E}_x, \mathbf{E}_y, \mathbf{E}_z)$$
$$= e^{j(k_{x,0}x + k_{y,0}y + k_z z)} \sum_{m=-M}^{M} \sum_{n=-N}^{N} (E_{x,m,n}, E_{y,m,n}, E_{z,m,n}) e^{j\left(\frac{2\pi m}{T_x}x + \frac{2\pi n}{T_y}y \right)}$$

(3.5a)

$$\mathbf{H} = (\mathbf{H}_x, \mathbf{H}_y, \mathbf{H}_z)$$
$$= j\sqrt{\frac{\varepsilon_0}{\mu_0}} e^{j(k_{x,0}x + k_{y,0}y + k_z z)} \sum_{m=-M}^{M} \sum_{n=-N}^{N} (H_{x,m,n}, H_{y,m,n}, H_{z,m,n}) e^{j\left(\frac{2\pi m}{T_x}x + \frac{2\pi n}{T_y}y \right)}$$

(3.5b)

where the propagation wavevector is denoted by $\mathbf{k} = (k_{x,0}, k_{y,0}, k_z)$. In this field representation, the transversal wavevector component, $(k_{x,0}, k_{y,0})$, is set to be given parameters, and the third wavevector component, k_z, is considered to be an eigenvalue that must be extracted from the Maxwell eigenvalue equation, i.e., dispersion relation. Substituting Equations (3.5a) and (3.5b) into the curl equations (3.2a and 3.2b), we can obtain the algebraic form of Maxwell's equations in the spatial frequency domain. Some manipulation gives the following results:

$$k_0 \sum_s \sum_t \mu_{m-s,n-t} H_{x,s,t,} = jk_z E_{y,m,n} - jk_{y,m,n} E_{z,m,n} \tag{3.6a}$$

$$k_0 \sum_s \sum_t \mu_{m-s,n-t} H_{y,s,t} = jk_{x,m,n} E_{z,m,n} - jk_z E_{x,m,n} \tag{3.6b}$$

$$k_0 \sum_s \sum_t \mu_{m-s,n-t} H_{z,s,t} = jk_{y,m,n} E_{x,m,n} - jk_{x,m,n} E_{y,m,n} \tag{3.6c}$$

$$k_0 \sum_s \sum_t \varepsilon_{m-s,n-t} E_{x,s,t} = jk_z H_{y,m,n} - jk_{y,m,n} H_{z,m,n} \tag{3.6d}$$

$$k_0 \sum_s \sum_t \varepsilon_{m-s,n-t} E_{y,s,t} = jk_{x,m,n} H_{z,m,n} - jk_z H_{x,m,n} \tag{3.6e}$$

$$k_0 \sum_s \sum_t \varepsilon_{m-s,n-t} E_{z,s,t} = jk_{y,m,n} H_{x,m,n} - jk_{x,m,n} H_{y,m,n} \tag{3.6f}$$

where $k_{x,m,n}$ and $k_{y,m,n}$ are defined by

$$(k_{x,m,n}, k_{y,m,n}) = (k_{x,m}, k_{y,n}) = \left(k_{x,0} + \frac{2\pi m}{T_x}, k_{y,0} + \frac{2\pi n}{T_y} \right) \tag{3.7}$$

It is interesting to see how to reflect the structural generalization in Equation (3.6) and compare the pattern of Equations (3.6a) to (3.6f) with the pattern of Equations (2.4a) to (2.4f). The matrix representations of the above equations can be written as

$$k_0 \underline{\underline{\mu}} H_x = j\underline{\underline{K}}_z E_y - j\underline{\underline{K}}_y E_z \tag{3.8a}$$

$$k_0 \underline{\underline{\mu}} H_y = j\underline{\underline{K}}_x E_z - j\underline{\underline{K}}_z E_x \tag{3.8b}$$

$$k_0 \underline{\underline{\mu}} \underline{H}_z = j\underline{\underline{K}}_y \underline{E}_x - j\underline{\underline{K}}_x \underline{E}_y \tag{3.8c}$$

$$k_0 \underline{\underline{\varepsilon}} \underline{E}_x = j\underline{\underline{K}}_z \underline{H}_y - j\underline{\underline{K}}_y \underline{H}_z \tag{3.8d}$$

$$k_0 \underline{\underline{\varepsilon}} \underline{E}_y = j\underline{\underline{K}}_x \underline{H}_z - j\underline{\underline{K}}_z \underline{H}_x \tag{3.8e}$$

$$k_0 \underline{\underline{\varepsilon}} \underline{E}_z = j\underline{\underline{K}}_y \underline{H}_x - j\underline{\underline{K}}_x \underline{H}_y \tag{3.8f}$$

Here, $\underline{\underline{\varepsilon}}$ is the Toeplitz matrix of Fourier coefficients of permittivity profile $\varepsilon(x,y)$ defined by

$$\underline{\underline{\varepsilon}} = \begin{pmatrix} \overline{\varepsilon}_0 & \overline{\varepsilon}_{-1} & \cdots & \overline{\varepsilon}_{-2M} \\ \overline{\varepsilon}_1 & \overline{\varepsilon}_0 & & \overline{\varepsilon}_{-2M+1} \\ \vdots & \vdots & & \\ \overline{\varepsilon}_{2M} & \overline{\varepsilon}_{2M-1} & \cdots & \overline{\varepsilon}_0 \end{pmatrix} \tag{3.9}$$

where

$$\overline{\varepsilon}_k = \begin{pmatrix} \varepsilon_{k,0} & \varepsilon_{k,-1} & \cdots & \varepsilon_{k,-2N} \\ \varepsilon_{k,1} & \varepsilon_{k,0} & & \varepsilon_{k,-2N+1} \\ \vdots & \vdots & & \\ \varepsilon_{k,2N} & \varepsilon_{k,2N-1} & \cdots & \varepsilon_{k,0} \end{pmatrix},$$

and $\varepsilon_{m,n}$ is the Fourier coefficients of $\varepsilon(x,y)$. The Toeplitz matrix, $\underline{\underline{\mu}}$, $\underline{\underline{\alpha}}$, and $\underline{\underline{\beta}}$ are defined in the same way. $\underline{\underline{K}}_x$ is the Toeplitz matrix of $k_{x,m,n}$ defined by

$$\underline{\underline{K}}_x = \begin{pmatrix} [k_x]_{-M} & 0 & \cdots & 0 \\ 0 & [k_x]_{-M+1} & 0 & 0 \\ \vdots & \vdots & \ddots & 0 \\ 0 & 0 & \cdots & [k_x]_M \end{pmatrix} \tag{3.10}$$

where

$$[k_x]_k = \begin{pmatrix} k_{x,k,-N} & 0 & \cdots & 0 \\ 0 & k_{x,k,-N+1} & 0 & 0 \\ \vdots & \vdots & \ddots & 0 \\ 0 & 0 & \cdots & k_{x,k,N} \end{pmatrix}$$

$\underline{K}_y$ is defined by the same way. The remained matrices, $\underline{\underline{I}}$ and $\underline{K}_z$, are the identity matrix and $k_z\underline{\underline{I}}$, respectively. The column vectors of the Fourier coefficients, $\underline{E}_y$, $\underline{E}_x$, $\underline{H}_y$, and $\underline{H}_x$, are expressed, respectively, by

$$
\underline{E}_y = \begin{pmatrix} \begin{pmatrix} E_{y,-M,-N} \\ \vdots \\ E_{y,-M,N} \end{pmatrix} \\ \begin{pmatrix} E_{y,-M+1,-N} \\ \vdots \\ E_{y,-M+1,N} \end{pmatrix} \\ \vdots \\ \begin{pmatrix} E_{y,M,-N} \\ \vdots \\ E_{y,M,N} \end{pmatrix} \end{pmatrix}, \quad
\underline{E}_x = \begin{pmatrix} \begin{pmatrix} E_{x,-M,-N} \\ \vdots \\ E_{x,-M,N} \end{pmatrix} \\ \begin{pmatrix} E_{x,-M+1,-N} \\ \vdots \\ E_{x,-M+1,N} \end{pmatrix} \\ \vdots \\ \begin{pmatrix} E_{x,M,-N} \\ \vdots \\ E_{x,M,N} \end{pmatrix} \end{pmatrix},
$$

$$
\underline{H}_y = \begin{pmatrix} \begin{pmatrix} H_{y,-M,-N} \\ \vdots \\ H_{y,-M,N} \end{pmatrix} \\ \begin{pmatrix} H_{y,-M+1,-N} \\ \vdots \\ H_{y,-M+1,N} \end{pmatrix} \\ \vdots \\ \begin{pmatrix} H_{y,M,-N} \\ \vdots \\ H_{y,M,N} \end{pmatrix} \end{pmatrix}, \quad
\underline{H}_x = \begin{pmatrix} \begin{pmatrix} H_{x,-M,-N} \\ \vdots \\ H_{x,-M,N} \end{pmatrix} \\ \begin{pmatrix} H_{x,-M+1,-N} \\ \vdots \\ H_{x,-M+1,N} \end{pmatrix} \\ \vdots \\ \begin{pmatrix} H_{x,M,-N} \\ \vdots \\ H_{x,M,N} \end{pmatrix} \end{pmatrix} \tag{3.11}
$$

$\underline{E}_y$, $\underline{H}_y$, and $\underline{H}_x$ are defined in the same way. The six matrices in Equations (3.8a) to (3.8f) can be rearranged by the following matrix form:

$$k_0 \begin{pmatrix} \underline{\underline{\varepsilon}} & 0 & 0 & 0 & 0 & 0 \\ 0 & \underline{\underline{\varepsilon}} & 0 & 0 & 0 & 0 \\ 0 & 0 & \underline{\underline{\varepsilon}} & 0 & 0 & 0 \\ 0 & 0 & 0 & \underline{\underline{\mu}} & 0 & 0 \\ 0 & 0 & 0 & 0 & \underline{\underline{\mu}} & 0 \\ 0 & 0 & 0 & 0 & 0 & \underline{\underline{\mu}} \end{pmatrix} \begin{pmatrix} \underline{E}_x \\ \underline{E}_y \\ \underline{E}_z \\ \underline{H}_x \\ \underline{H}_y \\ \underline{H}_z \end{pmatrix}$$

$$= \begin{pmatrix} 0 & 0 & 0 & 0 & j\underline{K}_z & -j\underline{K}_y \\ 0 & 0 & 0 & -j\underline{K}_z & 0 & j\underline{K}_x \\ 0 & 0 & 0 & j\underline{K}_y & -j\underline{K}_x & 0 \\ 0 & j\underline{K}_z & -j\underline{K}_y & 0 & 0 & 0 \\ -j\underline{K}_z & 0 & j\underline{K}_x & 0 & 0 & 0 \\ j\underline{K}_y & -j\underline{K}_x & 0 & 0 & 0 & 0 \end{pmatrix} \begin{pmatrix} \underline{E}_x \\ \underline{E}_y \\ \underline{E}_z \\ \underline{H}_x \\ \underline{H}_y \\ \underline{H}_z \end{pmatrix} \quad (3.12a)$$

Meanwhile, Equation (3.12a) can be further generalized to the form with the anisotropic tensor profiles of permittivity and permeability given as

$$k_0 \begin{pmatrix} \underline{\underline{\varepsilon}}_x & 0 & 0 & 0 & 0 & 0 \\ 0 & \underline{\underline{\varepsilon}}_y & 0 & 0 & 0 & 0 \\ 0 & 0 & \underline{\underline{\varepsilon}}_z & 0 & 0 & 0 \\ 0 & 0 & 0 & \underline{\underline{\mu}}_x & 0 & 0 \\ 0 & 0 & 0 & 0 & \underline{\underline{\mu}}_y & 0 \\ 0 & 0 & 0 & 0 & 0 & \underline{\underline{\mu}}_z \end{pmatrix} \begin{pmatrix} \underline{E}_x \\ \underline{E}_y \\ \underline{E}_z \\ \underline{H}_x \\ \underline{H}_y \\ \underline{H}_z \end{pmatrix}$$

$$= \begin{pmatrix} 0 & 0 & 0 & 0 & j\underline{K}_z & -j\underline{K}_y \\ 0 & 0 & 0 & -j\underline{K}_z & 0 & j\underline{K}_x \\ 0 & 0 & 0 & j\underline{K}_y & -j\underline{K}_x & 0 \\ 0 & j\underline{K}_z & -j\underline{K}_y & 0 & 0 & 0 \\ -j\underline{K}_z & 0 & j\underline{K}_x & 0 & 0 & 0 \\ j\underline{K}_y & -j\underline{K}_x & 0 & 0 & 0 & 0 \end{pmatrix} \begin{pmatrix} \underline{E}_x \\ \underline{E}_y \\ \underline{E}_z \\ \underline{H}_x \\ \underline{H}_y \\ \underline{H}_z \end{pmatrix} \quad (3.12b)$$

Moreover, Equation (3.12b) can be fully generalized to the form with the anisotropic tensor profiles of permittivity and permeability as

$$
k_0
\begin{pmatrix}
\underline{\underline{\varepsilon}}_{xx} & \underline{\underline{\varepsilon}}_{xy} & \underline{\underline{\varepsilon}}_{xz} & 0 & 0 & 0 \\
\underline{\underline{\varepsilon}}_{yx} & \underline{\underline{\varepsilon}}_{yy} & \underline{\underline{\varepsilon}}_{yz} & 0 & 0 & 0 \\
\underline{\underline{\varepsilon}}_{zx} & \underline{\underline{\varepsilon}}_{zy} & \underline{\underline{\varepsilon}}_{zz} & 0 & 0 & 0 \\
0 & 0 & 0 & \underline{\underline{\mu}}_{xx} & \underline{\underline{\mu}}_{xy} & \underline{\underline{\mu}}_{xz} \\
0 & 0 & 0 & \underline{\underline{\mu}}_{yx} & \underline{\underline{\mu}}_{yy} & \underline{\underline{\mu}}_{yz} \\
0 & 0 & 0 & \underline{\underline{\mu}}_{zx} & \underline{\underline{\mu}}_{zy} & \underline{\underline{\mu}}_{zz}
\end{pmatrix}
\begin{pmatrix}
\underline{E}_x \\ \underline{E}_y \\ \underline{E}_z \\ \underline{H}_x \\ \underline{H}_y \\ \underline{H}_z
\end{pmatrix}
$$

$$
=
\begin{pmatrix}
0 & 0 & 0 & 0 & j\underline{K}_z & -j\underline{K}_y \\
0 & 0 & 0 & -j\underline{K}_z & 0 & j\underline{K}_x \\
0 & 0 & 0 & j\underline{K}_y & -j\underline{K}_x & 0 \\
0 & j\underline{K}_z & -j\underline{K}_y & 0 & 0 & 0 \\
-j\underline{K}_z & 0 & j\underline{K}_x & 0 & 0 & 0 \\
j\underline{K}_y & -j\underline{K}_x & 0 & 0 & 0 & 0
\end{pmatrix}
\begin{pmatrix}
\underline{E}_x \\ \underline{E}_y \\ \underline{E}_z \\ \underline{H}_x \\ \underline{H}_y \\ \underline{H}_z
\end{pmatrix}
\qquad (3.12c)
$$

Figure 3.3 shows the principle of FMM with the above-derived results. FMM is the Fourier analysis of Maxwell's equations in the spatial frequency domain. The above transformation process is equivalent to 3D Fourier transform of Maxwell's equations. The next step is to formulate the eigenvalue equation.

From these matrix forms of Maxwell's equations, Equations (3.12a) to (3.12c), it is possible to obtain an eigenvalue equation that gives Bloch eigenmodes and eigenvalues for the single-block structure. In FMM, the z-directional wavevector component, k_z is taken as the eigenvalue, $k_z = k_z(k_{x,0}, k_{y,0}; \omega)$, that is a function of $k_{x,0}$, $k_{y,0}$, and ω. The Fourier coefficients of $(\underline{E}_x, \underline{E}_y, \underline{E}_z)$ or $(\underline{H}_x, \underline{H}_y, \underline{H}_z)$ are taken as the eigenvectors. Equation (3.12b) for diagonal anisotropic material is rearranged as

$$
j(k_z/k_0)\underline{E}_y = \underline{\underline{\mu}}_x \underline{H}_x + j\left(\underline{K}_y/k_0\right)\underline{E}_z
\qquad (3.13a)
$$

$$
j(k_z/k_0)\underline{E}_x = -\underline{\underline{\mu}}_y \underline{H}_y + j\left(\underline{K}_x/k_0\right)\underline{E}_z
\qquad (3.13b)
$$

$$
\underline{H}_z = \underline{\underline{\mu}}_z^{-1}\left(j\left(\underline{K}_y/k_0\right)\underline{E}_x - j\left(\underline{K}_x/k_0\right)\underline{E}_y\right)
\qquad (3.13c)
$$

$$
j(k_z/k_0)\underline{H}_y = \underline{\underline{\varepsilon}}_x \underline{E}_x + j\left(\underline{K}_y/k_0\right)\underline{H}_z
\qquad (3.13d)
$$

$$
j(k_z/k_0)\underline{H}_x = -\underline{\underline{\varepsilon}}_y \underline{E}_y + j\left(\underline{K}_x/k_0\right)\underline{H}_z
\qquad (3.13e)
$$

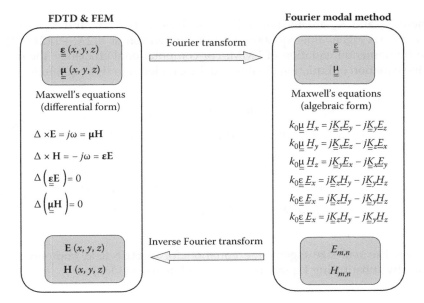

FIGURE 3.3
Principle of Fourier modal method.

$$\underline{E}_z = \underline{\underline{\varepsilon}}_z^{-1}\left(j\left(\underline{K}_y/k_0\right)\underline{H}_x - j\left(\underline{K}_x/k_0\right)\underline{H}_y \right) \qquad (3.13f)$$

By substituting Equations (3.13c) and (3.13f) into Equations (3.13a), (3.13b), (3.13d), and (3.13e), we can obtain the eigenvalue equation with the eigenvalue of the z-directional wavevector component k_z:

$$\begin{pmatrix} 0 & 0 & \underline{K}_y\underline{\underline{\varepsilon}}_z^{-1}\underline{K}_x & \underline{\underline{\mu}}_x - \underline{K}_y\underline{\underline{\varepsilon}}_z^{-1}\underline{K}_y \\ 0 & 0 & -\underline{\underline{\mu}}_y + \underline{K}_x\underline{\underline{\varepsilon}}_z^{-1}\underline{K}_x & -\underline{K}_x\underline{\underline{\varepsilon}}_z^{-1}\underline{K}_y \\ \underline{K}_y\underline{\underline{\mu}}_z^{-1}\underline{K}_x & \underline{\underline{\varepsilon}}_x - \underline{K}_y\underline{\underline{\mu}}_z^{-1}\underline{K}_y & 0 & 0 \\ -\underline{\underline{\varepsilon}}_y + \underline{K}_x\underline{\underline{\mu}}_z^{-1}\underline{K}_x & -\underline{K}_x\underline{\underline{\mu}}_z^{-1}\underline{K}_y & 0 & 0 \end{pmatrix}\begin{pmatrix} \underline{E}_y \\ \underline{E}_x \\ \underline{H}_y \\ \underline{H}_x \end{pmatrix}$$

$$= \frac{jk_z}{k_0}\begin{pmatrix} \underline{E}_y \\ \underline{E}_x \\ \underline{H}_y \\ \underline{H}_x \end{pmatrix} \qquad (3.14)$$

where $\underline{\underline{K}}_x$ and $\underline{\underline{K}}_y$ are normalized by k_0 such as $\underline{\underline{K}}_x \leftarrow \underline{\underline{K}}_x/k_0$ and $\underline{\underline{K}}_y \leftarrow \underline{\underline{K}}_y/k_0$. This matrix eigenvalue equation with the dimension of $4(2M+1)(2N+1)$ is the fundamental equation of FMM. For computational efficiency, the eigenvalue equation of Equation (3.14) is usually divided into two equations;

$$
\begin{pmatrix}
\underline{\underline{K}}_y \underline{\underline{\varepsilon}}_z^{-1} \underline{\underline{K}}_x & \underline{\underline{\mu}}_x - \underline{\underline{K}}_y \underline{\underline{\varepsilon}}_z^{-1} \underline{\underline{K}}_y \\
-\underline{\underline{\mu}}_y + \underline{\underline{K}}_x \underline{\underline{\varepsilon}}_z^{-1} \underline{\underline{K}}_x & -\underline{\underline{K}}_x \underline{\underline{\varepsilon}}_z^{-1} \underline{\underline{K}}_y
\end{pmatrix}
\begin{pmatrix} \underline{H}_y \\ \underline{H}_x \end{pmatrix}
= \frac{jk_z}{k_0}
\begin{pmatrix} \underline{E}_y \\ \underline{E}_x \end{pmatrix}
\tag{3.15a}
$$

$$
\begin{pmatrix}
\underline{\underline{K}}_y \underline{\underline{\mu}}_z^{-1} \underline{\underline{K}}_x & \underline{\underline{\varepsilon}}_x - \underline{\underline{K}}_y \underline{\underline{\mu}}_z^{-1} \underline{\underline{K}}_y \\
-\underline{\underline{\varepsilon}}_y + \underline{\underline{K}}_x \underline{\underline{\mu}}_z^{-1} \underline{\underline{K}}_x & -\underline{\underline{K}}_x \underline{\underline{\mu}}_z^{-1} \underline{\underline{K}}_y
\end{pmatrix}
\begin{pmatrix} \underline{E}_y \\ \underline{E}_x \end{pmatrix}
= \frac{jk_z}{k_0}
\begin{pmatrix} \underline{H}_y \\ \underline{H}_x \end{pmatrix}
\tag{3.15b}
$$

Then the half-size eigenvalue equation for electric field components is made by substituting Equation (3.15b) into Equation (3.15a) as

$$
k_0^2
\begin{pmatrix}
\underline{\underline{K}}_y \underline{\underline{\varepsilon}}_z^{-1} \underline{\underline{K}}_x & \underline{\underline{\mu}}_x - \underline{\underline{K}}_y \underline{\underline{\varepsilon}}_z^{-1} \underline{\underline{K}}_y \\
-\underline{\underline{\mu}}_y + \underline{\underline{K}}_x \underline{\underline{\varepsilon}}_z^{-1} \underline{\underline{K}}_x & -\underline{\underline{K}}_x \underline{\underline{\varepsilon}}_z^{-1} \underline{\underline{K}}_y
\end{pmatrix}
\begin{pmatrix}
\underline{\underline{K}}_y \underline{\underline{\mu}}_z^{-1} \underline{\underline{K}}_x & \underline{\underline{\varepsilon}}_x - \underline{\underline{K}}_y \underline{\underline{\mu}}_z^{-1} \underline{\underline{K}}_y \\
-\underline{\underline{\varepsilon}}_y + \underline{\underline{K}}_x \underline{\underline{\mu}}_z^{-1} \underline{\underline{K}}_x & -\underline{\underline{K}}_x \underline{\underline{\mu}}_z^{-1} \underline{\underline{K}}_y
\end{pmatrix}
\begin{pmatrix} \underline{E}_y \\ \underline{E}_x \end{pmatrix}
$$

$$
= (jk_z)^2
\begin{pmatrix} \underline{E}_y \\ \underline{E}_x \end{pmatrix}
\tag{3.16}
$$

In the case of the diagonal anisotropic material, the magnetic components, $\underline{H}_y$ and $\underline{H}_x$, are given by

$$
\begin{pmatrix} \underline{H}_y \\ \underline{H}_x \end{pmatrix}
= \left(\frac{jk_z}{k_0} \right)
\begin{pmatrix}
\underline{\underline{K}}_y \underline{\underline{\varepsilon}}_z^{-1} \underline{\underline{K}}_x & \underline{\underline{\mu}}_x - \underline{\underline{K}}_y \underline{\underline{\varepsilon}}_z^{-1} \underline{\underline{K}}_y \\
-\underline{\underline{\mu}}_y + \underline{\underline{K}}_x \underline{\underline{\varepsilon}}_z^{-1} \underline{\underline{K}}_x & -\underline{\underline{K}}_x \underline{\underline{\varepsilon}}_z^{-1} \underline{\underline{K}}_y
\end{pmatrix}^{-1}
\begin{pmatrix} \underline{E}_y \\ \underline{E}_x \end{pmatrix}
\tag{3.17a}
$$

Hence the z-directional electric and magnetic field components are obtained as

$$
\underline{E}_z = \underline{\underline{\varepsilon}}_z^{-1} \left[j\left(\underline{\underline{K}}_y/k_0 \right) \underline{H}_x - j\left(\underline{\underline{K}}_x/k_0 \right) \underline{H}_y \right]
\tag{3.17b}
$$

$$
\underline{H}_z = \underline{\underline{\mu}}_z^{-1} \left[j\left(\underline{\underline{K}}_y/k_0 \right) \underline{E}_x - j\left(\underline{\underline{K}}_x/k_0 \right) \underline{E}_y \right]
\tag{3.17c}
$$

For the case of general anisotropic materials, Equation (3.12c) is rearranged as

$$j(k_z/k_0)\underline{E}_y = \underline{\underline{\mu}}_{xx}\underline{H}_x + \underline{\underline{\mu}}_{xy}\underline{H}_y + \underline{\underline{\mu}}_{xz}\underline{H}_z + j\left(\underline{\underline{K}}_y/k_0\right)\underline{E}_z \qquad (3.18a)$$

$$j(k_z/k_0)\underline{E}_x = -\underline{\underline{\mu}}_{yx}\underline{H}_x - \underline{\underline{\mu}}_{yy}\underline{H}_y - \underline{\underline{\mu}}_{yz}\underline{H}_z + j\left(\underline{\underline{K}}_x/k_0\right)\underline{E}_z \qquad (3.18b)$$

$$\underline{H}_z = \underline{\underline{\mu}}_{zz}^{-1}\left[j\left(\underline{\underline{K}}_y/k_0\right)\underline{E}_x - j\left(\underline{\underline{K}}_x/k_0\right)\underline{E}_y - \left(\underline{\underline{\mu}}_{zx}\underline{H}_x + \underline{\underline{\mu}}_{zy}\underline{H}_y\right)\right] \qquad (3.18c)$$

$$j\left(\underline{\underline{K}}_z/k_0\right)\underline{H}_y = \underline{\underline{\varepsilon}}_{xx}\underline{E}_x + \underline{\underline{\varepsilon}}_{xy}\underline{E}_y + \underline{\underline{\varepsilon}}_{xz}\underline{E}_z + j\left(\underline{\underline{K}}_y/k_0\right)\underline{H}_z \qquad (3.18d)$$

$$j\left(\underline{\underline{K}}_z/k_0\right)\underline{H}_x = -\underline{\underline{\varepsilon}}_{yx}\underline{E}_x - \underline{\underline{\varepsilon}}_{yy}\underline{E}_y - \underline{\underline{\varepsilon}}_{yz}\underline{E}_z + j\left(\underline{\underline{K}}_x/k_0\right)\underline{H}_z \qquad (3.18e)$$

$$\underline{E}_z = \underline{\underline{\varepsilon}}_{zz}^{-1}\left[j\left(\underline{\underline{K}}_y/k_0\right)\underline{H}_x - j\left(\underline{\underline{K}}_x/k_0\right)\underline{H}_y - \left(\underline{\underline{\varepsilon}}_{zx}\underline{E}_x + \underline{\underline{\varepsilon}}_{zy}\underline{E}_y\right)\right] \qquad (3.18f)$$

Then, by substituting Equations (3.18c) and (3.18f) into Equations (3.18a), (3.18b), (3.18d), and (3.18e) and with some lengthy manipulation, we have the generalized version of the Maxwell eigenvalue equation:

$$k_0 \begin{bmatrix} -j\underline{\underline{\mu}}_{xz}\underline{\underline{\mu}}_{zz}^{-1}\underline{\underline{K}}_x - j\underline{\underline{K}}_y\underline{\underline{\varepsilon}}_{zz}^{-1}\underline{\underline{\varepsilon}}_{zy} & j\underline{\underline{\mu}}_{xz}\underline{\underline{\mu}}_{zz}^{-1}\underline{\underline{K}}_y - j\underline{\underline{K}}_y\underline{\underline{\varepsilon}}_{zz}^{-1}\underline{\underline{\varepsilon}}_{zx} & \underline{\underline{\mu}}_{xy} - \underline{\underline{\mu}}_{xz}\underline{\underline{\mu}}_{zz}^{-1}\underline{\underline{\mu}}_{zy} + \underline{\underline{K}}_y\underline{\underline{\varepsilon}}_{zz}^{-1}\underline{\underline{K}}_x & \underline{\underline{\mu}}_{xx} - \underline{\underline{\mu}}_{xz}\underline{\underline{\mu}}_{zz}^{-1}\underline{\underline{\mu}}_{zx} - \underline{\underline{K}}_y\underline{\underline{\varepsilon}}_{zz}^{-1}\underline{\underline{K}}_y \\ j\underline{\underline{\mu}}_{yz}\underline{\underline{\mu}}_{zz}^{-1}\underline{\underline{K}}_x - j\underline{\underline{K}}_x\underline{\underline{\varepsilon}}_{zz}^{-1}\underline{\underline{\varepsilon}}_{zy} & -j\underline{\underline{\mu}}_{yz}\underline{\underline{\mu}}_{zz}^{-1}\underline{\underline{K}}_y - j\underline{\underline{K}}_x\underline{\underline{\varepsilon}}_{zz}^{-1}\underline{\underline{\varepsilon}}_{zx} & -\underline{\underline{\mu}}_{yy} + \underline{\underline{\mu}}_{yz}\underline{\underline{\mu}}_{zz}^{-1}\underline{\underline{\mu}}_{zy} + \underline{\underline{K}}_x\underline{\underline{\varepsilon}}_{zz}^{-1}\underline{\underline{K}}_x & -\underline{\underline{\mu}}_{yx} + \underline{\underline{\mu}}_{yz}\underline{\underline{\mu}}_{zz}^{-1}\underline{\underline{\mu}}_{zx} - \underline{\underline{K}}_x\underline{\underline{\varepsilon}}_{zz}^{-1}\underline{\underline{K}}_y \\ \underline{\underline{\varepsilon}}_{xy} - \underline{\underline{\varepsilon}}_{xz}\underline{\underline{\varepsilon}}_{zz}^{-1}\underline{\underline{\varepsilon}}_{zy} + \underline{\underline{K}}_y\underline{\underline{\mu}}_{zz}^{-1}\underline{\underline{K}}_x & \underline{\underline{\varepsilon}}_{xx} - \underline{\underline{\varepsilon}}_{xz}\underline{\underline{\varepsilon}}_{zz}^{-1}\underline{\underline{\varepsilon}}_{zx} - \underline{\underline{K}}_y\underline{\underline{\mu}}_{zz}^{-1}\underline{\underline{K}}_y & -j\underline{\underline{\varepsilon}}_{xz}\underline{\underline{\varepsilon}}_{zz}^{-1}\underline{\underline{K}}_x - j\underline{\underline{K}}_y\underline{\underline{\mu}}_{zz}^{-1}\underline{\underline{\mu}}_{zy} & j\underline{\underline{\varepsilon}}_{xz}\underline{\underline{\varepsilon}}_{zz}^{-1}\underline{\underline{K}}_y - j\underline{\underline{K}}_y\underline{\underline{\mu}}_{zz}^{-1}\underline{\underline{\mu}}_{zx} \\ -\underline{\underline{\varepsilon}}_{yy} - \underline{\underline{\varepsilon}}_{yz}\underline{\underline{\varepsilon}}_{zz}^{-1}\underline{\underline{\varepsilon}}_{zy} + \underline{\underline{K}}_x\underline{\underline{\mu}}_{zz}^{-1}\underline{\underline{K}}_x & -\underline{\underline{\varepsilon}}_{yx} - \underline{\underline{\varepsilon}}_{yz}\underline{\underline{\varepsilon}}_{zz}^{-1}\underline{\underline{\varepsilon}}_{zx} + \underline{\underline{K}}_x\underline{\underline{\mu}}_{zz}^{-1}\underline{\underline{K}}_y & j\underline{\underline{\varepsilon}}_{yz}\underline{\underline{\varepsilon}}_{zz}^{-1}\underline{\underline{K}}_x - j\underline{\underline{K}}_x\underline{\underline{\mu}}_{zz}^{-1}\underline{\underline{\mu}}_{zy} & -j\underline{\underline{\varepsilon}}_{yz}\underline{\underline{\varepsilon}}_{zz}^{-1}\underline{\underline{K}}_y - j\underline{\underline{K}}_x\underline{\underline{\mu}}_{zz}^{-1}\underline{\underline{\mu}}_{zx} \end{bmatrix}$$

$$\times \begin{bmatrix} \underline{E}_y \\ \underline{E}_x \\ \underline{H}_y \\ \underline{H}_x \end{bmatrix} = jk_z \begin{bmatrix} \underline{E}_y \\ \underline{E}_x \\ \underline{H}_y \\ \underline{H}_x \end{bmatrix}. \qquad (3.19)$$

In the case of the general anisotropic material, the z-directional electric and magnetic field components are obtained by Equations (3.18c) and (3.18f);

$$\underline{E}_z = \underline{\underline{\varepsilon}}_{zz}^{-1}\left[j\left(\underline{\underline{K}}_y/k_0\right)\underline{H}_x - j\left(\underline{\underline{K}}_x/k_0\right)\underline{H}_y - \left(\underline{\underline{\varepsilon}}_{zx}\underline{E}_x + \underline{\underline{\varepsilon}}_{zy}\underline{E}_y\right)\right] \qquad (3.20a)$$

$$\underline{H}_z = \underline{\underline{\mu}}_{zz}^{-1}\left[j\left(\underline{\underline{K}}_y/k_0\right)\underline{E}_x - j\left(\underline{\underline{K}}_x/k_0\right)\underline{E}_y - \left(\underline{\underline{\mu}}_{zx}\underline{H}_x + \underline{\underline{\mu}}_{zy}\underline{H}_y\right)\right] \qquad (3.20b)$$

Recent findings in the field of transformation optics and metamaterials can be modeled by the general anisotropic Maxwell eigenvalue equation of Equation (3.19). As a consequence, we have the fundamental eigenvalue equation of Maxwell's equations in the spatial frequency domain. According to the rule of classification of eigenvalues proposed in the previous chapter, the Bloch eigenmodes can be classified into two groups of positive mode and negative mode. (The eigenmodes with eigenvalues of $jk_z = a + jb$ or $jk_z = a - jb$ ($a, b > 0$) are referred to as the negative mode, and the notation $k_{z,0}^-$ with the minus superscript is used to indicate the negative mode. Eigenmodes with eigenvalues of $jk_z = -a + jb$ and $jk_z = -a - jb$ are referred to as positive mode, and the notation $k_{z,0}^+$ with the plus superscript is used to indicate the positive mode. In particular, the eigenmodes with pure real eigenvalues of $jk_z = jb$ and $jk_z = -jb$ with $a = 0$ can be classified into the positive mode.) The number of the positive modes and that of the negative modes are denoted by M^+ and M^-, respectively. The sum of M^+ and M^- is $\varepsilon(x, y, z)$. The magnetic field is renormalized by $(H_x, H_y, H_z) \leftarrow j\sqrt{\frac{\varepsilon_0}{\mu_0}}(H_x, H_y, H_z)$, while the internal electric and magnetic field distributions are expressed, respectively, as the following symmetrically truncated two-dimensional pseudo-Fourier series:

$$
\begin{pmatrix} \mathbf{E}^{(g)} \\ \mathbf{H}^{(g)} \end{pmatrix} = \begin{pmatrix} \mathbf{E}_x^{(g)}, \mathbf{E}_y^{(g)}, \mathbf{E}_z^{(g)} \\ \mathbf{H}_x^{(g)}, \mathbf{H}_y^{(g)}, \mathbf{H}_z^{(g)} \end{pmatrix}
$$

$$
= e^{jk_z^{(g)}z} \sum_{m=-M}^{M} \sum_{n=-N}^{N} \begin{pmatrix} E_{x,m,n}^{(g)}, E_{y,m,n}^{(g)}, E_{z,m,n}^{(g)} \\ H_{x,m,n}^{(g)}, H_{y,m,n}^{(g)}, H_{z,m,n}^{(g)} \end{pmatrix} e^{j(k_{x,m,n}x + k_{y,m,n}y)} \quad (3.21)
$$

It is interesting to compare the patterns of Equation (3.21) and Equation (2.18).

3.1.2 S-Matrix and Coupling Coefficient Operator Calculation of Single Block

In the previous step, the Bloch eigenmodes in a single block with transversal periodic structure were analyzed in the general fashion from Maxwell's equations. The object of an S-matrix analysis is to find four reflection and transmission coefficient vectors by the bidirectional characterization as shown in Figure 3.4. The internal electromagnetic field is represented by the superposition of the Bloch eigenmodes with their own specific coupling coefficients as

$$
\begin{pmatrix} \mathbf{E} \\ \mathbf{H} \end{pmatrix} = \sum_{g=1}^{M^+} C_{a,g}^+ \begin{pmatrix} \mathbf{E}^{(g)+} \\ \mathbf{H}^{(g)+} \end{pmatrix} + \sum_{g=1}^{M^-} C_{a,g}^- \begin{pmatrix} \mathbf{E}^{(g)-} \\ \mathbf{H}^{(g)-} \end{pmatrix} \quad (3.22a)
$$

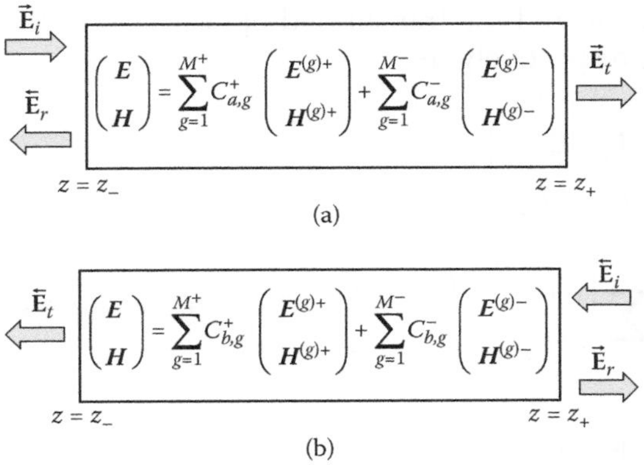

FIGURE 3.4
Bidirectional characterization: (a) left-to-right directional characterization and (b) right-to-left directional characterization.

where M^+ and M^- are equal to $2(2M+1)(2N+1)$ simultaneously. For numerical stability, the representation of the positive Bloch eigenmode and the negative Bloch eigenmode are slightly modified from Equation (3.21), with the Bloch phase terms of $e^{j(k_x x + k_y y + k_z^{(g)+}(z-z_-))}$ and $e^{j(k_x x + k_y y + k_z^{(g)+}(z-z_+))}$, respectively, as

$$
\begin{pmatrix} \mathbf{E}^{(g)+} \\ \mathbf{H}^{(g)+} \end{pmatrix} = \begin{pmatrix} \mathbf{E}_x^{(g)+}, \mathbf{E}_y^{(g)+}, \mathbf{E}_z^{(g)+} \\ \mathbf{H}_x^{(g)+}, \mathbf{H}_y^{(g)+}, \mathbf{H}_z^{(g)+} \end{pmatrix}
$$

$$
= e^{jk_z^{(g)+}(z-z_-)} \sum_{m=-M}^{M} \sum_{n=-N}^{N} \begin{pmatrix} E_{x,m,n}^{(g)+}, E_{y,m,n}^{(g)+}, E_{z,m,n}^{(g)+} \\ H_{x,m,n}^{(g)+}, H_{y,m,n}^{(g)+}, H_{z,m,n}^{(g)+} \end{pmatrix} e^{j(k_{x,m,n}x + k_{y,m,n}y)} \tag{3.22b}
$$

$$
\begin{pmatrix} \mathbf{E}^{(g)-} \\ \mathbf{H}^{(g)-} \end{pmatrix} = \begin{pmatrix} \mathbf{E}_x^{(g)-}, \mathbf{E}_y^{(g)-}, \mathbf{E}_z^{(g)-} \\ \mathbf{H}_x^{(g)-}, \mathbf{H}_y^{(g)-}, \mathbf{H}_z^{(g)-} \end{pmatrix}
$$

$$
= e^{jk_z^{(g)-}(z-z_+)} \sum_{m=-M}^{M} \sum_{n=-N}^{N} \begin{pmatrix} E_{x,m,n}^{(g)-}, E_{y,m,n}^{(g)-}, E_{z,m,n}^{(g)-} \\ H_{x,m,n}^{(g)-}, H_{y,m,n}^{(g)-}, H_{z,m,n}^{(g)-} \end{pmatrix} e^{j(k_{x,m,n}x + k_{y,m,n}y)} \tag{3.22c}
$$

First, the left-to-right directional characterization is conducted. The representations of the incident optical field, the reflected optical field in the lefthand free space, and the transmitted optical field in the righthand free space are given, respectively, as

$$
\begin{pmatrix} \vec{\mathbf{E}}_i \\ \vec{\mathbf{H}}_i \end{pmatrix} = \begin{pmatrix} \vec{\mathbf{E}}_{i,x}, \vec{\mathbf{E}}_{i,y}, \vec{\mathbf{E}}_{i,z} \\ \vec{\mathbf{H}}_{i,x}, \vec{\mathbf{H}}_{i,y}, \vec{\mathbf{H}}_{i,z} \end{pmatrix}
$$

$$
= \sum_{m=-M}^{M} \sum_{n=-N}^{N} \begin{pmatrix} \vec{E}_{i,x,m,n}, \vec{E}_{i,y,m,n}, \vec{E}_{i,z,m,n} \\ \vec{H}_{i,x,m,n}, \vec{H}_{i,y,m,n}, \vec{H}_{i,z,m,n} \end{pmatrix} e^{j(k_{x,m,n}x + k_{y,m,n}y + k_{z,m,n}(z-z_-))} \quad (3.23a)
$$

$$
\begin{pmatrix} \vec{\mathbf{E}}_r \\ \vec{\mathbf{H}}_r \end{pmatrix} = \begin{pmatrix} \vec{\mathbf{E}}_{r,x}, \vec{\mathbf{E}}_{r,y}, \vec{\mathbf{E}}_{r,z} \\ \vec{\mathbf{H}}_{r,x}, \vec{\mathbf{H}}_{r,y}, \vec{\mathbf{H}}_{r,z} \end{pmatrix}
$$

$$
= \sum_{m=-M}^{M} \sum_{n=-N}^{N} \begin{pmatrix} \vec{E}_{r,x,m,n}, \vec{E}_{r,y,m,n}, \vec{E}_{r,z,m,n} \\ \vec{H}_{r,x,m,n}, \vec{H}_{r,y,m,n}, \vec{H}_{r,z,m,n} \end{pmatrix} e^{j(k_{x,m,n}x + k_{y,m,n}y - k_{z,m,n}(z-z_-))} \quad (3.23b)
$$

$$
\begin{pmatrix} \vec{\mathbf{E}}_t \\ \vec{\mathbf{H}}_t \end{pmatrix} = \begin{pmatrix} \vec{\mathbf{E}}_{t,x}, \vec{\mathbf{E}}_{t,y}, \vec{\mathbf{E}}_{t,z} \\ \vec{\mathbf{H}}_{t,x}, \vec{\mathbf{H}}_{t,y}, \vec{\mathbf{H}}_{t,z} \end{pmatrix}
$$

$$
= \sum_{m=-M}^{M} \sum_{n=-N}^{N} \begin{pmatrix} \vec{E}_{t,x,m,n}, \vec{E}_{t,y,m,n}, \vec{E}_{t,z,m,n} \\ \vec{H}_{t,x,m,n}, \vec{H}_{t,y,m,n}, \vec{H}_{t,z,m,n} \end{pmatrix} e^{j(k_{x,m,n}x + k_{y,m,n}y + k_{z,m,n}(z-z_+))} \quad (3.23c)
$$

The tangential components of the incidence, reflection, and transmission magnetic fields are solved in Maxwell's equations by the electric field components as follows:

$$
\vec{\mathbf{H}}_{i,y} = \frac{1}{j\omega\mu_0}\left(\frac{\partial \vec{\mathbf{E}}_{i,x}}{\partial z} - \frac{\partial \vec{\mathbf{E}}_{i,z}}{\partial x} \right) \Leftrightarrow \vec{H}_{i,y,m,n} = \frac{1}{\omega\mu_0}(k_{z,m,n}\vec{E}_{i,x,m,n} - k_{x,m,n}\vec{E}_{i,z,m,n}) \quad (3.24a)
$$

$$
\vec{\mathbf{H}}_{i,x} = \frac{1}{j\omega\mu_0}\left(\frac{\partial \vec{\mathbf{E}}_{i,z}}{\partial y} - \frac{\partial \vec{\mathbf{E}}_{i,y}}{\partial z} \right) \Leftrightarrow \vec{H}_{i,x,m,n} = \frac{1}{\omega\mu_0}(k_{y,m,n}\vec{E}_{i,z,m,n} - k_{z,m,n}\vec{E}_{i,y,m,n}) \quad (3.24b)
$$

$$
\vec{\mathbf{H}}_{r,y} = \frac{1}{j\omega\mu_0}\left(\frac{\partial \vec{\mathbf{E}}_{r,x}}{\partial z} - \frac{\partial \vec{\mathbf{E}}_{r,z}}{\partial x} \right) \Leftrightarrow \vec{H}_{r,y,m,n} = \frac{1}{\omega\mu_0}(-k_{z,m,n}\vec{E}_{r,x,m,n} - k_{x,m,n}\vec{E}_{r,z,m,n}) \quad (3.24c)
$$

$$\vec{\mathbf{H}}_{r,x} = \frac{1}{j\omega\mu_0}\left(\frac{\partial \vec{E}_{r,z}}{\partial y} - \frac{\partial \vec{E}_{r,y}}{\partial z}\right) \Leftrightarrow \vec{H}_{r,x,m,n} = \frac{1}{\omega\mu_0}(k_{y,m,n}\vec{E}_{r,z,m,n} + k_{z,m,n}\vec{E}_{r,y,m,n})$$

$$(3.24d)$$

$$\vec{\mathbf{H}}_{t,y} = \frac{1}{j\omega\mu_0}\left(\frac{\partial \vec{E}_{t,x}}{\partial z} - \frac{\partial \vec{E}_{t,z}}{\partial x}\right) \Leftrightarrow \vec{H}_{t,y,m,n} = \frac{1}{\omega\mu_0}(k_{z,m,n}\vec{E}_{t,x,m,n} - k_{x,m,n}\vec{E}_{t,z,m,n}) \quad (3.24e)$$

$$\vec{\mathbf{H}}_{t,x} = \frac{1}{j\omega\mu_0}\left(\frac{\partial \vec{E}_{t,z}}{\partial y} - \frac{\partial \vec{E}_{t,y}}{\partial z}\right) \Leftrightarrow \vec{H}_{t,x,m,n} = \frac{1}{\omega\mu_0}(k_{y,m,n}\vec{E}_{t,z,m,n} - k_{z,m,n}\vec{E}_{t,y,m,n}) \quad (3.24f)$$

The tangential H-field coefficients, $\vec{H}_{i,y,m,n}$, $\vec{H}_{i,x,m,n}$, are represented by the three E-field coefficients, $\vec{E}_{i,x,m,n}$, $\vec{E}_{i,y,m,n}$, and $\vec{E}_{i,z,m,n}$. The pairs of $\vec{H}_{r,y,m,n}$ and $\vec{H}_{r,x,m,n}$, and $\vec{H}_{t,y,m,n}$ and $\vec{H}_{t,x,m,n}$, are represented by the corresponding three electric field coefficients. However, considering the following plane wave condition in free space,

$$k_{x,m,n}\vec{E}_{i,x,m,n} + k_{y,m,n}\vec{E}_{i,y,m,n} + k_{z,m,n}\vec{E}_{i,z,m,n} = 0 \Leftrightarrow \vec{E}_{i,z,m,n}$$

$$= -\frac{k_{x,m,n}\vec{E}_{i,x,m,n} + k_{y,m,n}\vec{E}_{i,y,m,n}}{k_{z,m,n}} \quad (3.25a)$$

$$k_{x,m,n}\vec{E}_{r,x,m,n} + k_{y,m,n}\vec{E}_{r,y,m,n} - k_{z,m,n}\vec{E}_{r,z,m,n} = 0 \Leftrightarrow \vec{E}_{r,z,m,n}$$

$$= \frac{k_{x,m,n}\vec{E}_{r,x,m,n} + k_{y,m,n}\vec{E}_{r,y,m,n}}{k_{z,m,n}} \quad (3.25b)$$

$$k_{x,m,n}\vec{E}_{t,x,m,n} + k_{y,m,n}\vec{E}_{t,y,m,n} + k_{z,m,n}\vec{E}_{t,z,m,n} = 0 \Leftrightarrow \vec{E}_{t,z,m,n}$$

$$= -\frac{k_{x,m,n}\vec{E}_{t,x,m,n} + k_{y,m,n}\vec{E}_{t,y,m,n}}{k_{z,m,n}} \quad (3.25c)$$

The z-directional components, $\vec{E}_{i,z,m,n}$, $\vec{E}_{r,z,m,n}$, and $\vec{E}_{t,z,m,n}$, can be eliminated by substituting these terms by the tangential electric field coefficients. According to the transverse field continuation condition of the harmonic term, $\exp(j(k_{x,m,n}x + k_{y,m,n}y))$, the boundary condition at the left boundary of $z = z_-$ can be expressed by the matrix equations for $-M \le m \le M$ and $-N \le n \le N$,

$$
\begin{pmatrix}
I & 0 & I & 0 \\
0 & I & 0 & I \\
\dfrac{1}{\omega\mu_0}\dfrac{k_{x,m}k_{y,n}}{k_{z,m,n}} & \dfrac{1}{\omega\mu_0}\dfrac{\left(k_{z,m,n}^2+k_{x,m}^2\right)}{k_{z,m,n}} & -\dfrac{1}{\omega\mu_0}\dfrac{k_{x,m}k_{y,n}}{k_{z,m,n}} & -\dfrac{1}{\omega\mu_0}\dfrac{\left(k_{z,m,n}^2+k_{x,m}^2\right)}{k_{z,m,n}} \\
-\dfrac{1}{\omega\mu_0}\dfrac{\left(k_{y,n}^2+k_{z,m,n}^2\right)}{k_{z,m,n}} & -\dfrac{1}{\omega\mu_0}\dfrac{k_{y,n}k_{x,m}}{k_{z,m,n}} & \dfrac{1}{\omega\mu_0}\dfrac{\left(k_{y,n}^2+k_{z,m,n}^2\right)}{k_{z,m,n}} & \dfrac{1}{\omega\mu_0}\dfrac{k_{y,n}k_{x,m}}{k_{z,m,n}}
\end{pmatrix}
$$

$$
\times
\begin{pmatrix}
\vec{E}_{i,y,m,n} \\
\vec{E}_{i,x,m,n} \\
\vec{E}_{r,y,m,n} \\
\vec{E}_{r,x,m,n}
\end{pmatrix}
=
\begin{pmatrix}
E_{y,m,n}^{(1)+} & \cdots & E_{y,m,n}^{(M^+)+} & E_{y,m,n}^{(1)-}e^{jk_z^{(1)-}(z_--z_+)} & \cdots & E_{y,m,n}^{(M^-)-}e^{jk_z^{(M^-)-}(z_--z_+)} \\
E_{x,m,n}^{(1)+} & \cdots & E_{x,m,n}^{(M^+)+} & E_{x,m,n}^{(1)-}e^{jk_z^{(1)-}(z_--z_+)} & \cdots & E_{x,m,n}^{(M^-)-}e^{jk_z^{(M^-)-}(z_--z_+)} \\
H_{y,m,n}^{(1)+} & \cdots & H_{y,m,n}^{(M^+)+} & H_{y,m,n}^{(1)-}e^{jk_z^{(1)-}(z_--z_+)} & \cdots & H_{y,m,n}^{(M^-)-}e^{jk_z^{(M^-)-}(z_--z_+)} \\
H_{x,m,n}^{(1)+} & \cdots & H_{x,m,n}^{(M^+)+} & H_{x,m,n}^{(1)-}e^{jk_z^{(1)-}(z_--z_+)} & \cdots & H_{x,m,n}^{(M^-)-}e^{jk_z^{(M^-)-}(z_--z_+)}
\end{pmatrix}
\begin{pmatrix}
C_{a,1}^+ \\
\vdots \\
C_{a,M^+}^+ \\
C_{a,1}^- \\
\vdots \\
C_{a,M^-}^-
\end{pmatrix}.
$$

$$(3.26a)$$

Similarly the boundary condition at the right boundary at $z = z_+$ is given by

$$
\begin{pmatrix}
E_{y,m,n}^{(1)+}e^{jk_z^{(1)+}(z_+-z_-)} & \cdots & E_{y,m,n}^{(M^+)+}e^{jk_z^{(M^+)+}(z_+-z_-)} & E_{y,m,n}^{(1)-} & \cdots & E_{y,m,n}^{(M^-)-} \\
E_{x,m,n}^{(1)+}e^{jk_z^{(1)+}(z_+-z_-)} & \cdots & E_{x,m,n}^{(M^+)+}e^{jk_z^{(M^+)+}(z_+-z_-)} & E_{x,m,n}^{(1)-} & \cdots & E_{x,m,n}^{(M^-)-} \\
H_{y,m,n}^{(1)+}e^{jk_z^{(1)+}(z_+-z_-)} & \cdots & H_{y,m,n}^{(M^+)+}e^{jk_z^{(M^+)+}(z_+-z_-)} & H_{y,m,n}^{(1)-} & \cdots & H_{y,m,n}^{(M^-)-} \\
H_{x,m,n}^{(1)+}e^{jk_z^{(1)+}(z_+-z_-)} & \cdots & H_{x,m,n}^{(M^+)+}e^{jk_z^{(M^+)+}(z_+-z_-)} & H_{x,m,n}^{(1)-} & \cdots & H_{x,m,n}^{(M^-)-}
\end{pmatrix}
\begin{pmatrix}
C_{a,1}^+ \\
\vdots \\
C_{a,M^+}^+ \\
C_{a,1}^- \\
\vdots \\
C_{a,M^-}^-
\end{pmatrix}
$$

$$
=
\begin{pmatrix}
I & 0 & I & 0 \\
0 & I & 0 & I \\
\dfrac{1}{\omega\mu_0}\dfrac{k_{x,m}k_{y,n}}{k_{z,m,n}} & \dfrac{1}{\omega\mu_0}\dfrac{\left(k_{z,m,n}^2+k_{x,m}^2\right)}{k_{z,m,n}} & -\dfrac{1}{\omega\mu_0}\dfrac{k_{x,m}k_{y,n}}{k_{z,m,n}} & -\dfrac{1}{\omega\mu_0}\dfrac{\left(k_{z,m,n}^2+k_{x,m}^2\right)}{k_{z,m,n}} \\
-\dfrac{1}{\omega\mu_0}\dfrac{\left(k_{y,n}^2+k_{z,m,n}^2\right)}{k_{z,m,n}} & -\dfrac{1}{\omega\mu_0}\dfrac{k_{y,n}k_{x,m}}{k_{z,m,n}} & \dfrac{1}{\omega\mu_0}\dfrac{\left(k_{y,n}^2+k_{z,m,n}^2\right)}{k_{z,m,n}} & \dfrac{1}{\omega\mu_0}\dfrac{k_{y,n}k_{x,m}}{k_{z,m,n}}
\end{pmatrix}
\begin{pmatrix}
\vec{E}_{t,y,m,n} \\
\vec{E}_{t,x,m,n} \\
0 \\
0
\end{pmatrix}.
$$

$$(3.26b)$$

The boundary condition matching Equations (3.26a) and (3.26b) can be collectively rewritten by the more compact matrix expression as

$$
\begin{pmatrix} \underline{\underline{W}}_h & \underline{\underline{W}}_h \\ \underline{\underline{V}}_h & -\underline{\underline{V}}_h \end{pmatrix} \begin{pmatrix} \vec{E}_i \\ \vec{E}_r \end{pmatrix} = \begin{pmatrix} \underline{\underline{W}}^+(0) & \underline{\underline{W}}^-(z_- - z_+) \\ \underline{\underline{V}}^+(0) & \underline{\underline{V}}^-(z_- - z_+) \end{pmatrix} \begin{pmatrix} \underline{C}_a^+ \\ \underline{C}_a^- \end{pmatrix} \tag{3.27a}
$$

$$
\begin{pmatrix} \underline{\underline{W}}^+(z_+ - z_-) & \underline{\underline{W}}^-(0) \\ \underline{\underline{V}}^+(z_+ - z_-) & \underline{\underline{V}}^-(0) \end{pmatrix} \begin{pmatrix} \underline{C}_a^+ \\ \underline{C}_a^- \end{pmatrix} = \begin{pmatrix} \underline{\underline{W}}_h & \underline{\underline{W}}_h \\ \underline{\underline{V}}_h & -\underline{\underline{V}}_h \end{pmatrix} \begin{pmatrix} \vec{E}_t \\ 0 \end{pmatrix} \tag{3.27b}
$$

where $\underline{\underline{W}}_h$ and $\underline{\underline{V}}_h$ are $[2(2M+1)(2N+1)] \times [2(2M+1)(2N+1)]$ matrices given, respectively, by

$$
\underline{\underline{W}}_h = \begin{pmatrix} \underline{\underline{I}} & 0 \\ 0 & \underline{\underline{I}} \end{pmatrix} \tag{3.27c}
$$

$$
\underline{\underline{V}}_h = \begin{pmatrix} \left[\dfrac{1}{\omega\mu_0} \dfrac{k_{x,m}k_{y,n}}{k_{z,m,n}} \right] & \left[\dfrac{1}{\omega\mu_0} \dfrac{\left(k_{z,m,n}^2 + k_{x,m}^2 \right)}{k_{z,m,n}} \right] \\ \left[-\dfrac{1}{\omega\mu_0} \dfrac{\left(k_{y,n}^2 + k_{z,m,n}^2 \right)}{k_{z,m,n}} \right] & \left[-\dfrac{1}{\omega\mu_0} \dfrac{k_{y,n}k_{x,m}}{k_{z,m,n}} \right] \end{pmatrix} \tag{3.27d}
$$

$\underline{\underline{W}}^+(z)$ and $\underline{\underline{V}}^+(z)$ are $[2(2M + 1)(2N + 1)] \times M^+$ matrices indicating the part of the positive modes in Equations (3.27a) and (3.27b), given, respectively, by

$$
\underline{\underline{W}}^+(z) = \begin{pmatrix} \left[E_{y,m,n}^{(1)+} e^{jk_z^{(1)+}z} \right] & \cdots & \left[E_{y,m,n}^{(M^+)+} e^{jk_z^{(M^+)+}z} \right] \\ \left[E_{x,m,n}^{(1)+} e^{jk_z^{(1)+}z} \right] & \cdots & \left[E_{x,m,n}^{(M^+)+} e^{jk_z^{(M^+)+}z} \right] \end{pmatrix} \tag{3.27e}
$$

$$
\underline{\underline{V}}^+(z) = \begin{pmatrix} \left[H_{y,m,n}^{(1)+} e^{jk_z^{(1)+}z} \right] & \cdots & \left[H_{y,m,n}^{(M^+)+} e^{jk_z^{(M^+)+}z} \right] \\ \left[H_{x,m,n}^{(1)+} e^{jk_z^{(1)+}z} \right] & \cdots & \left[H_{x,m,n}^{(M^+)+} e^{jk_z^{(M^+)+}z} \right] \end{pmatrix} \tag{3.27f}
$$

$\underline{\underline{W}}^-(z)$, and $\underline{\underline{V}}^-(z)$ are $[2(2M+1)(2N+1)] \times M^-$ matrices indicating the part of the negative modes given, respectively, by

$$\underline{\underline{W}}^-(z) = \begin{pmatrix} \left[E^{(1)-}_{y,m,n} e^{jk^{(1)-}_z z} \right] & \cdots & \left[E^{(M-)-}_{y,m,n} e^{jk^{(M-)-}_z z} \right] \\ \left[E^{(1)-}_{x,m,n} e^{jk^{(1)-}_z z} \right] & \cdots & \left[E^{(M-)-}_{x,m,n} e^{jk^{(M-)-}_z z} \right] \end{pmatrix} \qquad (3.27g)$$

$$\underline{\underline{V}}^-(z) = \begin{pmatrix} \left[H^{(1)-}_{y,m,n} e^{jk^{(1)-}_z z} \right] & \cdots & \left[H^{(M-)-}_{y,m,n} e^{jk^{(M-)-}_z z} \right] \\ \left[H^{(1)-}_{x,m,n} e^{jk^{(1)-}_z z} \right] & \cdots & \left[H^{(M-)-}_{x,m,n} e^{jk^{(M-)-}_z z} \right] \end{pmatrix} \qquad (3.27h)$$

The meaning of the notation [ ] forms a matrix according to the following indexing scheme. The diffraction order indices, m and n, are integers in the range of $-M \leq m \leq M$ and $-N \leq n \leq N$, respectively. In this case, the total number of Fourier harmonics used in the representation of the field is $(2M+1)(2N+1)$. It should be noted that there are two independent diffraction channels with the same wavevector. For the (m,n)th harmonics with $(k_{x,m,n}, k_{y,m,n}, k_{z,m,n})$, two independently polarized modes are excited. Therefore, in FMM, the total number of the diffraction channels is considered to be $2(2M+1)(2N+1)$, which is twice the number of the retained Fourier harmonics.

For convenience, using the 1D raw leading ordering in the notation [ ] we let the index pair (m,n) be equivalently indicated by a single index f given in the range of $1 \leq f \leq 2H$, where H is set to $H = (2M+1)(2N+1)$. The relationship between the index pair (m,n) and the index f is defined by

$$f = (m+M)(2N+1) + n + N + 1 \quad \text{for} \quad 1 \leq f \leq H$$

$$f = (m+M)(2N+1) + n + M + N + 1 \quad \text{for} \quad H+1 \leq f \leq 2H$$

through which we can extract the index pair (m, n) from the index f, using the above relationship. In addition, let the index f in the range of $1 \leq f \leq H$ and the index f in the range of $H+1 \leq f \leq 2H$ be allocated to the y-directional field component and x-directional field component, respectively. Set $\underline{\underline{U}}$ to be the input operator, an $[2(2M+1)(2N+1)] \times [2(2M+1)(2N+1)]$ identity matrix. $\underline{\underline{R}}$ and $\underline{\underline{T}}$ are the reflection coefficient matrix operator and transmission coefficient matrix operator, respectively. The coupling coefficient matrix operators are denoted by $\underline{\underline{C}}^+_a$ and $\underline{\underline{C}}^-_a$.

With these linear operators, we can write Equations (3.27a) and (3.27b) by the operator mathematics form

$$
\begin{pmatrix} \underline{\underline{W}}_h & \underline{\underline{W}}_h \\ \underline{\underline{V}}_h & -\underline{\underline{V}}_h \end{pmatrix} \begin{pmatrix} \vec{\underline{U}} \\ \overleftarrow{\underline{R}} \end{pmatrix} = \begin{pmatrix} \underline{\underline{W}}^+(0) & \underline{\underline{W}}^-(z_- - z_+) \\ \underline{\underline{V}}^+(0) & \underline{\underline{V}}^-(z_- - z_+) \end{pmatrix} \begin{pmatrix} \underline{\underline{C}}_a^+ \\ \underline{\underline{C}}_a^- \end{pmatrix}
\tag{3.28a}
$$

$$
\begin{pmatrix} \underline{\underline{W}}^+(z_+ - z_-) & \underline{\underline{W}}^-(0) \\ \underline{\underline{V}}^+(z_+ - z_-) & \underline{\underline{V}}^-(0) \end{pmatrix} \begin{pmatrix} \underline{\underline{C}}_a^+ \\ \underline{\underline{C}}_a^- \end{pmatrix} = \begin{pmatrix} \underline{\underline{W}}_h & \underline{\underline{W}}_h \\ \underline{\underline{V}}_h & -\underline{\underline{V}}_h \end{pmatrix} \begin{pmatrix} \vec{\underline{T}} \\ \underline{\underline{0}} \end{pmatrix}
\tag{3.28b}
$$

The solution is obtained as

$$
\begin{pmatrix} \underline{\underline{C}}_a^+ \\ \underline{\underline{C}}_a^- \end{pmatrix} = \begin{pmatrix} \underline{\underline{W}}_h^{-1}\underline{\underline{W}}^+(0) + \underline{\underline{V}}_h^{-1}\underline{\underline{V}}^+(0) & \underline{\underline{W}}_h^{-1}\underline{\underline{W}}^-(z_- - z_+) + \underline{\underline{V}}_h^{-1}\underline{\underline{V}}^-(z_- - z_+) \\ \underline{\underline{W}}_h^{-1}\underline{\underline{W}}^+(z_+ - z_-) - \underline{\underline{V}}_h^{-1}\underline{\underline{V}}^+(z_+ - z_-) & \underline{\underline{W}}_h^{-1}\underline{\underline{W}}^-(0) - \underline{\underline{V}}_h^{-1}\underline{\underline{V}}^-(0) \end{pmatrix}^{-1}
$$

$$
\times \begin{pmatrix} 2\vec{\underline{U}} \\ \underline{\underline{0}} \end{pmatrix}
\tag{3.29a}
$$

The reflection and transmission operators $\overleftarrow{\underline{R}}$ and $\vec{\underline{T}}$ are obtained, respectively, by

$$
\overleftarrow{\underline{R}} = \underline{\underline{W}}_h^{-1} \left[\underline{\underline{W}}^+(0)\underline{\underline{C}}_a^+ + \underline{\underline{W}}^-(z_- - z_+)\underline{\underline{C}}_a^- - \underline{\underline{W}}_h\vec{\underline{U}} \right]
\tag{3.29b}
$$

$$
\vec{\underline{T}} = \underline{\underline{W}}_h^{-1} \left[\underline{\underline{W}}^+(z_+ - z_-)\underline{\underline{C}}_a^+ + \underline{\underline{W}}^-(0)\underline{\underline{C}}_a^- \right]
\tag{3.29c}
$$

The obtained matrix operators, $\overleftarrow{\underline{R}}$, $\vec{\underline{T}}$, $\underline{\underline{C}}_a^+$, and $\underline{\underline{C}}_a^-$, provide complete information on the left-to-right directional characteristics of the single block. The coupling coefficient matrix operator $\underline{\underline{C}}_a$ is defined by

$$
\underline{\underline{C}}_a = \begin{pmatrix} \underline{\underline{C}}_a^+ \\ \underline{\underline{C}}_a^- \end{pmatrix}
\tag{3.29d}
$$

This operator mathematics of the left-to-right directional characterization is symbolically represented in Figure 3.5(a).

The right-to-left directional characterization illustrated in Figure 3.5(b) proceeds similarly. In the case of the right-to-left characterization, the excitation

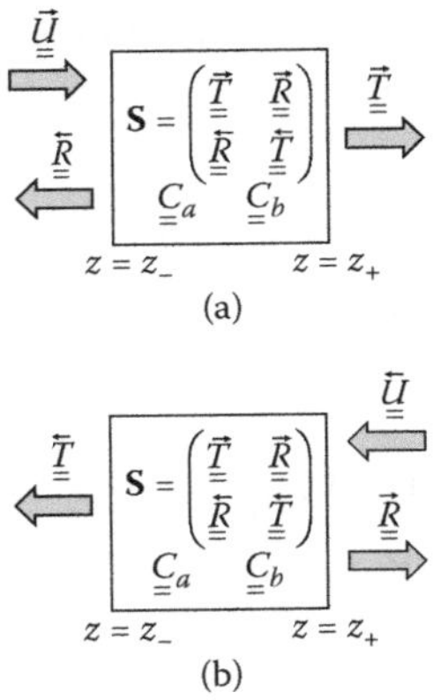

FIGURE 3.5
Operator form description of bidirectional characterization of a single block.

field of the right boundary $\overleftarrow{U}$, the reflection field $\overrightarrow{R}$, and the transmission field $\overleftarrow{T}$ are given, respectively, by

$$
\begin{pmatrix} \overleftarrow{\mathbf{E}}_i \\ \overleftarrow{\mathbf{H}}_i \end{pmatrix} = \begin{pmatrix} \overleftarrow{\mathbf{E}}_{i,x}, \overleftarrow{\mathbf{E}}_{i,y}, \overleftarrow{\mathbf{E}}_{i,z} \\ \overleftarrow{\mathbf{H}}_{i,x}, \overleftarrow{\mathbf{H}}_{i,y}, \overleftarrow{\mathbf{H}}_{i,z} \end{pmatrix}
$$
$$
= \sum_{m=-M}^{M} \sum_{n=-N}^{N} \begin{pmatrix} \overleftarrow{E}_{i,x,m,n}, \overleftarrow{E}_{i,y,m,n}, \overleftarrow{E}_{i,z,m,n} \\ \overleftarrow{H}_{i,x,m,n}, \overleftarrow{H}_{i,y,m,n}, \overleftarrow{H}_{i,z,m,n} \end{pmatrix} e^{j(k_{x,m,n}x + k_{y,m,n}y - k_{z,m,n}(z - z_+))} \quad (3.30a)
$$

$$
\begin{pmatrix} \overrightarrow{\mathbf{E}}_r \\ \overleftarrow{\mathbf{H}}_r \end{pmatrix} = \begin{pmatrix} \overrightarrow{\mathbf{E}}_{r,x}, \overrightarrow{\mathbf{E}}_{r,y}, \overrightarrow{\mathbf{E}}_{r,z} \\ \overrightarrow{\mathbf{H}}_{r,x}, \overrightarrow{\mathbf{H}}_{r,y}, \overrightarrow{\mathbf{H}}_{r,z} \end{pmatrix}
$$
$$
= \sum_{m=-M}^{M} \sum_{n=-N}^{N} \begin{pmatrix} \overrightarrow{E}_{r,x,m,n}, \overrightarrow{E}_{r,y,m,n}, \overrightarrow{E}_{r,z,m,n} \\ \overrightarrow{H}_{r,x,m,n}, \overrightarrow{H}_{r,y,m,n}, \overrightarrow{H}_{r,z,m,n} \end{pmatrix} e^{j(k_{x,m,n}x + k_{y,m,n}y + k_{z,m,n}(z - z_+))} \quad (3.30b)
$$

$$
\begin{pmatrix} \overleftarrow{\mathbf{E}}_t \\ \overleftarrow{\mathbf{H}}_t \end{pmatrix} = \begin{pmatrix} \overleftarrow{\mathbf{E}}_{t,x}, \overleftarrow{\mathbf{E}}_{t,y}, \overleftarrow{\mathbf{E}}_{t,z} \\ \overleftarrow{\mathbf{H}}_{t,x}, \overleftarrow{\mathbf{H}}_{t,y}, \overleftarrow{\mathbf{H}}_{t,z} \end{pmatrix}
$$
$$
= \sum_{m=-M}^{M} \sum_{n=-N}^{N} \begin{pmatrix} \overleftarrow{E}_{t,x,m,n}, \overleftarrow{E}_{t,y,m,n}, \overleftarrow{E}_{t,z,m,n} \\ \overleftarrow{H}_{t,x,m,n}, \overleftarrow{H}_{t,y,m,n}, \overleftarrow{H}_{t,z,m,n} \end{pmatrix} e^{j(k_{x,m,n}x + k_{y,m,n}y - k_{z,m,n}(z - z_-))} \quad (3.30c)
$$

The boundary conditions at the left and right boundaries are described, respectively, by the following matrix operator equations:

$$
\begin{pmatrix} \underline{\underline{W}}_h & \underline{\underline{W}}_h \\ \underline{\underline{V}}_h & -\underline{\underline{V}}_h \end{pmatrix} \begin{pmatrix} \underline{\underline{0}} \\ \vec{\underline{T}} \end{pmatrix} = \begin{pmatrix} \underline{\underline{W}}^+(0) & \underline{\underline{W}}^-(z_- - z_+) \\ \underline{\underline{V}}^+(0) & \underline{\underline{V}}^-(z_- - z_+) \end{pmatrix} \begin{pmatrix} \underline{\underline{C}}_b^+ \\ \underline{\underline{C}}_b^- \end{pmatrix}
\tag{3.31a}
$$

$$
\begin{pmatrix} \underline{\underline{W}}^+(z_+ - z_-) & \underline{\underline{W}}^-(0) \\ \underline{\underline{V}}^+(z_+ - z_-) & \underline{\underline{V}}^-(0) \end{pmatrix} \begin{pmatrix} \underline{\underline{C}}_b^+ \\ \underline{\underline{C}}_b^- \end{pmatrix} = \begin{pmatrix} \underline{\underline{W}}_h & \underline{\underline{W}}_h \\ \underline{\underline{V}}_h & -\underline{\underline{V}}_h \end{pmatrix} \begin{pmatrix} \vec{\underline{R}} \\ \vec{\underline{U}} \end{pmatrix}
\tag{3.31b}
$$

The coupling coefficient matrix operators $\underline{\underline{C}}_b^+$ and $\underline{\underline{C}}_b^-$ are obtained by

$$
\begin{pmatrix} \underline{\underline{C}}_b^+ \\ \underline{\underline{C}}_b^- \end{pmatrix} = \begin{pmatrix} \underline{\underline{W}}_h^{-1}\underline{\underline{W}}^+(0) + \underline{\underline{V}}_h^{-1}\underline{\underline{V}}^+(0) & \underline{\underline{W}}_h^{-1}\underline{\underline{W}}^-(z_- - z_+) + \underline{\underline{V}}_h^{-1}\underline{\underline{V}}^-(z_- - z_+) \\ \underline{\underline{W}}_h^{-1}\underline{\underline{W}}^+(z_+ - z_-) - \underline{\underline{V}}_h^{-1}\underline{\underline{V}}^+(z_+ - z_-) & \underline{\underline{W}}_h^{-1}\underline{\underline{W}}^-(0) - \underline{\underline{V}}_h^{-1}\underline{\underline{V}}^-(0) \end{pmatrix}^{-1}
$$
$$
\times \begin{pmatrix} \underline{\underline{0}} \\ 2\vec{\underline{U}} \end{pmatrix}
\tag{3.32a}
$$

The reflection and transmission coefficient matrix operators $\vec{\underline{R}}$ and $\vec{\underline{T}}$ are obtained, respectively, as

$$
\vec{\underline{R}} = \underline{\underline{W}}_h^{-1}\left[\underline{\underline{W}}^+(z_+ - z_-)\underline{\underline{C}}_b^+ + \underline{\underline{W}}^-(0)\underline{\underline{C}}_b^- - \underline{\underline{W}}_h\vec{\underline{U}} \right]
\tag{3.32b}
$$

$$
\vec{\underline{T}} = W_h^{-1}\left[\underline{\underline{W}}^+(0)\underline{\underline{C}}_b^+ + \underline{\underline{W}}^-(z_- - z_+)\underline{\underline{C}}_b^- \right]
\tag{3.32c}
$$

The obtained matrix operators, $\vec{\underline{R}}$, $\vec{\underline{T}}$, $\underline{\underline{C}}_b^+$, and $\underline{\underline{C}}_b^-$, provide complete information on the right-to-left directional characteristics of the single block. The coupling coefficient matrix operator $\underline{\underline{C}}_b$ is defined by

$$
\underline{\underline{C}}_b = \begin{pmatrix} \underline{\underline{C}}_b^+ \\ \underline{\underline{C}}_b^- \end{pmatrix}
\tag{3.32d}
$$

This operator mathematics of the right-to-left characterization is symbolically represented in Figure 3.5(b).

3.1.3 Field Visualization

The total electromagnetic field distributions have the following mathematical expressions.

1. Left-to-right field visualization. When the incident field, $\vec{\mathbf{E}}_i$, in the left half-infinite free space, strikes on the single-block structure at the boundary $z = z_-$, the reflection field coefficient inside the left half-infinite free space and the transmission field coefficient inside the right half-infinite free space are given, respectively, as

$$\begin{pmatrix} \vec{\bar{E}}_{r,y} \\ \vec{\bar{E}}_{r,x} \end{pmatrix} = \underline{\underline{\bar{R}}} \begin{pmatrix} \vec{\bar{E}}_{i,y} \\ \vec{\bar{E}}_{i,x} \end{pmatrix} \text{ and } \begin{pmatrix} \vec{\bar{E}}_{t,y} \\ \vec{\bar{E}}_{t,x} \end{pmatrix} = \underline{\underline{\bar{T}}} \begin{pmatrix} \vec{\bar{E}}_{i,y} \\ \vec{\bar{E}}_{i,x} \end{pmatrix}.$$

And the coupling coefficients of the internal optical field distribution are represented by

$$\underline{C}_a^+ = \underline{C}_a^+ \begin{pmatrix} \vec{\bar{E}}_{i,y} \\ \vec{\bar{E}}_{i,x} \end{pmatrix} \text{ and } \underline{C}_a^- = \underline{C}_a^- \begin{pmatrix} \vec{\bar{E}}_{i,y} \\ \vec{\bar{E}}_{i,x} \end{pmatrix}.$$

The field distributions take the form (Figure 3.6):

a. For $z < z_-$

$$\begin{pmatrix} \mathbf{E} \\ \mathbf{H} \end{pmatrix} = \begin{pmatrix} \vec{\mathbf{E}}_i \\ \vec{\mathbf{H}}_i \end{pmatrix} + \begin{pmatrix} \vec{\mathbf{E}}_r \\ \vec{\mathbf{H}}_r \end{pmatrix}$$

$$= \sum_{m=-M}^{M} \sum_{n=-N}^{N} \begin{pmatrix} \vec{E}_{i,x,m,n}, \vec{E}_{i,y,m,n}, \vec{E}_{i,z,m,n} \\ \vec{H}_{i,x,m,n}, \vec{H}_{i,y,m,n}, \vec{H}_{i,z,m,n} \end{pmatrix} e^{j(k_{x,m}x + k_{y,m,n}y + k_{z,m,n}(z-z_-))}$$

$$+ \sum_{m=-M}^{M} \sum_{n=-N}^{N} \begin{pmatrix} \vec{E}_{r,x,m,n}, \vec{E}_{r,y,m,n}, \vec{E}_{r,z,m,n} \\ \vec{H}_{r,x,m,n}, \vec{H}_{r,y,m,n}, \vec{H}_{r,z,m,n} \end{pmatrix} e^{j(k_{x,m}x + k_{y,m,n}y + k_{z,m,n}(z-z_-))} \quad (3.33)$$

b. For $z_- \leq z < z_+$

$$\begin{pmatrix} \mathbf{E} \\ \mathbf{H} \end{pmatrix} = \sum_{g=1}^{M^+} C_{a,g}^+ \begin{pmatrix} \mathbf{E}^{(g)+} \\ \mathbf{H}^{(g)+} \end{pmatrix} + \sum_{g=1}^{M^-} C_{a,g}^- \begin{pmatrix} \mathbf{E}^{(g)-} \\ \mathbf{H}^{(g)-} \end{pmatrix}, \quad (3.34a)$$

where $\mathbf{E}^{(g)+}$, $\mathbf{E}^{(g)-}$, $\mathbf{H}^{(g)+}$, and $\mathbf{H}^{(g)-}$ are given, respectively, by

$$\mathbf{E}^{(g)+}(x,y,z) = \sum_{m=-M}^{M} \sum_{n=-N}^{N} \left(E_{x,m,n}^{(g)+}, E_{y,m,n}^{(g)+}, E_{z,m,n}^{(g)+} \right) e^{j\left(k_{x,m}x + k_{y,n}y + k_z^{(g)+}(z-z_-)\right)} \quad (3.34b)$$

$$\mathbf{E}^{(g)-}(x,y,z) = \sum_{m=-M}^{M} \sum_{n=-N}^{N} \left(E_{x,m,n}^{(g)-}, E_{y,m,n}^{(g)-}, E_{z,m,n}^{(g)-} \right) e^{j\left(k_{x,m}x + k_{y,n}y + k_z^{(g)-}(z-z_+)\right)} \quad (3.34c)$$

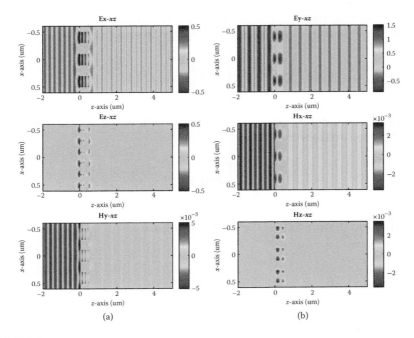

FIGURE 3.6

(a) E_x, E_z, and H_y field distributions for TM mode and (b) E_y, H_x, and H_z field distributions for TE mode diffracted by a metallic binary grating. (MATLAB code: *FMM_singleblock*\ *FMM_singleblock_analysis.m*.)

$$\mathbf{H}^{(g)+}(x,y,z) = \sum_{m=-M}^{M} \sum_{n=-N}^{N} \left(H_{x,m,n}^{(g)+}, H_{y,m,n}^{(g)+}, H_{z,m,n}^{(g)+} \right) e^{j\left(k_{x,m}x + k_{y,n}y + k_z^{(g)+}(z-z_-)\right)} \quad (3.34d)$$

$$\mathbf{H}^{(g)-}(x,y,z) = \sum_{m=-M}^{M} \sum_{n=-N}^{N} \left(H_{x,m,n}^{(g)-}, H_{y,m,n}^{(g)-}, H_{z,m,n}^{(g)-} \right) e^{j\left(k_{x,m}x + k_{y,n}y + k_z^{(g)-}(z-z_+)\right)} \quad (3.34e)$$

c. For $z_+ \leq z$

$$\begin{pmatrix} \vec{\mathbf{E}}_t \\ \vec{\mathbf{H}}_t \end{pmatrix} = \begin{pmatrix} \vec{\mathbf{E}}_{t,x}, \vec{\mathbf{E}}_{t,y}, \vec{\mathbf{E}}_{t,z} \\ \vec{\mathbf{H}}_{t,x}, \vec{\mathbf{H}}_{t,y}, \vec{\mathbf{H}}_{t,z} \end{pmatrix}$$

$$= \sum_{m=-M}^{M} \sum_{n=-N}^{N} \begin{pmatrix} \vec{E}_{t,x,m,n}, \vec{E}_{t,y,m,n}, \vec{E}_{t,z,m,n} \\ \vec{H}_{t,x,m,n}, \vec{H}_{t,y,m,n}, \vec{H}_{t,z,m,n} \end{pmatrix} e^{j(k_{x,m,n}x + k_{y,m,n}y + k_{z,m,n}(z-z_+))} \quad (3.35)$$

2. Right-to-left field visualization. When the incident field, $\vec{\mathbf{E}}_i$, in the right half-infinite free space, goes through the single-block structure at the boundary $z = z_+$, the reflection field coefficient inside the left

half-infinite free space and the transmission field coefficient inside the right half-infinite free space are given, respectively, as

$$
\begin{pmatrix} \vec{E}_{r,y} \\ \vec{E}_{r,x} \end{pmatrix} = \vec{\bar{R}} \begin{pmatrix} \vec{E}_{i,y} \\ \vec{E}_{i,x} \end{pmatrix} \quad \text{and} \quad \begin{pmatrix} \vec{E}_{t,y} \\ \vec{E}_{t,x} \end{pmatrix} = \vec{\bar{T}} \begin{pmatrix} \vec{E}_{i,y} \\ \vec{E}_{i,x} \end{pmatrix}.
$$

In addition, the coupling coefficients of the internal optical field distribution are represented by

$$
\underline{C}_b^+ = \underline{\underline{C}}_b^+ \begin{pmatrix} \vec{E}_{i,y} \\ \vec{E}_{i,x} \end{pmatrix} \quad \text{and} \quad \underline{C}_b^- = \underline{\underline{C}}_b^- \begin{pmatrix} \vec{E}_{i,y} \\ \vec{E}_{i,x} \end{pmatrix}.
$$

The field distributions take the form (Figure 3.6):

a. For $z < z_-$

$$
\begin{pmatrix} \vec{E}_t \\ \vec{H}_t \end{pmatrix} = \begin{pmatrix} \vec{E}_{t,x}, \vec{E}_{t,y}, \vec{E}_{t,z} \\ \vec{H}_{t,x}, \vec{H}_{t,y}, \vec{H}_{t,z} \end{pmatrix}
$$

$$
= \sum_{m=-M}^{M} \sum_{n=-N}^{N} \begin{pmatrix} \vec{E}_{t,x,m,n}, \vec{E}_{t,y,m,n}, \vec{E}_{t,z,m,n} \\ \vec{H}_{t,x,m,n}, \vec{H}_{t,y,m,n}, \vec{H}_{t,z,m,n} \end{pmatrix} e^{j(k_{x,m,n}x + k_{y,m,n}y + k_{z,m,n}(z-z_-))} \quad (3.36)
$$

b. For $z_- \le z < z_+$

$$
\begin{pmatrix} \mathbf{E} \\ \mathbf{H} \end{pmatrix} = \sum_{g=1}^{M^+} C_{b,g}^+ \begin{pmatrix} \mathbf{E}^{(g)+} \\ \mathbf{H}^{(g)+} \end{pmatrix} + \sum_{g=1}^{M^-} C_{b,g}^- \begin{pmatrix} \mathbf{E}^{(g)-} \\ \mathbf{H}^{(g)-} \end{pmatrix} \quad (3.37)
$$

where $\mathbf{E}^{(g)+}$, $\mathbf{E}^{(g)-}$, $\mathbf{H}^{(g)+}$, and $\mathbf{H}^{(g)-}$ are given in Equations (3.34b) to (3.34e).

c. For $z_+ \le z$

$$
\begin{pmatrix} \mathbf{E} \\ \mathbf{H} \end{pmatrix} = \begin{pmatrix} \vec{E}_i \\ \vec{H}_i \end{pmatrix} + \begin{pmatrix} \vec{E}_r \\ \vec{H}_r \end{pmatrix}
$$

$$
= \sum_{m=-M}^{M} \sum_{n=-N}^{N} \begin{pmatrix} \vec{E}_{i,x,m,n}, \vec{E}_{i,y,m,n}, \vec{E}_{i,z,m,n} \\ \vec{H}_{i,x,m,n}, \vec{H}_{i,y,m,n}, \vec{H}_{i,z,m,n} \end{pmatrix} e^{j(k_{x,m,n}x + k_{y,m,n}y - k_{z,m,n}(z-z_+))}
$$

$$
+ \sum_{m=-M}^{M} \sum_{n=-N}^{N} \begin{pmatrix} \vec{E}_{r,x,m,n}, \vec{E}_{r,y,m,n}, \vec{E}_{r,z,m,n} \\ \vec{H}_{r,x,m,n}, \vec{H}_{r,y,m,n}, \vec{H}_{r,z,m,n} \end{pmatrix} e^{j(k_{x,m,n}x + k_{y,m,n}y + k_{z,m,n}(z-z_+))} \quad (3.38)
$$

3.2 Fourier Modal Analysis of Collinear Multiblock Structures

This section provides the principles and mathematical framework for S-matrix analyses of collinear multiblock structures. A multiblock structure is the stack of collinearly cascaded single blocks as shown in Figure 3.7. The pseudo-Fourier representation provided by FMM and the S-matrix method are the main concepts of the whole mathematical framework in this book. The linear optical single block can be completely characterized by the Fourier modal analysis of internal Bloch eigenmodes. Multiblock structures can be completely analyzed using the S-matrix method. The algorithm of

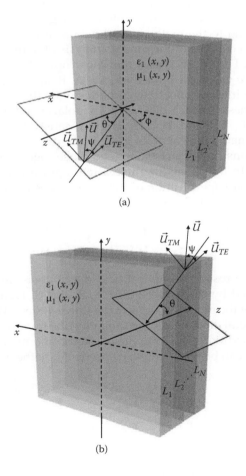

(a)

(b)

FIGURE 3.7
Bidirectional characterization: (a) left-to-right directional characterization and (b) right-to-left directional characterization.

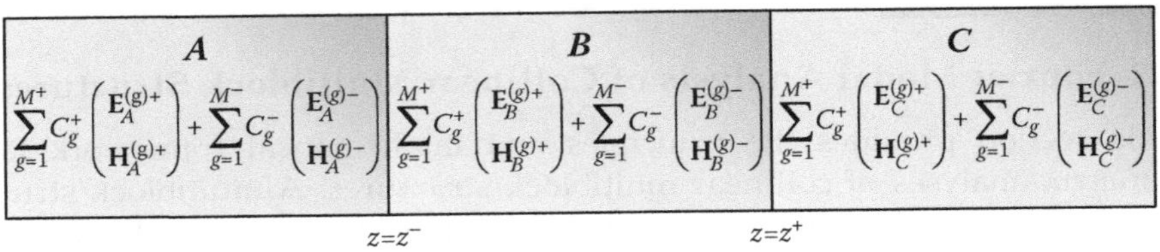

$$\underbrace{\sum_{g=1}^{M^+} C_g^+ \begin{pmatrix} \mathbf{E}_A^{(g)+} \\ \mathbf{H}_A^{(g)+} \end{pmatrix} + \sum_{g=1}^{M^-} C_g^- \begin{pmatrix} \mathbf{E}_A^{(g)-} \\ \mathbf{H}_A^{(g)-} \end{pmatrix}}_{A} \quad \underbrace{\sum_{g=1}^{M^+} C_g^+ \begin{pmatrix} \mathbf{E}_B^{(g)+} \\ \mathbf{H}_B^{(g)+} \end{pmatrix} + \sum_{g=1}^{M^-} C_g^- \begin{pmatrix} \mathbf{E}_B^{(g)-} \\ \mathbf{H}_B^{(g)-} \end{pmatrix}}_{B} \quad \underbrace{\sum_{g=1}^{M^+} C_g^+ \begin{pmatrix} \mathbf{E}_C^{(g)+} \\ \mathbf{H}_C^{(g)+} \end{pmatrix} + \sum_{g=1}^{M^-} C_g^- \begin{pmatrix} \mathbf{E}_C^{(g)-} \\ \mathbf{H}_C^{(g)-} \end{pmatrix}}_{C}$$

$$z = z^- \qquad\qquad\qquad\qquad z = z^+$$

FIGURE 3.8
Linear optical multiblock structure and modal representation of optical fields.

obtaining the electromagnetic field distribution in the multiblock structure is constructed with the block S-matrix and boundary S-matrix of the multiblock structure. The block S-matrix of the block structures is obtained by the bidirectional characterization presented in Figure 3.7.

Before proceeding to the mathematical details, it is neccessary to understand the general concepts of the modal analysis of a linear system and the S-matrix method. Figure 3.8 shows the linear optical multiblock structure composed of three cascaded optical blocks: A, B, and C. Each single-block structure has the internal Bloch eigenmodes with the property of mathematical completeness. In each block, the six polarization terms of the optical field, $\mathbf{E} = (\mathbf{E}_x, \mathbf{E}_y, \mathbf{E}_z)$ and $\mathbf{H} = (\mathbf{H}_x, \mathbf{H}_y, \mathbf{H}_z)$, satisfy Maxwell's equations. According to the mathematical proof given in the previous section, the linear Maxwell's equations have electromagnetic Bloch eigenmodes $(\mathbf{E}^{(g)+}, \mathbf{H}^{(g)+})$ and $(\mathbf{E}^{(g)-}, \mathbf{H}^{(g)-})$. In linear media, the existent optical field $(\mathbf{E}, \mathbf{H})$ is represented by the linear superposition of the eigenmodes with appropriate coupling coefficients C_g^+ and C_g^- as

$$\begin{pmatrix} \mathbf{E} \\ \mathbf{H} \end{pmatrix} = \sum_{g=1}^{M^+} C_g^+ \begin{pmatrix} \mathbf{E}^{(g)+} \\ \mathbf{H}^{(g)+} \end{pmatrix} + \sum_{g=1}^{M^-} C_g^- \begin{pmatrix} \mathbf{E}^{(g)-} \\ \mathbf{H}^{(g)-} \end{pmatrix} \tag{3.39}$$

In Figure 3.8, the respective optical field distributions in the three single blocks are indicated by the subscripts of A, B, and C. Figure 3.8 shows the basic framework of the modal analysis of the linear system. We have to get the concept of the general mode matching satisfying the transversal field continuation boundary condition. The transversal electric and magnetic fields in the blocks A, B, and C can be represented, respectively, by the matrix forms

$$\begin{pmatrix} \mathbf{E}_{A,y} \\ \mathbf{E}_{A,x} \\ \mathbf{H}_{A,y} \\ \mathbf{H}_{A,x} \end{pmatrix} = \begin{pmatrix} \mathbf{E}_{A,y}^{(1)+} & \cdots & \mathbf{E}_{A,y}^{(M^+)+} & \mathbf{E}_{A,y}^{(1)-} & \cdots & \mathbf{E}_{A,y}^{(M^-)-} \\ \mathbf{E}_{A,x}^{(1)+} & \cdots & \mathbf{E}_{A,x}^{(M^+)+} & \mathbf{E}_{A,x}^{(1)-} & \cdots & \mathbf{E}_{A,x}^{(M^-)-} \\ \mathbf{H}_{A,y}^{(1)+} & \cdots & \mathbf{H}_{A,y}^{(M^+)+} & \mathbf{H}_{A,y}^{(1)-} & \cdots & \mathbf{H}_{A,y}^{(M^-)-} \\ \mathbf{H}_{A,x}^{(1)+} & \cdots & \mathbf{H}_{A,x}^{(M^+)+} & \mathbf{H}_{A,x}^{(1)-} & \cdots & \mathbf{H}_{A,x}^{(M^-)-} \end{pmatrix} \begin{pmatrix} C_{A,1}^+ \\ \vdots \\ C_{A,M^+}^+ \\ C_{A,1}^- \\ \vdots \\ C_{A,M^-}^- \end{pmatrix} \tag{3.40a}$$

$$
\begin{pmatrix} \mathbf{E}_{B,y} \\ \mathbf{E}_{B,x} \\ \mathbf{H}_{B,y} \\ \mathbf{H}_{B,x} \end{pmatrix} = \begin{pmatrix} \mathbf{E}_{B,y}^{(1)+} & \cdots & \mathbf{E}_{B,y}^{(M^+)+} & \mathbf{E}_{B,y}^{(1)-} & \cdots & \mathbf{E}_{B,y}^{(M^-)-} \\ \mathbf{E}_{B,x}^{(1)+} & \cdots & \mathbf{E}_{B,x}^{(M^+)+} & \mathbf{E}_{B,x}^{(1)-} & \cdots & \mathbf{E}_{B,x}^{(M^-)-} \\ \mathbf{H}_{B,y}^{(1)+} & \cdots & \mathbf{H}_{B,y}^{(M^+)+} & \mathbf{H}_{B,y}^{(1)-} & \cdots & \mathbf{H}_{B,y}^{(M^-)-} \\ \mathbf{H}_{B,x}^{(1)+} & \cdots & \mathbf{H}_{B,x}^{(M^+)+} & \mathbf{H}_{B,x}^{(1)-} & \cdots & \mathbf{H}_{B,x}^{(M^-)-} \end{pmatrix} \begin{pmatrix} C_{B,1}^+ \\ \vdots \\ C_{B,M^+}^+ \\ C_{B,1}^- \\ \vdots \\ C_{B,M^-}^- \end{pmatrix} \tag{3.40b}
$$

$$
\begin{pmatrix} \mathbf{E}_{C,y} \\ \mathbf{E}_{C,x} \\ \mathbf{H}_{C,y} \\ \mathbf{H}_{.C,x} \end{pmatrix} = \begin{pmatrix} \mathbf{E}_{C,y}^{(1)+} & \cdots & \mathbf{E}_{C,y}^{(M^+)+} & \mathbf{E}_{C,y}^{(1)-} & \cdots & \mathbf{E}_{C,y}^{(M^-)-} \\ \mathbf{E}_{C,x}^{(1)+} & \cdots & \mathbf{E}_{C,x}^{(M^+)+} & \mathbf{E}_{C,x}^{(1)-} & \cdots & \mathbf{E}_{C,x}^{(M^-)-} \\ \mathbf{H}_{C,y}^{(1)+} & \cdots & \mathbf{H}_{C,y}^{(M^+)+} & \mathbf{H}_{C,y}^{(1)-} & \cdots & \mathbf{H}_{C,y}^{(M^-)-} \\ \mathbf{H}_{C,x}^{(1)+} & \cdots & \mathbf{H}_{C,x}^{(M^+)+} & \mathbf{H}_{C,x}^{(1)-} & \cdots & \mathbf{H}_{C,x}^{(M^-)-} \end{pmatrix} \begin{pmatrix} C_{C,1}^+ \\ \vdots \\ C_{C,M^+}^+ \\ C_{C,1}^- \\ \vdots \\ C_{C,M^-}^- \end{pmatrix} \tag{3.40c}
$$

The multiblock structure depicted in Figure 3.8 has two interfaces at $z = z_-$ and $z = z_+$. The boundary conditions of transversal field continuation are expressed as:

1. At $z = z_-$

$$
\begin{pmatrix} \mathbf{E}_{A,y}^{(1)+}(x,y,z_-) & \cdots & \mathbf{E}_{A,y}^{(M^+)+}(x,y,z_-) & \mathbf{E}_{A,y}^{(1)-}(x,y,z_-) & \cdots & \mathbf{E}_{A,y}^{(M^-)-}(x,y,z_-) \\ \mathbf{E}_{A,x}^{(1)+}(x,y,z_-) & \cdots & \mathbf{E}_{A,x}^{(M^+)+}(x,y,z_-) & \mathbf{E}_{A,x}^{(1)-}(x,y,z_-) & \cdots & \mathbf{E}_{A,x}^{(M^-)-}(x,y,z_-) \\ \mathbf{H}_{A,y}^{(1)+}(x,y,z_-) & \cdots & \mathbf{H}_{A,y}^{(M^+)+}(x,y,z_-) & \mathbf{H}_{A,y}^{(1)-}(x,y,z_-) & \cdots & \mathbf{H}_{A,y}^{(M^-)-}(x,y,z_-) \\ \mathbf{H}_{A,x}^{(1)+}(x,y,z_-) & \cdots & \mathbf{H}_{A,x}^{(M^+)+}(x,y,z_-) & \mathbf{H}_{A,x}^{(1)-}(x,y,z_-) & \cdots & \mathbf{H}_{A,x}^{(M^-)-}(x,y,z_-) \end{pmatrix} \begin{pmatrix} C_{A,1}^+ \\ \vdots \\ C_{A,M^+}^+ \\ C_{A,1}^- \\ \vdots \\ C_{A,M^-}^- \end{pmatrix} =
$$

$$
\begin{pmatrix} \mathbf{E}_{B,y}^{(1)+}(x,y,z_-) & \cdots & \mathbf{E}_{B,y}^{(M^+)+}(x,y,z_-) & \mathbf{E}_{B,y}^{(1)-}(x,y,z_-) & \cdots & \mathbf{E}_{B,y}^{(M^-)-}(x,y,z_-) \\ \mathbf{E}_{B,x}^{(1)+}(x,y,z_-) & \cdots & \mathbf{E}_{B,x}^{(M^+)+}(x,y,z_-) & \mathbf{E}_{B,x}^{(1)-}(x,y,z_-) & \cdots & \mathbf{E}_{B,x}^{(M^-)-}(x,y,z_-) \\ \mathbf{H}_{B,y}^{(1)+}(x,y,z_-) & \cdots & \mathbf{H}_{B,y}^{(M^+)+}(x,y,z_-) & \mathbf{H}_{B,y}^{(1)-}(x,y,z_-) & \cdots & \mathbf{H}_{B,y}^{(M^-)-}(x,y,z_-) \\ \mathbf{H}_{B,x}^{(1)+}(x,y,z_-) & \cdots & \mathbf{H}_{B,x}^{(M^+)+}(x,y,z_-) & \mathbf{H}_{B,x}^{(1)-}(x,y,z_-) & \cdots & \mathbf{H}_{B,x}^{(M^-)-}(x,y,z_-) \end{pmatrix} \begin{pmatrix} C_{B,1}^+ \\ \vdots \\ C_{B,M^+}^+ \\ C_{B,1}^- \\ \vdots \\ C_{B,M^-}^- \end{pmatrix}
$$

$$\tag{3.41a}$$

2. At $z = z_+$

$$
\begin{pmatrix}
\mathbf{E}_{B,y}^{(1)+}(x,y,z_+) & \cdots & \mathbf{E}_{B,y}^{(M^+)+}(x,y,z_+) & \mathbf{E}_{B,y}^{(1)-}(x,y,z_+) & \cdots & \mathbf{E}_{B,y}^{(M^-)-}(x,y,z_+) \\
\mathbf{E}_{B,x}^{(1)+}(x,y,z_+) & \cdots & \mathbf{E}_{B,x}^{(M^+)+}(x,y,z_+) & \mathbf{E}_{B,x}^{(1)-}(x,y,z_+) & \cdots & \mathbf{E}_{B,x}^{(M^-)-}(x,y,z_+) \\
\mathbf{H}_{B,y}^{(1)+}(x,y,z_+) & \cdots & \mathbf{H}_{B,y}^{(M^+)+}(x,y,z_+) & \mathbf{H}_{B,y}^{(1)-}(x,y,z_+) & \cdots & \mathbf{H}_{B,y}^{(M^-)-}(x,y,z_+) \\
\mathbf{H}_{B,x}^{(1)+}(x,y,z_+) & \cdots & \mathbf{H}_{B,x}^{(M^+)+}(x,y,z_+) & \mathbf{H}_{B,x}^{(1)-}(x,y,z_+) & \cdots & \mathbf{H}_{B,x}^{(M^-)-}(x,y,z_+)
\end{pmatrix}
\begin{pmatrix}
C_{B,1}^+ \\ \vdots \\ C_{B,M^+}^+ \\ C_{B,1}^- \\ \vdots \\ C_{B,M^-}^-
\end{pmatrix}
$$

$$
=
\begin{pmatrix}
\mathbf{E}_{C,y}^{(1)+}(x,y,z_+) & \cdots & \mathbf{E}_{C,y}^{(M^+)+}(x,y,z_+) & \mathbf{E}_{C,y}^{(1)-}(x,y,z_+) & \cdots & \mathbf{E}_{C,y}^{(M^-)-}(x,y,z_+) \\
\mathbf{E}_{C,x}^{(1)+}(x,y,z_+) & \cdots & \mathbf{E}_{C,y}^{(M^+)+}(x,y,z_+) & \mathbf{E}_{C,x}^{(1)-}(x,y,z_+) & \cdots & \mathbf{E}_{C,x}^{(M^-)-}(x,y,z_+) \\
\mathbf{H}_{C,y}^{(1)+}(x,y,z_+) & \cdots & \mathbf{H}_{C,y}^{(M^+)+}(x,y,z_+) & \mathbf{H}_{C,y}^{(1)-}(x,y,z_+) & \cdots & \mathbf{H}_{C,y}^{(M^-)-}(x,y,z_+) \\
\mathbf{H}_{C,x}^{(1)+}(x,y,z_+) & \cdots & \mathbf{H}_{C,x}^{(M^+)+}(x,y,z_+) & \mathbf{H}_{C,x}^{(1)-}(x,y,z_+) & \cdots & \mathbf{H}_{C,x}^{(M^-)-}(x,y,z_+)
\end{pmatrix}
\begin{pmatrix}
C_{C,1}^+ \\ \vdots \\ C_{C,M^+}^+ \\ C_{C,1}^- \\ \vdots \\ C_{C,M^-}^-
\end{pmatrix}
$$

$$
\text{(3.41b)}
$$

The basic process of a modal analysis for obtaining the total field distribution within the cascaded multiblock structure is simple. First, prepare a complete set of eigenmodes for each block, and second, solve the coupling coefficients. This simple process is known to be mathematically correct but numerically unstable. With this simple process, some difficulties can be expected in finding the correct solution. For the last few decades, the stable numerical process of the modal analysis has been actively investigated. In particular, in FMM, eliminating the numerical instability occurring in algebraic calculations was a challenging problem. The final answer for this problem is provided by the S-matrix method. We can say that the S-matrix is a general recursive algorithm for solving these types of linear modal analyses with a confirmed numerical stability. In FMM, the S-matrix method can be applied fairly easily to the electromagnetic modal analysis of multiblock structures. As was identified in the previous chapter, numerical stability and parallelism are the most distinguishable advantages of the S-matrix method. The extended Redheffer star product defined in Chapter 2 suits fairly well a multiblock structure with transversal periodic permittivity and permeability profiles. It is necessary to prepare the complete set of eigenmodes $(\mathbf{E}^{(g)}, \mathbf{H}^{(g)})$ and the S-matrices of each individual single block by FMM as elucidated in Section 3.1, comprising multiblock structure for multiblock interconnection. The information on the internal field distribution in an optical single block is not pertained in the S-matrix operator. The coupling coefficient operators, $\mathbf{C}_a$ and $\mathbf{C}_b$, are required to derive the internal field distribution.

Consequently, the developed S-matrix interconnection algorithm developed in Chapter 2 is directly used without any structural modification. The summary of the S-matrix method for FMM is given below. However, it should be noted that the half-infinte space can also be a transversely inhomogeneous space with periodic transversal permittivity and peameability in general.

3.2.1 Two-Block Interconnection

The reflection and transmission matrix operators, $\underline{\underline{\tilde{R}}}^{(1,2)}$, $\underline{\underline{\tilde{T}}}^{(1,2)}$, $\underline{\underline{\vec{R}}}^{(1,2)}$, and $\underline{\underline{\vec{T}}}^{(1,2)}$ of $\mathbf{S}^{(1,2)}$ of the multiblock, are obtained as

$$\underline{\underline{\tilde{R}}}^{(1,2)} = \underline{\underline{\tilde{R}}}^{(1,1)} + \underline{\underline{\tilde{T}}}^{(1,1)}\left(\underline{\underline{I}} - \underline{\underline{\tilde{R}}}^{(2,2)}\underline{\underline{\vec{R}}}^{(1,1)}\right)^{-1}\underline{\underline{\tilde{R}}}^{(2,2)}\underline{\underline{\vec{T}}}^{(1,1)} \tag{3.42a}$$

$$\underline{\underline{\vec{T}}}^{(1,2)} = \underline{\underline{\vec{T}}}^{(2,2)}\left(\underline{\underline{I}} - \underline{\underline{\vec{R}}}^{(1,1)}\underline{\underline{\tilde{R}}}^{(2,2)}\right)^{-1}\underline{\underline{\vec{T}}}^{(1,1)} \tag{3.42b}$$

$$\underline{\underline{\vec{R}}}^{(1,2)} = \underline{\underline{\vec{R}}}^{(2,2)} + \underline{\underline{\vec{T}}}^{(2,2)}\left(\underline{\underline{I}} - \underline{\underline{\vec{R}}}^{(1,1)}\underline{\underline{\tilde{R}}}^{(2,2)}\right)^{-1}\underline{\underline{\vec{R}}}^{(1,1)}\underline{\underline{\tilde{T}}}^{(2,2)} \tag{3.42c}$$

$$\underline{\underline{\tilde{T}}}^{(1,2)} = \underline{\underline{\tilde{T}}}^{(1,1)}\left(\underline{\underline{I}} - \underline{\underline{\tilde{R}}}^{(2,2)}\underline{\underline{\vec{R}}}^{(1,1)}\right)^{-1}\underline{\underline{\tilde{T}}}^{(2,2)} \tag{3.42d}$$

This relationship is symbolized by the star product of the block S-matrices as

$$\mathbf{S}^{(1,2)} = \mathbf{S}^{(1,1)} * \mathbf{S}^{(2,2)} \tag{3.43}$$

The internal coupling coefficient matrix operators of the combined multi-block, $\mathbf{C}_{a,(1,2)}^{(1,2)}$ and $\mathbf{C}_{b,(1,2)}^{(1,2)}$, are the set of the coupling coefficient matrix operators of the first and second blocks,

$$\mathbf{C}_{a,(1,2)}^{(1,2)} = \left\{\underline{\underline{C}}_{a,(1)}^{(1,2)}, \underline{\underline{C}}_{a,(2)}^{(1,2)}\right\} \tag{3.44a}$$

$$\mathbf{C}_{b,(1,2)}^{(1,2)} = \left\{\underline{\underline{C}}_{b,(1)}^{(1,2)}, \underline{\underline{C}}_{b,(2)}^{(1,2)}\right\} \tag{3.44b}$$

where $\{\underline{\underline{C}}_{a,(1)}^{(1,2)}, \underline{\underline{C}}_{b,(1)}^{(1,2)}\}$ and $\{\underline{\underline{C}}_{a,(2)}^{(1,2)}, \underline{\underline{C}}_{b,(2)}^{(1,2)}\}$ are given, respectively, by

$$\underline{\underline{C}}_{a,(1)}^{(1,2)} = \underline{\underline{C}}_{a,(1)}^{(1,1)} + \underline{\underline{C}}_{b,(1)}^{(1,1)}\left(\underline{\underline{I}} - \underline{\underline{\tilde{R}}}^{(2,2)}\underline{\underline{\vec{R}}}^{(1,1)}\right)^{-1}\underline{\underline{\tilde{R}}}^{(2,2)}\underline{\underline{\vec{T}}}^{(1,1)} \tag{3.45a}$$

$$\underline{\underline{C}}_{b,(1)}^{(1,2)} = \underline{\underline{C}}_{b,(1)}^{(1,1)}\left(\underline{\underline{I}} - \underline{\underline{\tilde{R}}}^{(2,2)}\underline{\underline{\vec{R}}}^{(1,1)}\right)^{-1}\underline{\underline{\tilde{T}}}^{(2,2)} \tag{3.45b}$$

$$\underline{\underline{C}}_{a,(2)}^{(1,2)} = \underline{\underline{C}}_{a,(2)}^{(2,2)}\left(\underline{\underline{I}} - \underline{\underline{\vec{R}}}^{(1,1)}\underline{\underline{\tilde{R}}}^{(2,2)}\right)^{-1}\underline{\underline{\vec{T}}}^{(1,1)} \tag{3.45c}$$

$$\underline{\underline{C}}_{b,(2)}^{(1,2)} = \underline{\underline{C}}_{b,(2)}^{(2,2)} + \underline{\underline{C}}_{a,(2)}^{(2,2)}\left(\underline{\underline{I}} - \underline{\underline{\vec{R}}}^{(1,1)}\underline{\underline{\tilde{R}}}^{(2,2)}\right)^{-1}\underline{\underline{\vec{R}}}^{(1,1)}\underline{\underline{\tilde{T}}}^{(2,2)} \tag{3.45d}$$

3.2.2 N-Block Interconnection with Parallelism

The S-matrix components of the combined multiblocks of $M^{(n,n+m)}$ and $M^{(n+m+1,n+m+l)}$ are mathematically described by the Redheffer star product as

$$\tilde{\underline{\underline{R}}}^{(n,n+m+l)} = \tilde{\underline{\underline{R}}}^{(n,n+m)} + \tilde{\underline{\underline{T}}}^{(n,n+m)}\left(\underline{\underline{I}} - \tilde{\underline{\underline{R}}}^{(n+m+1,n+m+l)}\vec{\underline{\underline{R}}}^{(n,n+m)}\right)^{-1}\tilde{\underline{\underline{R}}}^{(n+m+1,n+m+l)}\vec{\underline{\underline{T}}}^{(n,n+m)} \quad (3.46a)$$

$$\vec{\underline{\underline{T}}}^{(n,n+m+l)} = \vec{\underline{\underline{T}}}^{(n+m+1,n+m+l)}\left(\underline{\underline{I}} - \vec{\underline{\underline{R}}}^{(n,n+m)}\tilde{\underline{\underline{R}}}^{(n+m+1,n+m+l)}\right)^{-1}\vec{\underline{\underline{T}}}^{(n,n+m)} \quad\quad (3.46b)$$

$$\vec{\underline{\underline{R}}}^{(n,n+m+l)} = \vec{\underline{\underline{R}}}^{(n+m+1,n+m+l)} + \vec{\underline{\underline{T}}}^{(n+m+1,n+m+l)}\left(\underline{\underline{I}} - \vec{\underline{\underline{R}}}^{(n,n+m)}\tilde{\underline{\underline{R}}}^{(n+m+1,n+m+l)}\right)^{-1}$$

$$\times \vec{\underline{\underline{R}}}^{(n,n+m)}\tilde{\underline{\underline{T}}}^{(n+m+1,n+m+l)} \quad\quad (3.46c)$$

$$\tilde{\underline{\underline{T}}}^{(n,n+m+l)} = \tilde{\underline{\underline{T}}}^{(n,n+m)}\left(\underline{\underline{I}} - \tilde{\underline{\underline{R}}}^{(n+m+1,n+m+l)}\vec{\underline{\underline{R}}}^{(n,n+m)}\right)^{-1}\tilde{\underline{\underline{T}}}^{(n+m+1,n+m+l)} \quad\quad (3.46d)$$

This relationship is symbolized by the star product of the block S-matrices as

$$\mathbf{S}^{(n,n+m+l)} = \mathbf{S}^{(n,n+m)} * \mathbf{S}^{(n+m+1,n+m+l)} \quad\quad (3.47)$$

From the associative rule of the S-matrices, the block S-matrix of the multiblock $M^{(1,N)}$ is obtained by

$$\mathbf{S}^{(1,N)} = \mathbf{S}^{(1,1)} * \mathbf{S}^{(2,2)} * \cdots * \mathbf{S}^{(N-1,N-1)} * \mathbf{S}^{(N,N)} \quad\quad (3.48)$$

The extended Redheffer star product of the internal coupling coefficient operator matrices of the combined multiblock of $M^{(n,n+m)}$ and $M^{(n+m+1,n+m+l)}$ produces the set of the coupling coefficient matrices $\mathbf{C}_{a,(n,n+m+l)}^{(n,n+m+l)}$ and $\mathbf{C}_{b,(n,n+m+l)}^{(n,n+m+l)}$ of the combined multiblock $M^{(n,n+m+l)}$, which are given, respectively, by

$$\mathbf{C}_{a,(n,n+m+l)}^{(n,n+m+l)} = \left\{\mathbf{C}_{a,(n,n+m)}^{(n,n+m+l)}, \mathbf{C}_{a,(n+m+1,n+m+l)}^{(n,n+m+l)}\right\} \quad\quad (3.49a)$$

$$\mathbf{C}_{b,(n,n+m+l)}^{(n,n+m+l)} = \left\{\mathbf{C}_{b,(n,n+m)}^{(n,n+m+l)}, \mathbf{C}_{b,(n+m+1,n+m+l)}^{(n,n+m+l)}\right\} \quad\quad (3.49b)$$

where the subsets $\mathbf{C}_{a,(n,n+m)}^{(n,n+m+l)}$, $\mathbf{C}_{b,(n,n+m)}^{(n,n+m+l)}$, $\mathbf{C}_{a,(n+m+1,n+m+l)}^{(n,n+m+l)}$, and $\mathbf{C}_{b,(n+m+1,n+m+l)}^{(n,n+m+l)}$ are given, respectively, by

$$\mathbf{C}_{a,(n,n+m)}^{(n,n+m+l)} = \left\{\underline{\underline{C}}_{a,(n)}^{(n,n+m+l)}, \underline{\underline{C}}_{a,(n+1)}^{(n,n+m+l)}, \dots, \underline{\underline{C}}_{a,(n+m)}^{(n,n+m+l)}\right\} \quad\quad (3.49c)$$

$$\mathbf{C}_{b,(n,n+m)}^{(n,n+m+l)} = \left\{\underline{\underline{C}}_{b,(n)}^{(n,n+m+l)}, \underline{\underline{C}}_{b,(n+1)}^{(n,n+m+l)}, \dots, \underline{\underline{C}}_{b,(n+m)}^{(n,n+m+l)}\right\} \quad\quad (3.49d)$$

$$C_{a,(n+m+1,n+m+l)}^{(n,n+m+l)} = \left\{ \underline{C}_{a,(n+m+1)}^{(n,n+m+l)}, \underline{C}_{a,(n+m+2)}^{(n,n+m+l)}, \cdots, \underline{C}_{a,(n+m+l)}^{(n,n+m+l)} \right\} \tag{3.49e}$$

$$C_{b,(n+m+1,n+m+l)}^{(n,n+m+l)} = \left\{ \underline{C}_{b,(n+m+1)}^{(n,n+m+l)}, \underline{C}_{b,(n+m+2)}^{(n,n+m+l)}, \cdots, \underline{C}_{b,(n+m+l)}^{(n,n+m+l)} \right\} \tag{3.49f}$$

The respective updated pairs of the set of the coupling coefficient matrices $\left(C_{a,(n,n+m)}^{(n,n+m+l)}, C_{b,(n,n+m)}^{(n,n+m+l)} \right)$ and $\left(C_{a,(n+m+1,n+m+l)}^{(n,n+m+l)}, C_{b,(n+m+1,n+m+l)}^{(n,n+m+l)} \right)$ in the part of blocks $L_n - L_{n+m}$ and in the part of blocks $L_{n+m+1} - L_{n+m+l}$ are obtained by the following relations. For k in the range of $n \le k \le n+m$, the coupling coefficient matrices, $\underline{C}_{a,(k)}^{(n,n+m+l)}$ and $\underline{C}_{b,(k)}^{(n,n+m+l)}$, are derived as

$$\underline{C}_{a,(k)}^{(n,n+m+l)} = \underline{C}_{a,(k)}^{(n,n+m)} + \underline{C}_{b,(k)}^{(n,n+m)} \left(\underline{\underline{I}} - \vec{\underline{R}}^{(n+m+1,n+m+l)} \tilde{\underline{R}}^{(n,n+m)} \right)^{-1} \vec{\underline{R}}^{(n+m+1,n+m+l)} \vec{\underline{T}}^{(n,n+m)} \tag{3.50a}$$

$$\underline{C}_{b,(k)}^{(n,n+m+l)} = \underline{C}_{b,(k)}^{(n,n+m)} \left(\underline{\underline{I}} - \vec{\underline{R}}^{(n+m+1,n+m+l)} \tilde{\underline{R}}^{(n,n+m)} \right)^{-1} \tilde{\underline{T}}^{(n+m+1,n+m+l)} \tag{3.50b}$$

For k in the range of $n+m+1 \le k \le n+m+l$, the coupling coefficient matrices, $\underline{C}_{a,(k)}^{(n,n+m+l)}$ and $\underline{C}_{b,(k)}^{(n,n+m+l)}$, are derived as

$$\underline{C}_{a,(k)}^{(n,n+m+l)} = \underline{C}_{a,(k)}^{(n+m+1,n+m+l)} \left(\underline{\underline{I}} - \vec{\underline{R}}^{(n,n+m)} \tilde{\underline{R}}^{(n+m+1,n+m+l)} \right)^{-1} \vec{\underline{T}}^{(n,n+m)} \tag{3.50c}$$

$$\underline{C}_{b,(k)}^{(n,n+m+l)} = \underline{C}_{b,(k)}^{(n+m+1,n+m+l)} + \underline{C}_{a,(k)}^{(n+m+1,n+m+l)} \left(\underline{\underline{I}} - \vec{\underline{R}}^{(n,n+m)} \tilde{\underline{R}}^{(n+m+1,n+m+l)} \right)^{-1}$$
$$\times \vec{\underline{R}}^{(n,n+m)} \tilde{\underline{T}}^{(n+m+1,n+m+l)} \tag{3.50d}$$

These relationships can be symbolized by the extended star product as

$$\left(C_{a,(n,n+m+l)}^{(n,n+m+l)}, C_{b,(n,n+m+l)}^{(n,n+m+l)} \right) = \left(C_{a,(n,n+m)}^{(n,n+m)}, C_{b,(n,n+m)}^{(n,n+m)} \right) * \left(C_{a,(n+m+1,n+m+l)}^{(n+m+1,n+m+l)}, C_{b,(n+m+1,n+m+l)}^{(n+m+1,n+m+l)} \right) \tag{3.51a}$$

The associative rules enable the parallel computation of the internal coupling coefficients:

$$\left(C_{a,(1,N)}^{(1,N)}, C_{b,(1,N)}^{(1,N)} \right) = \left(\underline{C}_{a,(1)}^{(1,1)}, \underline{C}_{b,(1)}^{(1,1)} \right) * \left(\underline{C}_{a,(2)}^{(2,2)}, \underline{C}_{b,(2)}^{(2,2)} \right)$$
$$* \cdots * \left(\underline{C}_{a,(N-1)}^{(N-1,N-1)}, \underline{C}_{b,(N-1)}^{(N-1,N-1)} \right) * \left(\underline{C}_{a,(N)}^{(N,N)}, \underline{C}_{b,(N)}^{(N,N)} \right) \tag{3.51b}$$

3.2.3 Half-Infinite Block Interconnection

The boundary S-matrix of the left half-infinite blocks is given by

$$\mathbf{S}^{(0,0)} = \begin{pmatrix} \vec{\underline{T}}^{(0,0)} & \vec{\underline{R}}^{(0,0)} \\ \tilde{\underline{R}}^{(0,0)} & \tilde{\underline{T}}^{(0,0)} \end{pmatrix} \tag{3.52}$$

The boundary condition at $z = z_c$ is described as

$$\begin{pmatrix} \underline{\underline{W}}^{(0)+}(z_c) & \underline{\underline{W}}^{(0)-}(z_c) \\ \underline{\underline{V}}^{(0)+}(z_c) & \underline{\underline{V}}^{(0)-}(z_c) \end{pmatrix} \begin{pmatrix} \vec{\underline{U}} \\ \tilde{\underline{\underline{R}}}^{(0,0)} \end{pmatrix} = \begin{pmatrix} \underline{\underline{W}}_h & \underline{\underline{W}}_h \\ \underline{\underline{V}}_h & -\underline{\underline{V}}_h \end{pmatrix} \begin{pmatrix} \vec{\underline{T}}^{(0,0)} \\ \underline{\underline{0}} \end{pmatrix} \tag{3.53a}$$

$$\begin{pmatrix} \underline{\underline{W}}^{(0)+}(z_c) & \underline{\underline{W}}^{(0)-}(z_c) \\ \underline{\underline{V}}^{(0)+}(z_c) & \underline{\underline{V}}^{(0)-}(z_c) \end{pmatrix} \begin{pmatrix} \underline{\underline{0}} \\ \tilde{\underline{\underline{T}}}^{(0,0)} \end{pmatrix} = \begin{pmatrix} \underline{\underline{W}}_h & \underline{\underline{W}}_h \\ \underline{\underline{V}}_h & -\underline{\underline{V}}_h \end{pmatrix} \begin{pmatrix} \vec{\underline{\underline{R}}}^{(0,0)} \\ \vec{\underline{U}} \end{pmatrix} \tag{3.53b}$$

The functions $\underline{\underline{W}}_h$, $\underline{\underline{V}}_h$, $\underline{\underline{W}}^{(0)+}(z_c)$, $\underline{\underline{W}}^{(0)-}(z_c)$, $\underline{\underline{V}}^{(0)+}(z_c)$, and $\underline{\underline{V}}^{(0)-}(z_c)$ are defined in Equation (3.27). The S-matrix elements $\vec{\underline{T}}^{(0,0)}$, $\vec{\underline{\underline{R}}}^{(0,0)}$, $\tilde{\underline{\underline{T}}}^{(0,0)}$, and $\tilde{\underline{\underline{R}}}^{(0,0)}$ are obtained, respectively, as

$$\tilde{\underline{\underline{R}}}^{(0,0)} = -\left[\left(\underline{\underline{W}}_h\right)^{-1}\underline{\underline{W}}^{(0)-}(z_c) - \left(\underline{\underline{V}}_h\right)^{-1}\underline{\underline{V}}^{(0)-}(z_c)\right]^{-1}$$
$$\times\left[\left(\underline{\underline{W}}_h\right)^{-1}\underline{\underline{W}}^{(0)+}(z_c) - \left(\underline{\underline{V}}_h\right)^{-1}\underline{\underline{V}}^{(0)+}(z_c)\right] \tag{3.53c}$$

$$\vec{\underline{T}}^{(0,0)} = \left[\left(\underline{\underline{W}}^{(0)-}(z_c)\right)^{-1}\underline{\underline{W}}_h - \left(\underline{\underline{V}}^{(0)-}(z_c)\right)^{-1}\underline{\underline{V}}_h\right]^{-1}$$
$$\times\left[\left(\underline{\underline{W}}^{(0)-}(z_c)\right)^{-1}\underline{\underline{W}}^{(0)+}(z_c) - \left(\underline{\underline{V}}^{(0)-}(z_c)\right)^{-1}\underline{\underline{V}}^{(0)+}(z_c)\right] \tag{3.53d}$$

$$\tilde{\underline{\underline{T}}}^{(0,0)} = 2\left[\left(\underline{\underline{W}}_h\right)^{-1}\underline{\underline{W}}^{(0)-}(z_c) - \left(\underline{\underline{V}}_h\right)^{-1}\underline{\underline{V}}^{(0)-}(z_c)\right]^{-1} \tag{3.53e}$$

$$\vec{\underline{\underline{R}}}^{(0,0)} = -\left[\left(\underline{\underline{W}}^{(0)-}(z_c)\right)^{-1}\underline{\underline{W}}_h - \left(\underline{\underline{V}}^{(0)-}(z_c)\right)^{-1}\underline{\underline{V}}_h\right]^{-1}$$
$$\left[\left(\underline{\underline{W}}^{(0)-}(z_c)\right)^{-1}\underline{\underline{W}}_h + \left(\underline{\underline{V}}^{(0)-}(z_c)\right)^{-1}\underline{\underline{V}}_h\right] \tag{3.53f}$$

The boundary S-matrix of the left half-infinite blocks is given by

$$\mathbf{S}^{(N+1,N+1)} = \begin{pmatrix} \vec{\underline{T}}^{(N+1,N+1)} & \vec{\underline{\underline{R}}}^{(N+1,N+1)} \\ \tilde{\underline{\underline{R}}}^{(N+1,N+1)} & \tilde{\underline{\underline{T}}}^{(N+1,N+1)} \end{pmatrix} \tag{3.54}$$

The boundary condition at $z = z_c$ is described as

$$\begin{pmatrix} \underline{\underline{W}}_h & \underline{\underline{W}}_h \\ \underline{\underline{V}}_h & -\underline{\underline{V}}_h \end{pmatrix} \begin{pmatrix} \vec{\underline{U}} \\ \tilde{\underline{\underline{R}}}^{(N+1,N+1)} \end{pmatrix} = \begin{pmatrix} \underline{\underline{W}}^{(N+1)+}(z_c) & \underline{\underline{W}}^{(N+1)-}(z_c) \\ \underline{\underline{V}}^{(N+1)+}(z_c) & \underline{\underline{V}}^{(N+1)-}(z_c) \end{pmatrix} \begin{pmatrix} \vec{\underline{T}}^{(N+1,N+1)} \\ \underline{\underline{0}} \end{pmatrix} \tag{3.55a}$$

$$\begin{pmatrix} \underline{\underline{W}}_h & \underline{\underline{W}}_h \\ \underline{\underline{V}}_h & -\underline{\underline{V}}_h \end{pmatrix}\begin{pmatrix} \underline{\underline{0}} \\ \vec{\underline{T}}^{(N+1,N+1)} \end{pmatrix} = \begin{pmatrix} \underline{\underline{W}}^{(N+1)+}(z_c) & \underline{\underline{W}}^{(N+1)-}(z_c) \\ \underline{\underline{V}}^{(N+1)+}(z_c) & \underline{\underline{V}}^{(N+1)-}(z_c) \end{pmatrix}\begin{pmatrix} \vec{\underline{R}}^{(N+1,N+1)} \\ \vec{\underline{U}} \end{pmatrix} \tag{3.55b}$$

The S-matrix elements $\vec{\underline{\underline{T}}}^{(N+1,N+1)}$, $\vec{\underline{\underline{R}}}^{(N+1,N+1)}$, $\vec{\underline{\underline{T}}}^{(N+1,N+1)}$, and $\vec{\underline{\underline{R}}}^{(N+1,N+1)}$ are given, respectively, as

$$\vec{\underline{\underline{T}}}^{(N+1,N+1)} = 2\left[\left(\underline{\underline{W}}_h\right)^{-1}\underline{\underline{W}}^{(N+1)+}(z_c) + \left(\underline{\underline{V}}_h\right)^{-1}\underline{\underline{V}}^{(N+1)+}(z_c)\right]^{-1} \tag{3.55c}$$

$$\vec{\underline{\underline{R}}}^{(N+1,N+1)} = -\left[\left(\underline{\underline{W}}^{(N+1)+}(z_c)\right)^{-1}\underline{\underline{W}}_h + \left(\underline{\underline{V}}^{(N+1)+}(z_c)\right)^{-1}\underline{\underline{V}}_h\right]^{-1}$$
$$\times\left[\left(\underline{\underline{W}}^{(N+1)+}(z_c)\right)^{-1}\underline{\underline{W}}_h - \left(\underline{\underline{V}}^{(N+1)+}(z_c)\right)^{-1}\underline{\underline{V}}_h\right] \tag{3.55d}$$

$$\vec{\underline{\underline{T}}}^{(N+1,N+1)} = \left[\left(\underline{\underline{W}}^{(N+1)+}(z_c)\right)^{-1}\underline{\underline{W}}_h + \left(\underline{\underline{V}}^{(N+1)+}(z_c)\right)^{-1}\underline{\underline{V}}_h\right]^{-1}$$
$$\times\left[\left(\underline{\underline{W}}^{(N+1)+}(z_c)\right)^{-1}\underline{\underline{W}}^{(N+1)-}(z_c) - \left(\underline{\underline{V}}^{(N+1)+}(z_c)\right)^{-1}\underline{\underline{V}}^{(N+1)-}(z_c)\right] \tag{3.55e}$$

$$\vec{\underline{\underline{R}}}^{(N+1,N+1)} = -\left[\left(\underline{\underline{W}}_h\right)^{-1}\underline{\underline{W}}^{(N+1)+}(z_c) + \left(\underline{\underline{V}}_h\right)^{-1}\underline{\underline{V}}^{(N+1)+}(z_c)\right]^{-1}$$
$$\times\left[\left(\underline{\underline{W}}_h\right)^{-1}\underline{\underline{W}}^{(N+1)-}(z_c) + \left(\underline{\underline{V}}_h\right)^{-1}\underline{\underline{V}}^{(N+1)-}(z_c)\right] \tag{3.55f}$$

With the above results, we can construct the total S-matrix of the whole multiblock $M^{(0,N+1)}$. At the first step, the S-matrix $\mathbf{S}^{(0,N)}$ of the multiblock $M^{(0,N)}$ is derived by the Redheffer star product of $\mathbf{S}^{(0,0)}$ of Equation (3.52) and $\mathbf{S}^{(1,N)}$ of Equation (3.48) as

$$\mathbf{S}^{(0,N)} = \mathbf{S}^{(0,0)} * \mathbf{S}^{(1,N)} \tag{3.56a}$$

The coupling coefficient matrices $\underline{\underline{C}}_{a,(k)}^{(0,N)}$ and $\underline{\underline{C}}_{b,(k)}^{(0,N)}$ ($1 \le k \le N$) of the blocks $L_1 \sim L_N$ in $M^{(0,N)}$ are obtained by, from Equations (3.48c) and (3.48d),

$$\left(\underline{\underline{C}}_{a,(k)}^{(0,N)}, \underline{\underline{C}}_{b,(k)}^{(0,N)}\right) = \left(\underline{\underline{C}}_{a,(k)}^{(1,N)}\left(\underline{\underline{I}} - \vec{\underline{\underline{R}}}^{(0,0)}\vec{\underline{\underline{R}}}^{(1,N)}\right)^{-1}\vec{\underline{\underline{T}}}^{(0,0)},\right.$$
$$\left.\underline{\underline{C}}_{b,(k)}^{(1,N)} + \underline{\underline{C}}_{a,(k)}^{(1,N)}\left(\underline{\underline{I}} - \vec{\underline{\underline{R}}}^{(0,0)}\vec{\underline{\underline{R}}}^{(1,N)}\right)^{-1}\vec{\underline{\underline{R}}}^{(0,0)}\vec{\underline{\underline{T}}}^{(1,N)}\right) \tag{3.56b}$$

Next, the total S-matrix $\mathbf{S}^{(0,N+1)}$ of the multiblock $M^{(0,N+1)}$ is obtained by the Redheffer star product of $\mathbf{S}^{(0,N)}$ of Equation (3.56a) and $\mathbf{S}^{(N+1,N+1)}$ of Equation (3.54) as

$$\mathbf{S}^{(0,N+1)} = \mathbf{S}^{(0,N)} * \mathbf{S}^{(N+1,N+1)} \tag{3.57a}$$

The final coupling coefficient matrix $\underline{\underline{C}}_{a,(k)}^{(0,N+1)}$ and $\underline{\underline{C}}_{b,(k)}^{(0,N+1)}$ ($1 \le k \le N$) of the blocks $L_1 - L_N$ in $M^{(0,N+1)}$ is obtained from Equations (3.50a) and (3.48b) as

$$\left(\underline{\underline{C}}_{a,(k)}^{(0,N+1)}, \underline{\underline{C}}_{b,(k)}^{(0,N+1)} \right) = \left(\underline{\underline{C}}_{a,(k)}^{(0,N)} + \underline{\underline{C}}_{b,(k)}^{(0,N)} \left(\underline{\underline{I}} - \underline{\underline{\tilde{R}}}^{(N+1,N+1)} \underline{\underline{\vec{R}}}^{(0,N)} \right)^{-1} \underline{\underline{\tilde{R}}}^{(N+1,N+1)} \underline{\underline{\vec{T}}}^{(0,N)} \right.,$$

$$\left. \underline{\underline{C}}_{b,(k)}^{(0,N)} \left(\underline{\underline{I}} - \underline{\underline{\tilde{R}}}^{(N+1,N+1)} \underline{\underline{\vec{R}}}^{(0,N)} \right)^{-1} \underline{\underline{\tilde{T}}}^{(N+1,N+1)} \right). \tag{3.57b}$$

The S-matrix $\mathbf{S}^{(0,N+1)}$ and the coupling coefficient matrices, $\underline{\underline{C}}_{a,(k)}^{(0,N+1)}$ and $\underline{\underline{C}}_{b,(k)}^{(0,N+1)}$, ($1 \le k \le N$) of the blocks $L_1 - L_N$ in $M^{(0,N+1)}$, provide the complete characterization of the multiblock $M^{(0,N+1)}$.

3.2.4 Field Visualization

Even though the theory of FMM can be presented in a unified manner, in practice, the analysis is classified into four cases with respect to four possible combinations of finite-sized multiblocks and half-infinite blocks: (A) half-infinite homogeneous space–multiblock structure–half-infinite homogeneous space, (B) half-infinite homogeneous space–multiblock structure–half-infinite inhomogeneous space, (C) half-infinite inhomogeneous space–multiblock structure–half-infinite homogeneous space, and (D) half-infinite inhomogeneous space–multiblock structure–half-infinite inhomogeneous space.

3.2.4.1 Case A: Half-Infinite Homogeneous Space–Multiblock Structure–Half-Infinite Homogeneous Space

3.2.4.1.1 Left-to-Right Field Visualization

When the incident field, $\vec{\mathbf{E}}_i$, in the left half-infinite block, strikes the multiblock structure $M^{(1,N)}$ at the boundary B_0, the reflection field coefficient inside the left half-infinite block and the transmission field coefficient inside the right half-infinite block are given, respectively, as

$$\begin{pmatrix} \vec{\underline{E}}_{r,y} \\ \vec{\underline{E}}_{r,x} \end{pmatrix} = \underline{\underline{\vec{R}}}^{(0,N+1)} \begin{pmatrix} \vec{\underline{E}}_{i,y} \\ \vec{\underline{E}}_{i,x} \end{pmatrix} \tag{3.58a}$$

$$\begin{pmatrix} \vec{\underline{E}}_{t,y} \\ \vec{\underline{E}}_{t,x} \end{pmatrix} = \underline{\underline{\vec{T}}}^{(0,N+1)} \begin{pmatrix} \vec{\underline{E}}_{i,y} \\ \vec{\underline{E}}_{i,x} \end{pmatrix} \tag{3.58b}$$

And the coupling coefficients of the internal optical field distribution are represented by

$$\left\{\underline{C}_{a,(1)}^{(1,N)}, \underline{C}_{a,(2)}^{(1,N)}, \ldots, \underline{C}_{a,(N)}^{(1,N)}\right\} = \left\{\underline{C}_{a,(1)}^{(1,N)}\begin{pmatrix} \vec{E}_{i,y} \\ \vec{E}_{i,x} \end{pmatrix}, \underline{C}_{a,(2)}^{(1,N)}\begin{pmatrix} \vec{E}_{i,y} \\ \vec{E}_{i,x} \end{pmatrix}, \ldots, \underline{C}_{a,(N)}^{(1,N)}\begin{pmatrix} \vec{E}_{i,y} \\ \vec{E}_{i,x} \end{pmatrix}\right\} \quad (3.58c)$$

The total field distribution **E** and **H** are obtained in the respective spatial regions (Figure 3.9a).

1. For $z < z_-$

$$\begin{pmatrix} \mathbf{E} \\ \mathbf{H} \end{pmatrix} = \begin{pmatrix} \vec{E}_i \\ \vec{H}_i \end{pmatrix} + \begin{pmatrix} \vec{E}_r \\ \vec{H}_r \end{pmatrix}$$

$$= \sum_{m=-M}^{M} \sum_{n=-N}^{N} \begin{pmatrix} \vec{E}_{i,x,m,n}, \vec{E}_{i,y,m,n}, \vec{E}_{i,z,m,n} \\ \vec{H}_{i,x,m,n}, \vec{H}_{i,y,m,n}, \vec{H}_{i,z,m,n} \end{pmatrix} e^{j(k_{x,m,n}x + k_{ym,n}y + k_{z,m,n}(z-z_-))}$$

$$+ \sum_{m=-M}^{M} \sum_{n=-N}^{N} \begin{pmatrix} \vec{E}_{r,x,m,n}, \vec{E}_{r,y,m,n}, \vec{E}_{r,z,m,n} \\ \vec{H}_{r,x,m,n}, \vec{H}_{r,y,m,n}, \vec{H}_{r,z,m,n} \end{pmatrix} e^{j(k_{x,m,n}x + k_{ym,n}y - k_{z,m,n}(z-z_-))} \quad (3.59)$$

2. For $z_- \leq z < z_+$

$$\begin{pmatrix} \mathbf{E} \\ \mathbf{H} \end{pmatrix} = \begin{cases} \displaystyle\sum_{g=1}^{M^+} C_{a,(1),g}^+ \begin{pmatrix} \mathbf{E}_{(1)}^{(g)+} \\ \mathbf{H}_{(1)}^{(g)+} \end{pmatrix} + \sum_{g=1}^{M^-} C_{a,(1),g}^- \begin{pmatrix} \mathbf{E}_{(1)}^{(g)-} \\ \mathbf{H}_{(1)}^{(g)-} \end{pmatrix} & \text{for } 0 \leq z \leq l_{1,1} \\[3ex] \displaystyle\sum_{g=1}^{M^+} C_{a,(2),g}^+ \begin{pmatrix} \mathbf{E}_{(2)}^{(g)+} \\ \mathbf{H}_{(2)}^{(g)+} \end{pmatrix} + \sum_{g=1}^{M^-} C_{a,(2),g}^- \begin{pmatrix} \mathbf{E}_{(2)}^{(g)-} \\ \mathbf{H}_{(2)}^{(g)-} \end{pmatrix} & \text{for } l_{1,1} \leq z \leq l_{1,2} \\[2ex] \qquad\qquad\qquad\vdots & \\[2ex] \displaystyle\sum_{g=1}^{M^+} C_{a,(N),g}^+ \begin{pmatrix} \mathbf{E}_{(N)}^{(g)+} \\ \mathbf{H}_{(N)}^{(g)+} \end{pmatrix} + \sum_{g=1}^{M^-} C_{a,(N),g}^- \begin{pmatrix} \mathbf{E}_{(N)}^{(g)-} \\ \mathbf{H}_{(N)}^{(g)-} \end{pmatrix} & \text{for } l_{1,N-1} \leq z \leq l_{1,N} \end{cases}$$

$$(3.60a)$$

where $\mathbf{E}^{(g)+}$, $\mathbf{E}^{(g)-}$, $\mathbf{H}^{(g)+}$, and $\mathbf{H}^{(g)-}$ are given, respectively, by

$$\mathbf{E}^{(g)+}(x,y,z) = \sum_{m=-M}^{M} \sum_{n=-N}^{N} \left(E_{x,m,n}^{(g)+}, E_{y,m,n}^{(g)+}, E_{z,m,n}^{(g)+}\right) e^{j\left(k_{x,m}x + k_{y,n}y + k_z^{(g)+}(z-z_-)\right)} \quad (3.60b)$$

$$\mathbf{E}^{(g)-}(x,y,z) = \sum_{m=-M}^{M} \sum_{n=-N}^{N} \left(E_{x,m,n}^{(g)-}, E_{y,m,n}^{(g)-}, E_{z,m,n}^{(g)-}\right) e^{j\left(k_{x,m}x + k_{y,n}y + k_z^{(g)-}(z-z_+)\right)} \quad (3.60c)$$

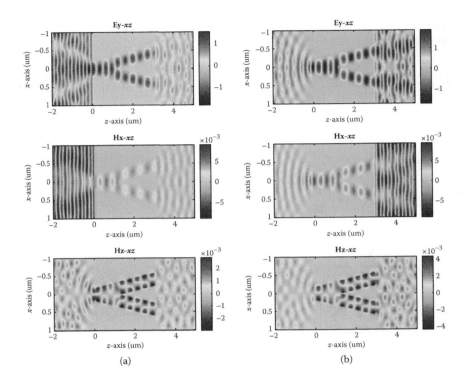

FIGURE 3.9
Case A: (a) Left-to-right directional characterization, E_y, H_x, and H_z field distributions for TE mode; and (b) right-to-left directional characterization, E_y, H_x, and H_z field distributions for TE mode. (MATLAB code: \FMM_Lwg_Rwg\FMM_Lwg_Rwg.m.)

$$\mathbf{H}^{(g)+}(x,y,z) = \sum_{m=-M}^{M} \sum_{n=-N}^{N} \left(H_{x,m,n}^{(g)+}, H_{y,m,n}^{(g)+}, H_{z,m,n}^{(g)+} \right) e^{j\left(k_{x,m}x+k_{y,n}y+k_z^{(g)+}(z-z_-)\right)} \quad (3.60d)$$

$$\mathbf{H}^{(g)-}(x,y,z) = \sum_{m=-M}^{M} \sum_{n=-N}^{N} \left(H_{x,m,n}^{(g)-}, H_{y,m,n}^{(g)-}, H_{z,m,n}^{(g)-} \right) e^{j\left(k_{x,m}x+k_{y,n}y+k_z^{(g)-}(z-z_+)\right)} \quad (3.60e)$$

3. For $z_+ \leq z$

$$\begin{pmatrix} \vec{\mathbf{E}}_t \\ \vec{\mathbf{H}}_t \end{pmatrix} = \begin{pmatrix} \vec{\mathbf{E}}_{t,x}, \vec{\mathbf{E}}_{t,y}, \vec{\mathbf{E}}_{t,z} \\ \vec{\mathbf{H}}_{t,x}, \vec{\mathbf{H}}_{t,y}, \vec{\mathbf{H}}_{t,z} \end{pmatrix}$$

$$= \sum_{m=-M}^{M} \sum_{n=-N}^{N} \begin{pmatrix} \vec{\mathbf{E}}_{t,x,m,n}, \vec{\mathbf{E}}_{t,y,m,n}, \vec{\mathbf{E}}_{t,z,m,n} \\ \vec{\mathbf{H}}_{t,x,m,n}, \vec{\mathbf{H}}_{t,y,m,n}, \vec{\mathbf{H}}_{t,z,m,n} \end{pmatrix} e^{j(k_{x,m,n}x+k_{y,m,n}y+k_{z,m,n}(z-z_+))} \quad (3.61a)$$

The incident electromagnetic power at the boundary B_0 is equal to the sum of the reflection power normal to the boundary B_0 and the transmission power normal to the boundary B_N.

$$\sum_{m=-M}^{M} \sum_{n=-N}^{N} \left(|\bar{E}_{r,x,m,n}|^2 + |\bar{E}_{r,y,m,n}|^2 + |\bar{E}_{r,z,m,n}|^2 \right) \mathrm{Re}\left(k_{z,m,n}^{(0)}/k_{z,0,0}^{(0)} \right)$$

$$+ \sum_{m=-M}^{M} \sum_{n=-N}^{N} \left(|\bar{E}_{t,x,m,n}|^2 + |\bar{E}_{t,y,m,n}|^2 + |\bar{E}_{t,z,m,n}|^2 \right) \mathrm{Re}\left(k_{z,m,n}^{(N+1)}/k_{z,0,0}^{(0)} \right) = 1 \qquad (3.61b)$$

3.2.4.1.2 Right-to-Left Field Visualization

When the incident field, $\bar{E}_i$, in the left half-infinite block impinges on the multiblock structure $M^{(1,N)}$ at the boundary B_N, the reflection field coefficient inside the left half-infinite block and the transmission field coefficient inside the right half-infinite block are given, respectively, as

$$\begin{pmatrix} \bar{E}_{r,y} \\ \bar{E}_{r,x} \end{pmatrix} = \bar{\bar{R}}^{(0,N+1)} \begin{pmatrix} \bar{E}_{i,y} \\ \bar{E}_{i,x} \end{pmatrix} \qquad (3.62a)$$

$$\begin{pmatrix} \bar{E}_{t,y} \\ \bar{E}_{t,x} \end{pmatrix} = \bar{\bar{T}}^{(0,N+1)} \begin{pmatrix} \bar{E}_{i,y} \\ \bar{E}_{i,x} \end{pmatrix} \qquad (3.62b)$$

And the coupling coefficients of the internal optical field distribution are represented by

$$\left\{ \underline{C}_{b,(1)}^{(1,N)}, \underline{C}_{b,(2)}^{(1,N)}, \ldots, \underline{C}_{b,(N)}^{(1,N)} \right\} = \left\{ \underline{\underline{C}}_{b,(1)}^{(1,N)} \begin{pmatrix} \bar{E}_{i,y} \\ \bar{E}_{i,x} \end{pmatrix}, \underline{\underline{C}}_{b,(2)}^{(1,N)} \begin{pmatrix} \bar{E}_{i,y} \\ \bar{E}_{i,x} \end{pmatrix}, \ldots, \underline{\underline{C}}_{b,(N)}^{(1,N)} \begin{pmatrix} \bar{E}_{i,y} \\ \bar{E}_{i,x} \end{pmatrix} \right\}$$

$$(3.62c)$$

The total field distribution $\mathbf{E}$ and $\mathbf{H}$ are obtained in the respective spatial regions (Figure 3.9b).

1. For $z < z_-$

$$\begin{pmatrix} \bar{E}_t \\ \bar{H}_t \end{pmatrix} = \begin{pmatrix} \bar{E}_{t,x}, \bar{E}_{t,y}, \bar{E}_{t,z} \\ \bar{H}_{t,x}, \bar{H}_{t,y}, \bar{H}_{t,z} \end{pmatrix}$$

$$= \sum_{m=-M}^{M} \sum_{n=-N}^{N} \begin{pmatrix} \bar{E}_{t,x,m,n}, \bar{E}_{t,y,m,n}, \bar{E}_{t,z,m,n} \\ \bar{H}_{t,x,m,n}, \bar{H}_{t,y,m,n}, \bar{H}_{t,z,m,n} \end{pmatrix} e^{j(k_{x,m,n}x + k_{y,m,n}y - k_{z,m,n}(z - z_-))} \qquad (3.63)$$

2. For $z_- \le z < z_+$

$$
\begin{pmatrix} \mathbf{E} \\ \mathbf{H} \end{pmatrix} = \begin{cases}
\displaystyle\sum_{g=1}^{M^+} C_{b,(1),g}^+ \begin{pmatrix} \mathbf{E}_{(1)}^{(g)+} \\ \mathbf{H}_{(1)}^{(g)+} \end{pmatrix} + \sum_{g=1}^{M^-} C_{b,(1),g}^- \begin{pmatrix} \mathbf{E}_{(1)}^{(g)-} \\ \mathbf{H}_{(1)}^{(g)-} \end{pmatrix} & \text{for } 0 \le z \le l_{1,1} \\[3mm]
\displaystyle\sum_{g=1}^{M^+} C_{b,(2),g}^+ \begin{pmatrix} \mathbf{E}_{(2)}^{(g)+} \\ \mathbf{H}_{(2)}^{(g)+} \end{pmatrix} + \sum_{g=1}^{M^-} C_{b,(2),g}^- \begin{pmatrix} \mathbf{E}_{(2)}^{(g)-} \\ \mathbf{H}_{(2)}^{(g)-} \end{pmatrix} & \text{for } l_{1,1} \le z \le l_{1,2} \\[3mm]
\qquad\qquad\qquad\qquad\vdots & \\[3mm]
\displaystyle\sum_{g=1}^{M^+} C_{b,(N),g}^+ \begin{pmatrix} \mathbf{E}_{(N)}^{(g)+} \\ \mathbf{H}_{(N)}^{(g)+} \end{pmatrix} + \sum_{g=1}^{M^-} C_{b,(N),g}^- \begin{pmatrix} \mathbf{E}_{(N)}^{(g)-} \\ \mathbf{H}_{(N)}^{(g)-} \end{pmatrix} & \text{for } l_{1,N-1} \le z \le l_{1,N}
\end{cases}
$$

$$(3.64)$$

where $\mathbf{E}^{(g)+}$, $\mathbf{E}^{(g)-}$, $\mathbf{H}^{(g)+}$, and $\mathbf{H}^{(g)-}$ are given in Equation (3.34).

3. For $z_+ \le z$

$$
\begin{pmatrix} \mathbf{E} \\ \mathbf{H} \end{pmatrix} = \begin{pmatrix} \vec{E}_i \\ \vec{H}_i \end{pmatrix} + \begin{pmatrix} \vec{E}_r \\ \vec{H}_r \end{pmatrix}
$$

$$
= \sum_{m=-M}^{M} \sum_{n=-N}^{N} \begin{pmatrix} \vec{E}_{i,x,m,n}, \vec{E}_{i,y,m,n}, \vec{E}_{i,z,m,n} \\ \vec{H}_{i,x,m,n}, \vec{H}_{i,y,m,n}, \vec{H}_{i,z,m,n} \end{pmatrix} e^{j(k_{x,m,n}x + k_{ym,n}y - k_{z,m,n}(z-z_+))}
$$

$$
+ \sum_{m=-M}^{M} \sum_{n=-N}^{N} \begin{pmatrix} \vec{E}_{r,x,m,n}, \vec{E}_{r,y,m,n}, \vec{E}_{r,z,m,n} \\ \vec{H}_{r,x,m,n}, \vec{H}_{r,y,m,n}, \vec{H}_{r,z,m,n} \end{pmatrix} e^{j(k_{x,m,n}x + k_{ym,n}y + k_{z,m,n}(z-z_+))} \qquad (3.65)
$$

The incident electromagnetic power at the boundary B_N is equal to the sum of the reflection power normal to the boundary B_N and the transmission power normal to the boundary B_0.

$$
\sum_{m=-M}^{M} \sum_{n=-N}^{N} \left(|\vec{E}_{r,x,m,n}|^2 + |\vec{E}_{r,y,m,n}|^2 + |\vec{E}_{r,z,m,n}|^2 \right) \mathrm{Re}\left(k_{z,m,n}^{(0)} / k_{z,0,0}^{(N+1)} \right)
$$

$$
+ \sum_{m=-M}^{M} \sum_{n=-N}^{N} \left(|\vec{E}_{t,x,m,n}|^2 + |\vec{E}_{t,y,m,n}|^2 + |\vec{E}_{t,z,m,n}|^2 \right) \mathrm{Re}\left(k_{z,m,n}^{(N+1)} / k_{z,0,0}^{(N+1)} \right) = 1. \quad (3.66)
$$

3.2.4.2 Case B: Half-Infinite Homogeneous Space–Multiblock Structure–Half-Infinite Inhomogeneous Space

3.2.4.2.1 Left-to-Right Field Visualization

$$
\begin{pmatrix} \vec{\bar{E}}_{r,y} \\ \vec{\bar{E}}_{r,x} \end{pmatrix} = \vec{\underline{\underline{R}}}^{(0,N+1)} \begin{pmatrix} \vec{E}_{i,y} \\ \vec{E}_{i,x} \end{pmatrix}
\tag{3.67a}
$$

$$
\vec{\underline{t}} = \vec{\underline{\underline{T}}}^{(0,N+1)} \begin{pmatrix} \vec{E}_{i,y} \\ \vec{E}_{i,x} \end{pmatrix}
\tag{3.67b}
$$

$$
\left\{ \underline{C}_{a,(1)}^{(1,N)}, \underline{C}_{a,(2)}^{(1,N)}, \cdots, \underline{C}_{a,(N)}^{(1,N)} \right\} = \left\{ \underline{\underline{C}}_{a,(1)}^{(1,N)} \begin{pmatrix} \vec{E}_{i,y} \\ \vec{E}_{i,x} \end{pmatrix}, \underline{C}_{a,(2)}^{(1,N)} \begin{pmatrix} \vec{E}_{i,y} \\ \vec{E}_{i,x} \end{pmatrix}, \cdots, \underline{C}_{a,(N)}^{(1,N)} \begin{pmatrix} \vec{E}_{i,y} \\ \vec{E}_{i,x} \end{pmatrix} \right\}
\tag{3.67c}
$$

The total field distributions are given as Figure 3.10(a).

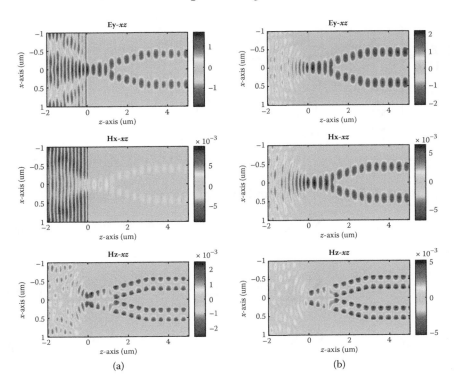

(a)   (b)

FIGURE 3.10
Case B: (a) Left-to-right directional characterization, E_y, H_x, and H_z field distributions for TE mode; and (b) right-to-left directional characterization, E_y, H_x, and H_z field distributions for TE mode. (MATLAB code: *FMM_Lwg_Rwg**FMM_Lwg_Rwg.m*.)

1. For $z < z_-$

$$\begin{pmatrix} \mathbf{E} \\ \mathbf{H} \end{pmatrix} = \begin{pmatrix} \vec{E}_i \\ \vec{H}_i \end{pmatrix} + \begin{pmatrix} \overleftarrow{E}_r \\ \overleftarrow{H}_r \end{pmatrix}$$

$$= \sum_{m=-M}^{M} \sum_{n=-N}^{N} \begin{pmatrix} \vec{E}_{i,x,m,n}, \vec{E}_{i,y,m,n}, \vec{E}_{i,z,m,n} \\ \vec{H}_{i,x,m,n}, \vec{H}_{i,y,m,n}, \vec{H}_{i,z,m,n} \end{pmatrix} e^{j(k_{x,m,n}x + k_{ym,n}y + k_{z,m,n}(z-z_-))}$$

$$+ \sum_{m=-M}^{M} \sum_{n=-N}^{N} \begin{pmatrix} \overleftarrow{E}_{r,x,m,n}, \overleftarrow{E}_{r,y,m,n}, \overleftarrow{E}_{r,z,m,n} \\ \overleftarrow{H}_{r,x,m,n}, \overleftarrow{H}_{r,y,m,n}, \overleftarrow{H}_{r,z,m,n} \end{pmatrix} e^{j(k_{x,m,n}x + k_{ym,n}y - k_{z,m,n}(z-z_-))} \quad (3.68)$$

2. For $z_- \leq z < z_+$

$$\begin{pmatrix} \mathbf{E} \\ \mathbf{H} \end{pmatrix} = \begin{cases} \displaystyle\sum_{g=1}^{M^+} C_{a,(1),g}^+ \begin{pmatrix} \mathbf{E}_{(1)}^{(g)+} \\ \mathbf{H}_{(1)}^{(g)+} \end{pmatrix} + \sum_{g=1}^{M^-} C_{a,(1),g}^- \begin{pmatrix} \mathbf{E}_{(1)}^{(g)-} \\ \mathbf{H}_{(1)}^{(g)-} \end{pmatrix} & \text{for } 0 \leq z \leq l_{1,1} \\[3ex] \displaystyle\sum_{g=1}^{M^+} C_{a,(2),g}^+ \begin{pmatrix} \mathbf{E}_{(2)}^{(g)+} \\ \mathbf{H}_{(2)}^{(g)+} \end{pmatrix} + \sum_{g=1}^{M^-} C_{a,(2),g}^- \begin{pmatrix} \mathbf{E}_{(2)}^{(g)-} \\ \mathbf{H}_{(2)}^{(g)-} \end{pmatrix} & \text{for } l_{1,1} \leq z \leq l_{1,2} \\[2ex] \qquad\qquad\qquad\qquad\vdots \\[2ex] \displaystyle\sum_{g=1}^{M^+} C_{a,(N),g}^+ \begin{pmatrix} \mathbf{E}_{(N)}^{(g)+} \\ \mathbf{H}_{(N)}^{(g)+} \end{pmatrix} + \sum_{g=1}^{M^-} C_{a,(N),g}^- \begin{pmatrix} \mathbf{E}_{(N)}^{(g)-} \\ \mathbf{H}_{(N)}^{(g)-} \end{pmatrix} & \text{for } l_{1,N-1} \leq z \leq l_{1,N} \end{cases}$$

$$(3.69\text{a})$$

where $\mathbf{E}^{(g)+}$, $\mathbf{E}^{(g)-}$, $\mathbf{H}^{(g)+}$, and $\mathbf{H}^{(g)-}$ are given, respectively, by

$$\mathbf{E}^{(g)+}(x,y,z) = \sum_{m=-M}^{M} \sum_{n=-N}^{N} \left(E_{x,m,n}^{(g)+}, E_{y,m,n}^{(g)+}, E_{z,m,n}^{(g)+} \right) e^{j\left(k_{x,m}x + k_{y,n}y + k_z^{(g)+}(z-z_-)\right)} \quad (3.69\text{b})$$

$$\mathbf{E}^{(g)-}(x,y,z) = \sum_{m=-M}^{M} \sum_{n=-N}^{N} \left(E_{x,m,n}^{(g)-}, E_{y,m,n}^{(g)-}, E_{z,m,n}^{(g)-} \right) e^{j\left(k_{x,m}x + k_{y,n}y + k_z^{(g)-}(z-z_+)\right)} \quad (3.69\text{c})$$

$$\mathbf{H}^{(g)+}(x,y,z) = \sum_{m=-M}^{M} \sum_{n=-N}^{N} \left(H_{x,m,n}^{(g)+}, H_{y,m,n}^{(g)+}, H_{z,m,n}^{(g)+} \right) e^{j\left(k_{x,m}x + k_{y,n}y + k_z^{(g)+}(z-z_-)\right)} \quad (3.69\text{d})$$

$$\mathbf{H}^{(g)-}(x,y,z) = \sum_{m=-M}^{M} \sum_{n=-N}^{N} \left(H_{x,m,n}^{(g)-}, H_{y,m,n}^{(g)-}, H_{z,m,n}^{(g)-} \right) e^{j\left(k_{x,m}x + k_{y,n}y + k_z^{(g)-}(z-z_+)\right)} \quad (3.69\text{e})$$

3. For $z_+ \leq z$

$$\begin{pmatrix} \vec{E}_t \\ \vec{H}_t \end{pmatrix} = \sum_{g=1}^{M^+} \vec{t}_g \begin{pmatrix} \mathbf{E}_{(N+1)}^{(g)+} \\ \mathbf{H}_{(N+1)}^{(g)+} \end{pmatrix} \tag{3.70}$$

3.2.4.2.2 Right-to-Left Field Visualization

$$\underline{\vec{r}} = \underline{\underline{\vec{R}}}^{(0,N+1)}\underline{\vec{u}} \tag{3.71a}$$

$$\begin{pmatrix} \underline{\bar{E}}_{t,y} \\ \underline{\bar{E}}_{t,x} \end{pmatrix} = \underline{\underline{\bar{T}}}^{(0,N+1)}\underline{\vec{u}} \tag{3.71b}$$

$$\left\{ \underline{C}_{b,(1)}^{(1,N)}, \underline{C}_{b,(2)}^{(1,N)}, \ldots, \underline{C}_{b,(N)}^{(1,N)} \right\} = \left\{ \underline{\underline{C}}_{b,(1)}^{(1,N)}\underline{\vec{u}}, \underline{\underline{C}}_{b,(2)}^{(1,N)}\underline{\vec{u}}, \ldots, \underline{\underline{C}}_{b,(N)}^{(1,N)}\underline{\vec{u}} \right\} \tag{3.71c}$$

The total field distributions are represented as Figure 3.10(b).

1. For $z < z_-$

$$\begin{pmatrix} \bar{\mathbf{E}}_t \\ \bar{\mathbf{H}}_t \end{pmatrix} = \begin{pmatrix} \bar{\mathbf{E}}_{t,x}, \bar{\mathbf{E}}_{t,y}, \bar{\mathbf{E}}_{t,z} \\ \bar{\mathbf{H}}_{t,x}, \bar{\mathbf{H}}_{t,y}, \bar{\mathbf{H}}_{t,z} \end{pmatrix}$$

$$= \sum_{m=-M}^{M} \sum_{n=-N}^{N} \begin{pmatrix} \bar{E}_{t,x,m,n}, \bar{E}_{t,y,m,n}, \bar{E}_{t,z,m,n} \\ \bar{H}_{t,x,m,n}, \bar{H}_{t,y,m,n}, \bar{H}_{t,z,m,n} \end{pmatrix} e^{j(k_{x,m,n}x + k_{y,m,n}y - k_{z,m,n}(z-z_-))} \tag{3.72}$$

2. For $z_- \leq z < z_+$

$$\begin{pmatrix} \mathbf{E} \\ \mathbf{H} \end{pmatrix} = \begin{cases} \displaystyle\sum_{g=1}^{M^+} C_{b,(1),g}^+ \begin{pmatrix} \mathbf{E}_{(1)}^{(g)+} \\ \mathbf{H}_{(1)}^{(g)+} \end{pmatrix} + \sum_{g=1}^{M^-} C_{b,(1),g}^- \begin{pmatrix} \mathbf{E}_{(1)}^{(g)-} \\ \mathbf{H}_{(1)}^{(g)-} \end{pmatrix} & \text{for } 0 \leq z \leq l_{1,1} \\[2em] \displaystyle\sum_{g=1}^{M^+} C_{b,(2),g}^+ \begin{pmatrix} \mathbf{E}_{(2)}^{(g)+} \\ \mathbf{H}_{(2)}^{(g)+} \end{pmatrix} + \sum_{g=1}^{M^-} C_{b,(2),g}^- \begin{pmatrix} \mathbf{E}_{(2)}^{(g)-} \\ \mathbf{H}_{(2)}^{(g)-} \end{pmatrix} & \text{for } l_{1,1} \leq z \leq l_{1,2} \\[1em] \qquad\qquad\qquad\vdots \\[1em] \displaystyle\sum_{g=1}^{M^+} C_{b,(N),g}^+ \begin{pmatrix} \mathbf{E}_{(N)}^{(g)+} \\ \mathbf{H}_{(N)}^{(g)+} \end{pmatrix} + \sum_{g=1}^{M^-} C_{b,(N),g}^- \begin{pmatrix} \mathbf{E}_{(N)}^{(g)-} \\ \mathbf{H}_{(N)}^{(g)-} \end{pmatrix} & \text{for } l_{1,N-1} \leq z \leq l_{1,N} \end{cases}$$

$$\tag{3.73}$$

where $\mathbf{E}^{(g)+}$, $\mathbf{E}^{(g)-}$, $\mathbf{H}^{(g)+}$, and $\mathbf{H}^{(g)-}$ are given in Equation (3.34).

3. For $z_+ \leq z$

$$\begin{pmatrix} \mathbf{E} \\ \mathbf{H} \end{pmatrix} = \sum_{g=1}^{M^-} \vec{u}_g \begin{pmatrix} \mathbf{E}_{(N+1)}^{(g)-} \\ \mathbf{H}_{(N+1)}^{(g)-} \end{pmatrix} + \sum_{g=1}^{M^+} \vec{r}_g \begin{pmatrix} \mathbf{E}_{(N+1)}^{(g)+} \\ \mathbf{H}_{(N+1)}^{(g)+} \end{pmatrix} \tag{3.74}$$

where $\mathbf{E}^{(g)+}$, $\mathbf{E}^{(g)-}$, $\mathbf{H}^{(g)+}$, and $\mathbf{H}^{(g)-}$ are given in Equation (3.34).

3.2.4.3 Case C: Half-Infinite Inhomogeneous Waveguide–Multiblock Structure–Half-Infinite Homogeneous Space

3.2.4.3.1 Left-to-Right Field Visualization

$$\vec{r} = \underline{\underline{\tilde{R}}}^{(0,N+1)}\vec{u} \tag{3.75a}$$

$$\begin{pmatrix} \vec{\underline{E}}_{t,y} \\ \vec{\underline{E}}_{t,x} \end{pmatrix} = \underline{\underline{\vec{T}}}^{(0,N+1)}\vec{u} \tag{3.75b}$$

$$\left\{ \underline{C}_{a,(1)}^{(1,N)}, \underline{C}_{a,(2)}^{(1,N)}, \ldots, \underline{C}_{a,(N)}^{(1,N)} \right\} = \left\{ \underline{\underline{C}}_{a,(1)}^{(1,N)}\vec{u}, \underline{\underline{C}}_{a,(2)}^{(1,N)}\vec{u}, \ldots, \underline{\underline{C}}_{a,(N)}^{(1,N)}\vec{u} \right\} \tag{3.75c}$$

The total field distributions are given as Figure 3.11(a).

1. For $z < z_-$

$$\begin{pmatrix} \mathbf{E} \\ \mathbf{H} \end{pmatrix} = \sum_{g=1}^{M^+} \vec{u}_g \begin{pmatrix} \mathbf{E}_{(0)}^{(g)+} \\ \mathbf{H}_{(0)}^{(g)+} \end{pmatrix} + \sum_{g=1}^{M^-} \vec{r}_g \begin{pmatrix} \mathbf{E}_{(0)}^{(g)-} \\ \mathbf{H}_{(0)}^{(g)-} \end{pmatrix} \tag{3.76}$$

2. For $z_- \leq z < z_+$

$$\begin{pmatrix} \mathbf{E} \\ \mathbf{H} \end{pmatrix} = \begin{cases} \sum_{g=1}^{M^+} C_{a,(1),g}^+ \begin{pmatrix} \mathbf{E}_{(1)}^{(g)+} \\ \mathbf{H}_{(1)}^{(g)+} \end{pmatrix} + \sum_{g=1}^{M^-} C_{a,(1),g}^- \begin{pmatrix} \mathbf{E}_{(1)}^{(g)-} \\ \mathbf{H}_{(1)}^{(g)-} \end{pmatrix} & \text{for } 0 \leq z \leq l_{1,1} \\[2em] \sum_{g=1}^{M^+} C_{a,(2),g}^+ \begin{pmatrix} \mathbf{E}_{(2)}^{(g)+} \\ \mathbf{H}_{(2)}^{(g)+} \end{pmatrix} + \sum_{g=1}^{M^-} C_{a,(2),g}^- \begin{pmatrix} \mathbf{E}_{(2)}^{(g)-} \\ \mathbf{H}_{(2)}^{(g)-} \end{pmatrix} & \text{for } l_{1,1} \leq z \leq l_{1,2} \\[2em] \qquad\qquad\qquad\qquad \vdots \\[1em] \sum_{g=1}^{M^+} C_{a,(N),g}^+ \begin{pmatrix} \mathbf{E}_{(N)}^{(g)+} \\ \mathbf{H}_{(N)}^{(g)+} \end{pmatrix} + \sum_{g=1}^{M^-} C_{a,(N),g}^- \begin{pmatrix} \mathbf{E}_{(N)}^{(g)-} \\ \mathbf{H}_{(N)}^{(g)-} \end{pmatrix} & \text{for } l_{1,N-1} \leq z \leq l_{1,N} \end{cases}$$

$$\tag{3.77a}$$

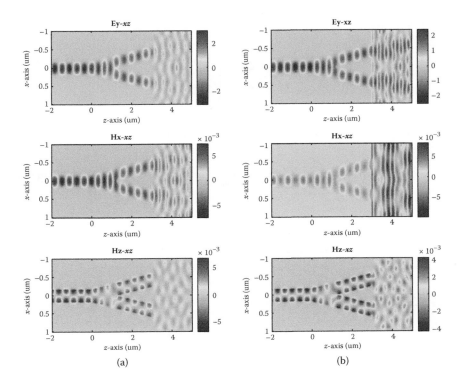

FIGURE 3.11
Case C: (a) Left-to-right directional characterization, E_y, H_x, and H_z field distributions for TE mode; and (b) right-to-left directional characterization, E_y, H_x, and H_z field distributions for TE mode. (MATLAB code: \$FMM_Lwg_Rwg\$FMM_Lwg_Rwg.m.)

where $\mathbf{E}^{(g)+}$, $\mathbf{E}^{(g)-}$, $\mathbf{H}^{(g)+}$, and $\mathbf{H}^{(g)-}$ are given, respectively, by

$$\mathbf{E}^{(g)+}(x,y,z) = \sum_{m=-M}^{M} \sum_{n=-N}^{N} \left(E_{x,m,n}^{(g)+}, E_{y,m,n}^{(g)+}, E_{z,m,n}^{(g)+} \right) e^{j\left(k_{x,m}x + k_{y,n}y + k_z^{(g)+}(z-z_-)\right)} \quad (3.77b)$$

$$\mathbf{E}^{(g)-}(x,y,z) = \sum_{m=-M}^{M} \sum_{n=-N}^{N} \left(E_{x,m,n}^{(g)-}, E_{y,m,n}^{(g)-}, E_{z,m,n}^{(g)-} \right) e^{j\left(k_{x,m}x + k_{y,n}y + k_z^{(g)-}(z-z_+)\right)} \quad (3.77c)$$

$$\mathbf{H}^{(g)+}(x,y,z) = \sum_{m=-M}^{M} \sum_{n=-N}^{N} \left(H_{x,m,n}^{(g)+}, H_{y,m,n}^{(g)+}, H_{z,m,n}^{(g)+} \right) e^{j\left(k_{x,m}x + k_{y,n}y + k_z^{(g)+}(z-z_-)\right)} \quad (3.77d)$$

$$\mathbf{H}^{(g)-}(x,y,z) = \sum_{m=-M}^{M} \sum_{n=-N}^{N} \left(H_{x,m,n}^{(g)-}, H_{y,m,n}^{(g)-}, H_{z,m,n}^{(g)-} \right) e^{j\left(k_{x,m}x + k_{y,n}y + k_z^{(g)-}(z-z_+)\right)} \quad (3.77e)$$

3. For $z_+ \leq z$

$$\begin{pmatrix} \mathbf{E} \\ \mathbf{H} \end{pmatrix} = \sum_{g=1}^{M^+} \vec{t}_g \begin{pmatrix} \mathbf{E}_{(N+1)}^{(g)+} \\ \mathbf{H}_{(N+1)}^{(g)+} \end{pmatrix} \tag{3.78a}$$

where $\mathbf{E}^{(g)+}$ and $\mathbf{H}^{(g)+}$ are given, respectively, by

$$\mathbf{E}^{(g)+}(x,y,z) = \sum_{m=-M}^{M} \sum_{n=-N}^{N} \left(E_{x,m,n}^{(g)+}, E_{y,m,n}^{(g)+}, E_{z,m,n}^{(g)+} \right) e^{j\left(k_{x,m}x + k_{y,n}y + k_z^{(g)+}(z-z_-) \right)} \tag{3.78b}$$

$$\mathbf{H}^{(g)+}(x,y,z) = \sum_{m=-M}^{M} \sum_{n=-N}^{N} \left(H_{x,m,n}^{(g)+}, H_{y,m,n}^{(g)+}, H_{z,m,n}^{(g)+} \right) e^{j\left(k_{x,m}x + k_{y,n}y + k_z^{(g)+}(z-z_-) \right)} \tag{3.78c}$$

3.2.4.3.2 Right-to-Left Field Visualization

$$\begin{pmatrix} \vec{\tilde{E}}_{r,y} \\ \vec{\tilde{E}}_{r,x} \end{pmatrix} = \vec{\underline{\underline{R}}}^{(0,N+1)} \begin{pmatrix} \vec{\tilde{E}}_{i,y} \\ \vec{\tilde{E}}_{i,x} \end{pmatrix} \tag{3.79a}$$

$$\vec{\tilde{t}} = \vec{\underline{\underline{T}}}^{(0,N+1)} \begin{pmatrix} \vec{\tilde{E}}_{i,y} \\ \vec{\tilde{E}}_{i,x} \end{pmatrix} \tag{3.79b}$$

And the coupling coefficients of the internal optical field distribution are represented by

$$\left\{ \underline{C}_{b,(1)}^{(1,N)}, \underline{C}_{b,(2)}^{(1,N)}, \ldots, \underline{C}_{b,(N)}^{(1,N)} \right\} = \left\{ \underline{\underline{C}}_{b,(1)}^{(1,N)} \begin{pmatrix} \vec{\tilde{E}}_{i,y} \\ \vec{\tilde{E}}_{i,x} \end{pmatrix}, \underline{\underline{C}}_{b,(2)}^{(1,N)} \begin{pmatrix} \vec{\tilde{E}}_{i,y} \\ \vec{\tilde{E}}_{i,x} \end{pmatrix}, \ldots, \underline{\underline{C}}_{b,(N)}^{(1,N)} \begin{pmatrix} \vec{\tilde{E}}_{i,y} \\ \vec{\tilde{E}}_{i,x} \end{pmatrix} \right\} \tag{3.79c}$$

The total field distributions are given as Figure 3.11(b).

1. For $z < z_-$

$$\begin{pmatrix} \mathbf{E} \\ \mathbf{H} \end{pmatrix} = \sum_{g=1}^{M^+} \vec{\tilde{t}}_g \begin{pmatrix} \mathbf{E}_{(0)}^{(g)-} \\ \mathbf{H}_{(0)}^{(g)-} \end{pmatrix} \tag{3.80}$$

2. For $z_- \leq z < z_+$

$$\begin{pmatrix} \mathbf{E} \\ \mathbf{H} \end{pmatrix} = \begin{cases} \displaystyle\sum_{g=1}^{M^+} C_{b,(1),g}^+ \begin{pmatrix} \mathbf{E}_{(1)}^{(g)+} \\ \mathbf{H}_{(1)}^{(g)+} \end{pmatrix} + \sum_{g=1}^{M^-} C_{b,(1),g}^- \begin{pmatrix} \mathbf{E}_{(1)}^{(g)-} \\ \mathbf{H}_{(1)}^{(g)-} \end{pmatrix} & \text{for } 0 \leq z \leq l_{1,1} \\[2em] \displaystyle\sum_{g=1}^{M^+} C_{b,(2),g}^+ \begin{pmatrix} \mathbf{E}_{(2)}^{(g)+} \\ \mathbf{H}_{(2)}^{(g)+} \end{pmatrix} + \sum_{g=1}^{M^-} C_{b,(2),g}^- \begin{pmatrix} \mathbf{E}_{(2)}^{(g)-} \\ \mathbf{H}_{(2)}^{(g)-} \end{pmatrix} & \text{for } l_{1,1} \leq z \leq l_{1,2} \\[2em] \quad\quad\quad\quad\quad\quad\quad \vdots & \\[1em] \displaystyle\sum_{g=1}^{M^+} C_{b,(N),g}^+ \begin{pmatrix} \mathbf{E}_{(N)}^{(g)+} \\ \mathbf{H}_{(N)}^{(g)+} \end{pmatrix} + \sum_{g=1}^{M^-} C_{b,(N),g}^- \begin{pmatrix} \mathbf{E}_{(N)}^{(g)-} \\ \mathbf{H}_{(N)}^{(g)-} \end{pmatrix} & \text{for } l_{1,N-1} \leq z \leq l_{1,N} \end{cases} \tag{3.81}$$

where $\mathbf{E}^{(g)+}$, $\mathbf{E}^{(g)-}$, $\mathbf{H}^{(g)+}$, and $\mathbf{H}^{(g)-}$ are given in Equation (3.34).

3. For $z_+ \leq z$

$$\begin{pmatrix} \mathbf{E} \\ \mathbf{H} \end{pmatrix} = \begin{pmatrix} \vec{\mathbf{E}}_i \\ \vec{\mathbf{H}}_i \end{pmatrix} + \begin{pmatrix} \vec{\mathbf{E}}_r \\ \vec{\mathbf{H}}_r \end{pmatrix}$$

$$= \sum_{m=-M}^{M} \sum_{n=-N}^{N} \begin{pmatrix} \vec{E}_{i,x,m,n}, \vec{E}_{i,y,m,n}, \vec{E}_{i,z,m,n} \\ \vec{H}_{i,x,m,n}, \vec{H}_{i,y,m,n}, \vec{H}_{i,z,m,n} \end{pmatrix} e^{j(k_{x,m,n}x + k_{ym,n}y - k_{z,m,n}(z-z_+))}$$

$$+ \sum_{m=-M}^{M} \sum_{n=-N}^{N} \begin{pmatrix} \vec{E}_{r,x,m,n}, \vec{E}_{r,y,m,n}, \vec{E}_{r,z,m,n} \\ \vec{H}_{r,x,m,n}, \vec{H}_{r,y,m,n}, \vec{H}_{r,z,m,n} \end{pmatrix} e^{j(k_{x,m,n}x + k_{ym,n}y + k_{z,m,n}(z-z_+))}. \tag{3.82}$$

3.2.4.4 Case D: Half-Infinite Inhomogeneous Waveguide–Multiblock Structure–Half-Infinite Inhomogeneous Space

3.2.4.4.1 Left-to-Right Field Visualization

$$\vec{\underline{r}} = \vec{\underline{R}}^{(0,N+1)} \vec{\underline{u}} \tag{3.83a}$$

$$\begin{pmatrix} \vec{\underline{E}}_{t,y} \\ \vec{\underline{E}}_{t,x} \end{pmatrix} = \vec{\underline{T}}^{(0,N+1)} \vec{\underline{u}} \tag{3.83b}$$

$$\left\{ \underline{C}_{a,(1)}^{(1,N)}, \underline{C}_{a,(2)}^{(1,N)}, \ldots, \underline{C}_{a,(N)}^{(1,N)} \right\} = \left\{ \underline{C}_{a,(1)}^{(1,N)} \vec{\underline{u}}, \underline{C}_{a,(2)}^{(1,N)} \vec{\underline{u}}, \ldots, \underline{C}_{a,(N)}^{(1,N)} \vec{\underline{u}} \right\} \tag{3.83c}$$

The total field distributions are given as Figure 3.12(a).

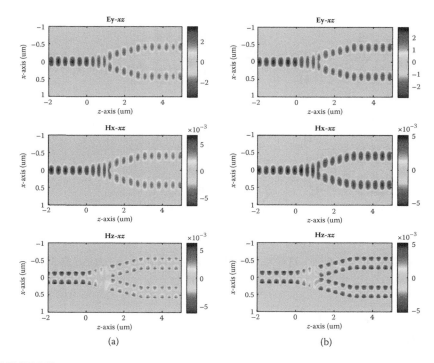

FIGURE 3.12
Case D: (a) Left-to-right directional characterization, E_y, H_x, and H_z field distributions for TE mode; and (b) right-to-left directional characterization, E_y, H_x, and H_z field distributions for TE mode. (MATLAB code: *FMM_Lwg_Rwg\FMM_Lwg_Rwg.m*.)

1. For $z < z_-$

$$
\begin{pmatrix} \mathbf{E} \\ \mathbf{H} \end{pmatrix} = \sum_{g=1}^{M^+} \vec{u}_g \begin{pmatrix} \mathbf{E}_{(0)}^{(g)+} \\ \mathbf{H}_{(0)}^{(g)+} \end{pmatrix} + \sum_{g=1}^{M^-} \vec{r}_g \begin{pmatrix} \mathbf{E}_{(0)}^{(g)-} \\ \mathbf{H}_{(0)}^{(g)-} \end{pmatrix}
\tag{3.84}
$$

2. For $z_- \leq z < z_+$

$$
\begin{pmatrix} \mathbf{E} \\ \mathbf{H} \end{pmatrix} = \begin{cases}
\displaystyle\sum_{g=1}^{M^+} C_{a,(1),g}^+ \begin{pmatrix} \mathbf{E}_{(1)}^{(g)+} \\ \mathbf{H}_{(1)}^{(g)+} \end{pmatrix} + \sum_{g=1}^{M^-} C_{a,(1),g}^- \begin{pmatrix} \mathbf{E}_{(1)}^{(g)-} \\ \mathbf{H}_{(1)}^{(g)-} \end{pmatrix} & \text{for } 0 \leq z \leq l_{1,1} \\[3ex]
\displaystyle\sum_{g=1}^{M^+} C_{a,(2),g}^+ \begin{pmatrix} \mathbf{E}_{(2)}^{(g)+} \\ \mathbf{H}_{(2)}^{(g)+} \end{pmatrix} + \sum_{g=1}^{M^-} C_{a,(2),g}^- \begin{pmatrix} \mathbf{E}_{(2)}^{(g)-} \\ \mathbf{H}_{(2)}^{(g)-} \end{pmatrix} & \text{for } l_{1,1} \leq z \leq l_{1,2} \\[3ex]
\qquad\qquad\qquad\vdots & \\[1ex]
\displaystyle\sum_{g=1}^{M^+} C_{a,(N),g}^+ \begin{pmatrix} \mathbf{E}_{(N)}^{(g)+} \\ \mathbf{H}_{(N)}^{(g)+} \end{pmatrix} + \sum_{g=1}^{M^-} C_{a,(N),g}^- \begin{pmatrix} \mathbf{E}_{(N)}^{(g)-} \\ \mathbf{H}_{(N)}^{(g)-} \end{pmatrix} & \text{for } l_{1,N-1} \leq z \leq l_{1,N}
\end{cases}
$$

$$\tag{3.85a}$$

where $\mathbf{E}^{(g)+}$, $\mathbf{E}^{(g)-}$, $\mathbf{H}^{(g)+}$, and $\mathbf{H}^{(g)-}$ are given, respectively, by

$$\mathbf{E}^{(g)+}(x,y,z) = \sum_{m=-M}^{M}\sum_{n=-N}^{N}\left(E_{x,m,n}^{(g)+}, E_{y,m,n}^{(g)+}, E_{z,m,n}^{(g)+}\right)e^{j\left(k_{x,m}x+k_{y,n}y+k_z^{(g)+}(z-z_-)\right)} \quad (3.85\text{b})$$

$$\mathbf{E}^{(g)-}(x,y,z) = \sum_{m=-M}^{M}\sum_{n=-N}^{N}\left(E_{x,m,n}^{(g)-}, E_{y,m,n}^{(g)-}, E_{z,m,n}^{(g)-}\right)e^{j\left(k_{x,m}x+k_{y,n}y+k_z^{(g)-}(z-z_+)\right)} \quad (3.85\text{c})$$

$$\mathbf{H}^{(g)+}(x,y,z) = \sum_{m=-M}^{M}\sum_{n=-N}^{N}\left(H_{x,m,n}^{(g)+}, H_{y,m,n}^{(g)+}, H_{z,m,n}^{(g)+}\right)e^{j\left(k_{x,m}x+k_{y,n}y+k_z^{(g)+}(z-z_-)\right)} \quad (3.85\text{d})$$

$$\mathbf{H}^{(g)-}(x,y,z) = \sum_{m=-M}^{M}\sum_{n=-N}^{N}\left(H_{x,m,n}^{(g)-}, H_{y,m,n}^{(g)-}, H_{z,m,n}^{(g)-}\right)e^{j\left(k_{x,m}x+k_{y,n}y+k_z^{(g)-}(z-z_+)\right)} \quad (3.85\text{e})$$

3. For $z_+ \leq z$

$$\left(\begin{array}{c}\mathbf{E}\\\mathbf{H}\end{array}\right) = \sum_{g=1}^{M^+}\vec{t}_g\left(\begin{array}{c}\mathbf{E}_{(N+1)}^{(g)+}\\\mathbf{H}_{(N+1)}^{(g)+}\end{array}\right) \quad (3.86\text{a})$$

where $\mathbf{E}^{(g)+}$ and $\mathbf{H}^{(g)+}$ are given, respectively, by

$$\mathbf{E}^{(g)+}(x,y,z) = \sum_{m=-M}^{M}\sum_{n=-N}^{N}\left(E_{x,m,n}^{(g)+}, E_{y,m,n}^{(g)+}, E_{z,m,n}^{(g)+}\right)e^{j\left(k_{x,m}x+k_{y,n}y+k_z^{(g)+}(z-z_-)\right)} \quad (3.86\text{b})$$

$$\mathbf{H}^{(g)+}(x,y,z) = \sum_{m=-M}^{M}\sum_{n=-N}^{N}\left(H_{x,m,n}^{(g)+}, H_{y,m,n}^{(g)+}, H_{z,m,n}^{(g)+}\right)e^{j\left(k_{x,m}x+k_{y,n}y+k_z^{(g)+}(z-z_-)\right)} \quad (3.86\text{c})$$

3.2.4.4.2 *Right-to-Left Field Visualization*

$$\vec{r} = \vec{\underline{R}}^{(0,N+1)}\vec{u} \quad (3.87\text{a})$$

$$\vec{t} = \vec{\underline{T}}^{(0,N+1)}\vec{u} \quad (3.87\text{b})$$

$$\left\{\underline{C}_{b,(1)}^{(1,N)}, \underline{C}_{b,(2)}^{(1,N)}, \ldots, \underline{C}_{b,(N)}^{(1,N)}\right\} = \left\{\underline{C}_{b,(1)}^{(1,N)}\vec{u}, \underline{C}_{b,(2)}^{(1,N)}\vec{u}, \ldots, \underline{C}_{b,(N)}^{(1,N)}\vec{u}\right\} \quad (3.87\text{c})$$

The total field distributions are given as Figure 3.12(b).

1. For $z < z_-$

$$\begin{pmatrix} \mathbf{E} \\ \mathbf{H} \end{pmatrix} = \sum_{g=1}^{M^+} \bar{t}_g \begin{pmatrix} \mathbf{E}_{(0)}^{(g)-} \\ \mathbf{H}_{(0)}^{(g)-} \end{pmatrix} \qquad\qquad (3.88)$$

2. For $z_- \leq z < z_+$

$$\begin{pmatrix} \mathbf{E} \\ \mathbf{H} \end{pmatrix} = \begin{cases} \displaystyle\sum_{g=1}^{M^+} C_{b,(1),g}^+ \begin{pmatrix} \mathbf{E}_{(1)}^{(g)+} \\ \mathbf{H}_{(1)}^{(g)+} \end{pmatrix} + \sum_{g=1}^{M^-} C_{b,(1),g}^- \begin{pmatrix} \mathbf{E}_{(1)}^{(g)-} \\ \mathbf{H}_{(1)}^{(g)-} \end{pmatrix} & \text{for } 0 \leq z \leq l_{1,1} \\[3ex] \displaystyle\sum_{g=1}^{M^+} C_{b,(2),g}^+ \begin{pmatrix} \mathbf{E}_{(2)}^{(g)+} \\ \mathbf{H}_{(2)}^{(g)+} \end{pmatrix} + \sum_{g=1}^{M^-} C_{b,(2),g}^- \begin{pmatrix} \mathbf{E}_{(2)}^{(g)-} \\ \mathbf{H}_{(2)}^{(g)-} \end{pmatrix} & \text{for } l_{1,1} \leq z \leq l_{1,2} \\[3ex] \qquad\qquad\qquad\vdots \\[2ex] \displaystyle\sum_{g=1}^{M^+} C_{b,(N),g}^+ \begin{pmatrix} \mathbf{E}_{(N)}^{(g)+} \\ \mathbf{H}_{(N)}^{(g)+} \end{pmatrix} + \sum_{g=1}^{M^-} C_{b,(N),g}^- \begin{pmatrix} \mathbf{E}_{(N)}^{(g)-} \\ \mathbf{H}_{(N)}^{(g)-} \end{pmatrix} & \text{for } l_{1,N-1} \leq z \leq l_{1,N} \end{cases}$$

$$(3.89)$$

where $\mathbf{E}^{(g)+}$, $\mathbf{E}^{(g)-}$, $\mathbf{H}^{(g)+}$, and $\mathbf{H}^{(g)-}$ are given in Equation (3.34).

3. For $z_+ \leq z$

$$\begin{pmatrix} \mathbf{E} \\ \mathbf{H} \end{pmatrix} = \sum_{g=1}^{M^-} \bar{u}_g \begin{pmatrix} \mathbf{E}_{(N+1)}^{(g)-} \\ \mathbf{H}_{(N+1)}^{(g)-} \end{pmatrix} + \sum_{g=1}^{M^+} \bar{r}_g \begin{pmatrix} \mathbf{E}_{(N+1)}^{(g)+} \\ \mathbf{H}_{(N+1)}^{(g)+} \end{pmatrix} \qquad (3.90)$$

where $\mathbf{E}^{(g)+}$, $\mathbf{E}^{(g)-}$, $\mathbf{H}^{(g)+}$, and $\mathbf{H}^{(g)-}$ are given in Equation (3.34).

3.3 MATLAB® Implementation

In this section, the numerical schemes for two main concepts will be discussed in detail with standard MATLAB codes. We prepared five field visualization MATLAB codes to test the S-matrix formulation and provide reference MATLAB codes. For analyzing multiblock structures, whole space is divided into three regions: left half-infinite space, finite-sized grating structure, and right half-infinite space. The MATLAB implementation of the above-described mathematical model is provided with MATLAB code *FMM_Lwg_Rwg.m*. The entire code is presented as follows:

MATLAB Code 3.1: *FMM_Lwg_Rwg.m*

```
1    % FMM standard written by H. Kim
2
3    % 0. free space - grating - free space
4    % 1. semi-infinite homogeneous space- grating - semi-
        infinite homogeneous space
5    % 2. semi-infinite homogeneous space - grating - semi-
        infinite inhomogeneous waveguide
6    % 3. semi-infinite inhomogeneous waveguide - grating
        - semi-infinite homogeneous space
7    % 4. semi-infinite inhomogeneous waveguide - grating
        - semi-infinite inhomogeneous waveguide
8
9    clear all;
10   close all;
11   clc;
12
13   addpath([pwd '\PRCWA_COM']);
14   addpath([pwd '\FIELD_VISUAL']);
15   addpath([pwd '\STRUCTURE']);
16
17
18   %% STEP 1 : wavevector setting and structure modeling
19
20   % length unit
21
22   global nm; % nano
23   global um; % micro
24   global mm; % mili
25
26   global k0;                              % wavenumber
27   global c0; global w0;
28   global eps0; global mu0;
29
30   % zero-thickness buffer refractive index, permittivity,
        permeability
31   global n0; global epr0; global mur0;
32   % refractive index, permittivity, permeability in free
        space I
33   global ni; global epri; global muri;
34   % refractive index, permittivity, permeability in free
        space II
```

```
35    global nf; global eprf; global murf;
36
37    % x-directional supercell period, y-directional supercell
         period
38    global Tx; global Ty;
39    global nx; global ny;
40    % # of x-direction Fourier harmonics, # of y-directional
         Fourier harmonics
41    global NBx; global NBy;
42    global num_hx; global num_hy;
43    global kx_vc; global ky_vc; global kz_vc;
44
45    % input output free space
46    global kix; global kiy; global kiz;
47    global kfz;
48    global kx_ref; global ky_ref; global kz_ref;
49    global kx_tra; global ky_tra; global kz_tra;
50
51    nm=1e-9;
52    lambda=532*nm;
53
54    % 0. free space - grating - free space : L-to-R, R-to-L
55    % 1. homogeneous space - grating - homogeneous space :
         L-to-R, R-to-L
56    % 2. homogeneous space - grating - inhomogeneous
         waveguide : L-to-R, R-to-L
57    % 3. inhomogeneous waveguide - grating - homogeneous
         space : L-to-R, R-to-L
58    % 4. inhomogeneous waveguide - grating - inhomogeneous
         waveguide : L-to-R, R-to-L
59
60    type_ =4;
61    direct_ =2; % 1 = left-to-right , 2 = right-to-left
62
63    PRCWA_basic;              % 3D structure
64    PRCWA_Gen_K;              % zero-thickness buffer
65
66    % The example structure is the triangle grating
         structure approximated by
67    % the stepwise multi-blocks. This structure modeling in
         the MATLAB code lines 23~68
68    % In PRCWA_Gen_Y_branch.m, open PRCWA_Gen_Y_branch.m and
         see the annotation
69
70    PRCWA_Gen_Y_branch; % SPP Y branch
71
```

```
72    %% STEP 2 Block S-matrix computation of single block
      structures
73
74    L=NBx*NBy;
75
76    Ta=zeros(2*L,2*L,Nlay); % left to rignt
77    Ra=zeros(2*L,2*L,Nlay); % left to right
78    Tb=zeros(2*L,2*L,Nlay); % right to left
79    Rb=zeros(2*L,2*L,Nlay); % right to left
80    Ca=zeros(4*L,2*L,Nlay); % left to right
81    Cb=zeros(4*L,2*L,Nlay); % right to left
82    tCa=zeros(4*L,2*L,Nlay); % left to right
83    tCb=zeros(4*L,2*L,Nlay); % right to left
84
85    Diagonal_SMM;
86    %Off_diagonal_tensor_SMM;
87
88    %% STEP3 S-matrix method
89    % The obtained S-matrix and coupling coefficient matrix
      operator of single
90    % blocks are combined to generate the S-matrix and
      coupling coefficient
91    % operator of interconnected multi-block structures by
      the Redheffer star
92    % product...
93
94    I=eye(2*L,2*L);
95    T_temp1a=Ta(:,:,1);
96    R_temp1a=Ra(:,:,1);
97    T_temp1b=Tb(:,:,1);
98    R_temp1b=Rb(:,:,1);
99
100   %% Important
101     tCa=Ca;
102     tCb=Cb;
103     %%
104
105   for laynt=2:Nlay
106
107     %laynt
108
109     T_temp2a=Ta(:,:,laynt);
110     R_temp2a=Ra(:,:,laynt);
111     T_temp2b=Tb(:,:,laynt);
112     R_temp2b=Rb(:,:,laynt);
113
```

```
114    RRa=(R_temp1a+T_temp1b*inv(I-R_temp2a*R_temp1b)*R_
       temp2a*T_temp1a);
115    TTa=T_temp2a*inv(I-R_temp1b*R_temp2a)*T_temp1a;
116
117    RRb=(R_temp2b+T_temp2a*inv(I-R_temp1b*R_temp2a)*R_
       temp1b*T_temp2b);
118    TTb=T_temp1b*inv(I-R_temp2a*R_temp1b)*T_temp2b;
119
120    for k=1:laynt-1
121
122    tCa(:,:,k)=Ca(:,:,k)+Cb(:,:,k)*inv(I-R_temp2a*R_
       temp1b)*R_temp2a*T_temp1a;
123    tCb(:,:,k)=Cb(:,:,k)*inv(I-R_temp2a*R_temp1b)*T_temp2b;
124
125    end; % for k
126
127    tCa(:,:,laynt)=Ca(:,:,laynt)*inv(I-R_temp1b*R_
       temp2a)*T_temp1a;
128    tCb(:,:,laynt)=Cb(:,:,laynt)+Ca(:,:,laynt)*inv(I-R_
       temp1b*R_temp2a)*R_temp1b*T_temp2b;
129
130    T_temp1a=TTa;
131    R_temp1a=RRa;
132    T_temp1b=TTb;
133    R_temp1b=RRb;
134
135    Ca=tCa;
136    Cb=tCb;
137    end; % laynt
138
139    TTa=T_temp1a; % left-to-right transmission operator
140    RRa=R_temp1a; % left-to-right reflection operator
141    TTb=T_temp1b; % right-to-left transmission operator
142    RRb=R_temp1b; % right-to-left reflection operator
143
144    %% STEP4 Half-infinite block interconnection & Field
       visualization
145    switch direct_
146
147    case 1 % left-to-right
148    % polarizatoin angle : TM=0, TE=pi/2
149    psi=0;
150    tm_Ux=cos(psi)*cos(theta)*cos(phi)-sin(psi)*sin(phi); %
       incident waveï¿½ï¿½ Ex
151    tm_Uy=cos(psi)*cos(theta)*sin(phi)+sin(psi)*cos(phi); %
       incident waveï¿½ï¿½ Ey
```

```
152    tm_Uz=-cos(psi)*sin(theta);
153
154    psi=pi/2;
155    te_Ux=cos(psi)*cos(theta)*cos(phi)-sin(psi)*sin(phi); %
       incident waveﻥﻥ Ex
156    te_Uy=cos(psi)*cos(theta)*sin(phi)+sin(psi)*cos(phi); %
       incident waveﻥﻥ Ey
157    te_Uz=-cos(psi)*sin(theta);
158
159    % Ux=cos(psi)*cos(theta)*cos(phi)-sin(psi)*sin(phi); %
       incident waveﻥﻥ Ex
160    % Uy=cos(psi)*cos(theta)*sin(phi)+sin(psi)*cos(phi); %
       incident waveﻥﻥ Ey
161    % Uz=-cos(psi)*sin(theta);
162
163    case 2 % right-to-left
164    % polarization angle : TM=0, TE=pi/2
165    psi=0;
166    tm_Ux=cos(psi)*cos(theta)*cos(phi)-sin(psi)*sin(phi); %
       incident waveﻥﻥ Ex
167    tm_Uy=cos(psi)*cos(theta)*sin(phi)+sin(psi)*cos(phi); %
       incident waveﻥﻥ Ey
168    tm_Uz=cos(psi)*sin(theta);
169
170    psi=pi/2;
171    te_Ux=cos(psi)*cos(theta)*cos(phi)-sin(psi)*sin(phi); %
       incident waveﻥﻥ Ex
       te_Uy=cos(psi)*cos(theta)*sin(phi)+sin(psi)*cos(phi); %
       incident waveﻥﻥ Ey
172    te_Uz=cos(psi)*sin(theta);
173
174    % Ux=cos(psi)*cos(theta)*cos(phi)-sin(psi)*sin(phi); %
       incident waveﻥﻥ Ex
175    % Uy=cos(psi)*cos(theta)*sin(phi)+sin(psi)*cos(phi); %
       incident waveﻥﻥ Ey
176    % Uz=cos(psi)*sin(theta);
177
178    end;
179
180    switch type_
181
182     case 0
183
184     Bdr_Smat_case0; % 0. free space - grating - free space
185
186     case 1
```

```
187
188    Bdr_Smat_case1; % 1. homogeneous space - grating -
       homogeneous space
189
190    case 2
191
192    Bdr_Smat_case2; % 2. homogeneous space - grating -
       inhomogeneous waveguide
193
194    case 3
195
196    Bdr_Smat_case3; % 3. inhomogeneous waveguide - grating
       - homogeneous space
197
198    case 4
199
200    Bdr_Smat_case4; % 4. inhomogeneous waveguide - grating
       - inhomogeneous waveguide
201
202    end;
203
```

The important constituents of the MATLAB functions are listed in the following table. In particular, the readers have to get the in-depth understanding of five MATLAB functions:

1. *FMM_single_block_tensor.m*
2. *FMM_single_block.m*
3. *Bdr_SMat_infr_outfr.m*
4. *Bdr_SMat_wg.m*
5. *Bdr_SMat_wg_tensor.m*

These MATLAB functions are the analysis functions of block and boundary S-matrices of the target structure.

File Name		Description
FMM_Lwg_Rwg.m		Main routine of S-matrix method with field visualization
PRCWA_COM/Directory		*Common Library*
FMM_single_block.m		S-matrix calculation of single layer with diagonal anisotropic material
	sWp_gen.m	Transversal boundary electric field distribution at z=z+
	sWm_gen.m	Transversal boundary electric field distribution at z=z−

	sVp_gen.m	Transversal boundary magnetic field distribution at z=z+
	sVm_gen.m	Transversal boundary magnetic field distribution at z=z–
FMM_single_block_tensor.m		S-matrix calculation of single block with general off-diagonal tensor anisotropic material
	Wp_gen.m	Transversal boundary electric field distribution at z=z+
	Wm_gen.m	Transversal boundary electric field distribution at z=z–
	Vp_gen.m	Transversal boundary magnetic field distribution at z=z+
	Vm_gen.m	Transversal boundary magnetic field distribution at z=z–
Bdr_SMat_infr_outfr.m		Boundary S-matrix calculation of left and right half-infinite isotropic homogeneous spaces
Bdr_SMat_wg.m		Boundary S-matrix calculation of left and right half-infinite inhomogeneous spaces with diagonal anisotropic material
Bdr_SMat_wg_tensor.m		Boundary S-matrix calculation of left and right half-infinite inhomogeneous spaces with general off-diagonal tensor anisotropic material
PRCWA_basic.m		Basic setting of Fourier harmonic orders, supercell period, buffer block material
PRCWA_Gen_K.m		Setting wavevector grid of Fourier harmonics
fun_*.m		Permittivity functions of Ag, Au, Si, PMMA for wavelength
Odd_*.m		FFT shift functions of odd number sampling signal for Fourier transform
FIELD_VISUAL/Directory		*Field Visualization*
Field_visualization_3D_xz_*.m		Cross-section profile of six field components at x-z plane
Field_visualization_3D_yz_*.m		Cross-section profile of six field components at y-z plane
Field_visualization_3D_xy.m		Cross-section profile of six field components at x-y plane
STRUCTURE/Directory		*Example Structures*
PRCWA_Gen_Y_branch.m		Setting the Toeplitz matrices of permittivity and permeability tensors
Grating_gen_Y_branch.m		Setting permittivity and permeability tensors

The code structure is divided into the following four steps:

Step 1: Wavevector setting and structure modeling (lines 18~70 in *FMM_Lwg_Rwg.m*). In *PRCWA_basic.m*, the permittivity and permeability of the left and right half-infinite spaces are determined. Equation (3.3) is coded in *PRCWA_Gen_K.m*. The material structures described in the lefthand side of Equations (3.12a) to (3.12c) are implemented in *PRCWA_Gen_Y_branch.m*. The material structures described in the righthand side of Equations (3.1a) and (3.1b) are written in *PRCWA_Gen_Y_branch.m*, *Grating_gen_Y_branch.m*, and *Grating_gen_Y_branch.m*.

Step 2: Block S-matrix computation of single-block structures (lines 72~87 in *FMM_Lwg_Rwg.m*). The block S-matrix analysis of a single-block structure composed of a diagonal anisotropic material is performed by *FMM_single_block.m* in *Diagonal_SMM.m*, while the block S-matrix for a single-block structure composed of a general off-diagonal anisotropic material is analyzed by *FMM_single_block_tensor.m* in *off_diagonal_tensor_SMM.m*. Both functions generate S-matrix, coupling coefficient operators, pseudo-Fourier series coefficients of internal eigenmodes, and eigenvalues for the input of the permittivity and permeability profiles. The material to *FMM_single_block_tensor.m* can be fully anisotropic with nonzero off-diagonal elements. Equations (3.15a) and (3.15b) correspond to *SA* and *SB* in *FMM_single_block.m* (see lines 100~112 in *FMM_single_block.m*), respectively. Equation (3.16) is implemented by *St* in *FMM_single_block.m*. Lalanne's experience rule for the use of reciprocal permittivity and permeability profiles (reference [31] in Chapter 1) is employed in both *FMM_single_block.m* and *FMM_single_block_tensor.m*. In these MATLAB codes, *Exx*, *Eyy*, *Ezz*, *Axx*, *Ayy*, and *Azz* correspond to ε_{xx}, ε_{yy}, ε_{zz}, α_{xx}, α_{yy}, α_{zz}, respectively. *Gxx*, *Gyy*, *Gzz*, *Bxx*, *Byy*, and *Bzz* correspond to μ_{xx}, μ_{yy}, μ_{zz}, β_{xx}, β_{yy}, and β_{zz}, respectively.

MATLAB Code 3.2: *FMM_single_block.m*

```
100    % System Matrix
101
102    SA=zeros(2*L, 2*L);
103    SB=zeros(2*L,2*L);
104
105    SA=[ (Ky)*inv(Ezz)*(Kx)  BG_x-(Ky)*inv(Ezz)*(Ky)
106      (Kx)*inv(Ezz)*(Kx)-GB_y -(Kx)*inv(Ezz)*(Ky)];
107
108    SB=[ (Ky)*inv(Gzz)*(Kx)  AE_x-(Ky)*inv(Gzz)*(Ky)
109      (Kx)*inv(Gzz)*(Kx)-EA_x -(Kx)*inv(Gzz)*(Ky)];
110
111    St=k0^2*SA*SB;
112    clear SB;
```

On the other hand, the Maxwell eigenvalue equation of Equation (3.19) for general anisotropic media is analyzed by *FMM_single_block_tensor.m* in *Off_diagonal_tensor_SMM.m*. The MATLAB code of Equation (3.19) is presented in lines 140~168 in *FMM_single_block_tensor.m*.

MATLAB Code 3.3: *FMM_single_block_tensor.m*

```
140     % System Matrix
141
142     St11= -j*Gxz*inv(Gzz)*Kx - j*Ky*inv(Ezz)*Ezy;
143     St12= j*Gxz*inv(Gzz)*Ky - j*Ky*inv(Ezz)*Ezx;
144     St13= Gxy - Gxz*inv(Gzz)*Gzy + Ky*inv(Ezz)*Kx;
145     St14= BG_x - Gxz*inv(Gzz)*Gzx - Ky*inv(Ezz)*Ky;
146
147
148     St21= j*Gyz*inv(Gzz)*Kx-j*Kx*inv(Ezz)*Ezy;
149     St22= -j*Gyz*inv(Gzz)*Ky-j*Kx*inv(Ezz)*Ezx;
150     St23= -GB_y+Gyz*inv(Gzz)*Gzy + Kx*inv(Ezz)*Kx;
151     St24= -Gyx+Gyz*inv(Gzz)*Gzx - Kx*inv(Ezz)*Ky;
152
153     St31= Exy - Exz*inv(Ezz)*Ezy+Ky*inv(Gzz)*Kx;
154     St32= AE_x - Exz*inv(Ezz)*Ezx-Ky*inv(Gzz)*Ky;
155     St33= -j*Exz*inv(Ezz)*Kx-j*Ky*inv(Gzz)*Gzy;
156     St34= j*Exz*inv(Ezz)*Ky-j*Ky*inv(Gzz)*Gzx;
157
158     St41= -EA_y+Eyz*inv(Ezz)*Ezy+Kx*inv(Gzz)*Kx;
159     St42= -Eyx+Eyz*inv(Ezz)*Ezx-Kx*inv(Gzz)*Ky;
160     St43= j*Eyz*inv(Ezz)*Kx-j*Kx*inv(Gzz)*Gzy;
161     St44= -j*Eyz*inv(Ezz)*Ky-j*Kx*inv(Gzz)*Gzx;
162
163     St=[ St11 St12 St13 St14
164        St21 St22 St23 St24
165        St31 St32 St33 St34
166        St41 St42 St43 St44];
167
168     St=k0*St;
```

The Maxwell eigenvalue equation (3.19) is numerically solved and the resulting eigenvalues and eigenvectors are saved to *W* and *Dt*, respectively, in *FMM_single_block.m* (see line 159) and *FMM_single_block_tensor.m* (see line 250). In the construction of Maxwell's eigenvalue equation for general off-diagonal anisotropic material as well as diagonal anisotropic material, Lalanne's empirical rule of the use of reciprocal permittivity and permeability is employed (reference [31] in Chapter 1), which is described in lines 175~212 of MATLAB code *FMM_single_block.m*.

MATLAB Code 3.4: *FMM_single_block_tensor.m*

```
175     alpha=alpha_tm; % convergence factor
176     beta=beta_tm;
177
```

```
178    AE_x=zeros(L,L);
179    EA_x=zeros(L,L);
180
181    AE_y=zeros(L,L);
182    EA_y=zeros(L,L);
183
184    AE_z=zeros(L,L);
185    EA_z=zeros(L,L);
186
187    AE_x=alpha*inv(Axx)+(1-alpha)*Exx;
188    EA_x=alpha*Exx+(1-alpha)*inv(Axx);
189
190    AE_y=alpha*inv(Ayy)+(1-alpha)*Eyy;
191    EA_y=alpha*Eyy+(1-alpha)*inv(Ayy);
192
193    AE_z=alpha*inv(Azz)+(1-alpha)*Ezz;
194    EA_z=alpha*Ezz+(1-alpha)*inv(Azz);
195
196    BG_x=zeros(L,L);
197    GB_x=zeros(L,L);
198
199    BG_y=zeros(L,L);
200    GB_y=zeros(L,L);
201
202    BG_z=zeros(L,L);
203    GB_z=zeros(L,L);
204
205    BG_x=beta*inv(Bxx)+(1-beta)*Gxx;
206    GB_x=beta*Gxx+(1-beta)*inv(Bxx);
207
208    BG_y=beta*inv(Byy)+(1-beta)*Gyy;
209    GB_y=beta*Gyy+(1-beta)*inv(Byy);
210
211    BG_z=beta*inv(Bzz)+(1-beta)*Gzz;
212    GB_z=beta*Gzz+(1-beta)*inv(Bzz);
```

| The system matrix (3.19) is implemented by *St* in the MATLAB code. This part is described in MATLAB code lines 177~250 in *FMM_single_block_tensor.m*. In the same way as the case of diagonal anisotropy material, Lalanne's experience rule for using reciprocal permittivity and permeability (reference [31] in Chapter 1) is employed in the implementation. The Maxwell eigenvalue equation (3.19) is numerically solved and the resulting eigenvalues and eigenvectors are saved to *W* and *Dt*, respectively. This part is in

lines 247~254 in *FMM_single_block_tensor.m*. Equations (3.29a) to (3.29c) related to the left-to-right characterization and Equations (3.32a) to (3.32c) related to the right-to-left characterization are coded in lines 292~334 in *FMM_single_block.m* and in lines 355~398 in *FMM_single_block_tensor.m*. The codes of the two files are exactly the same.

MATLAB Code 3.5: *FMM_single_block.m (FMM_single_block_tensor.m)*

```
292      % left-to-right
293      U=eye(2*L,2*L);
294
295      S11=inv(Wh)*Wp_zm+inv(Vh)*Vp_zm;
296      S12=inv(Wh)*Wm_zm+inv(Vh)*Vm_zm;
297      S21=inv(Wh)*Wp_zp-inv(Vh)*Vp_zp;
298      S22=inv(Wh)*Wm_zp-inv(Vh)*Vm_zp;
299
300      S=[S11 S12; S21 S22];
301
302      clear S11;
303      clear S12;
304      clear S21;
305      clear S22;
306      D=[2*U;zeros(2*L,2*L)];
307
308      CCa=inv(S)*D;
309      Cap=CCa(1:2*L,:);
310      Cam=CCa(2*L+1:4*L,:);
311      Ra=inv(Wh)*(Wp_zm*Cap+Wm_zm*Cam-Wh*U);
312      Ta=inv(Wh)*(Wp_zp*Cap+Wm_zp*Cam);
313
314      % right-to-left
315
316      S11=inv(Wh)*Wp_zm+inv(Vh)*Vp_zm;
317      S12=inv(Wh)*Wm_zm+inv(Vh)*Vm_zm;
318      S21=inv(Wh)*Wp_zp-inv(Vh)*Vp_zp;
319      S22=inv(Wh)*Wm_zp-inv(Vh)*Vm_zp;
320      S=[S11 S12 ; S21 S22];
321
322      clear S11;
323      clear S12;
324      clear S21;
325      clear S22;
326
327      D=[zeros(2*L,2*L);2*U];
328      CCb=inv(S)*D;
```

```
329      Cbp=CCb(1:2*L,:);
330      Cbm=CCb(2*L+1:4*L,:);
331
332      Rb=inv(Wh)*(Wp_zp*Cbp+Wm_zp*Cbm-Wh*U);
333      Tb=inv(Wh)*(Wp_zm*Cbp+Wm_zm*Cbm);
```

At lines 236~254 of *FMM_single_block.m*, the necessary variables to make the system matrices of Equations (3.28a) and (3.28b), and Equations (3.31a) and (3.31b) are prepared as follows.

$\underline{\underline{W}}^{+}(0)$	Wp_zm	sWp_gen(pW,pevalue,pcnt,L,zm-zm)
$\underline{\underline{W}}^{-}(z_- - z_+)$	Wm_zm	sWm_gen(mW,mevalue,mcnt,L,zm-zp)
$\underline{\underline{V}}^{+}(0)$	Vp_zm	sVp_gen(pV,pevalue,pcnt,L,zm-zm)
$\underline{\underline{V}}^{-}(z_- - z_+)$	Vm_zm	sVm_gen(mV,mevalue,mcnt,L,zm-zp)
$\underline{\underline{W}}^{+}(z_+ - z_-)$	Wp_zp	sWp_gen(pW,pevalue,pcnt,L,zm-zm)
$\underline{\underline{V}}^{+}(z_+ - z_-)$	Vp_zp	sWm_gen(mW,mevalue,mcnt,L,zm-zp)
$\underline{\underline{W}}^{-}(0)$	Wm_zp	sVp_gen(pV,pevalue,pcnt,L,zm-zm)
$\underline{\underline{V}}^{-}(0)$	Vm_zp	sVm_gen(mV,mevalue,mcnt,L,zm-zp)

$\underline{\underline{W}}_h$ and $\underline{\underline{V}}_h$ are implemented in lines 286 and 287 of *FMM_single_block.m*. It is noteworthy that the final sentence is KII divided by (w0*mu0) at line 287. Equation (3.29a) is solved and $\underline{C}_a$ is obtained at line 309 by *CCa*. Equations (3.29b) and (3.29c) are solved and $\vec{\underline{R}}$ and $\vec{\underline{T}}$ are obtained at lines 312 and 313, respectively, by *Ra* and *Ta*. In the MATLAB code of *FMM_single_block.m*, Equations (3.32a)~(3.33d) are solved and $\underline{C}_b$, $\vec{\underline{R}}$, and $\vec{\underline{T}}$ are obtained in lines 328~334. In the case of *FMM_single_block_tensor.m*, the exact same codes are used in lines 395~401. In *FMM_single_block.m* the calculation of Fourier coefficients of pseudo-Fourier series is performed at lines 210~233. In *FMM_single_block_tensor.m*, the same calculation is being done at lines 281~303.

MATLAB Code 3.6: *FMM_single_block.m*

```
210      % Fourier coefficients (pfEx,pfEy,pfEz,pfHx,pfHy,pfHz)
          , (mfEx,mfEy,mfEz,mfHx,mfHy,mfHz)
211
212      pfEy=pW(1:L,:);                          % pfEy
213      pfEx=pW(L+1:2*L,:);                       % pfEx
214      pfHy=pV(1:L,:);                           % pfHy
```

```
215    pfHx=pV(L+1:2*L,:);                        % pfHx
216    pfEz=inv(Ezz)*(j*Ky*pfHx-j*Kx*pfHy);       % pfEz
217    pfHz=inv(Gzz)*(j*Ky*pfEx-j*Kx*pfEy);       % pfHz
218
219    pfHy=j*(eps0/mu0)^0.5*pfHy;                 % pfHy
220    pfHx=j*(eps0/mu0)^0.5*pfHx;                 % pfHx
221    pfHz=j*(eps0/mu0)^0.5*pfHz;                 % pfHz
222
223
224    mfEy=mW(1:L,:);                             % mfEy
225    mfEx=mW(L+1:2*L,:);                         % mfEx
226    mfHy=mV(1:L,:);                             % mfHy
227    mfHx=mV(L+1:2*L,:);                         % mfHx
228    mfEz=inv(Ezz)*(j*Ky*mfHx-j*Kx*mfHy);       % mfEz
229    mfHz=inv(Gzz)*(j*Ky*mfEx-j*Kx*mfEy);       % mfHz
230
231    mfHy=j*(eps0/mu0)^0.5*mfHy;                 % mfHy
232    mfHx=j*(eps0/mu0)^0.5*mfHx;                 % mfHx
233    mfHz=j*(eps0/mu0)^0.5*mfHz;
```

MATLAB Code 3.7: *FMM_single_block.m*

```
281    % Fourier coefficients (pfEx,pfEy,pfEz,pfHx,pfHy,pfHz) ,
          (mfEx,mfEy,mfEz,mfHx,mfHy,mfHz)
282
283    pfEy=pW(1:L,:);                             % pfEy
284    pfEx=pW(L+1:2*L,:);                         % pfEx
285    pfHy=pW(2*L+1:3*L,:);                       % pfHy
286    pfHx=pW(3*L+1:4*L,:);                       % pfHx
287    pfEz=inv(Ezz)*(j*Ky*pfHx-j*Kx*pfHy-Ezx*pfEx-Ezy*pfEy);
          % pfEz
288    pfHz=inv(Gzz)*(j*Ky*pfEx-j*Kx*pfEy-Gzx*pfHx-Gzy*pfHy);
          % pfHz
289
290    pfHy=j*(eps0/mu0)^0.5*pfHy;
291    pfHx=j*(eps0/mu0)^0.5*pfHx;
292    pfHz=j*(eps0/mu0)^0.5*pfHz;
293
294    mfEy=mW(1:L,:);                             % mfEy
295    mfEx=mW(L+1:2*L,:);                         % mfEx
296    mfHy=mW(2*L+1:3*L,:);                       % mfHy
297    mfHx=mW(3*L+1:4*L,:);                       % mfHx
298    mfEz=inv(Ezz)*(j*Ky*mfHx-j*Kx*mfHy-Ezx*mfEx-Ezy*mfEy);
          % mfEz
299    mfHz=inv(Gzz)*(j*Ky*mfEx-j*Kx*mfEy-Gzx*mfHx-Gzy*mfHy);
          % mfHz
300
```

```
301    mfHy=j*(eps0/mu0)^0.5*mfHy;
302    mfHx=j*(eps0/mu0)^0.5*mfHx;
303    mfHz=j*(eps0/mu0)^0.5*mfHz;
```

The *FMM_single_block.m* and *FMM_single_block_tensor.m* have the forms of

$$[Ta, Ra, Tb, Rb, CCa, CCb, pfEx, pfEy, pfEz, pfHx, pfHy, pfHz,$$
$$pevalue, mfEx, mfEy, mfEz, mfHx, mfHy, mfHz, mevalue] =$$
$$FMM\_single\_block(lay\_thick, str\_tensor, alpha\_tm, beta\_tm)$$

$$[Ta, Ra, Tb, Rb, CCa, CCb, pfEx, pfEy, pfEz, pfHx, pfHy, pfHz,$$
$$pevalue, mfEx, mfEy, mfEz, mfHx, mfHy, mfHz, mevalue] =$$
$$FMM\_single\_block\_tensor(lay\_thick, str\_tensor, alpha\_tm, beta\_tm)$$

The outputs of *FMM_single_block_tensor.m* and *FMM_single_block.m* are given by:

S-Matrix and Coupling Coefficient Operator		Positive Eigenmodes		Negative Eigenmodes	
$\overrightarrow{\underline{\underline{T}}}$	Ta	$E_{x,m,n}^{(g)+}$	pfEx	$E_{x,m,n}^{(g)-}$	mfEx
$\overrightarrow{\underline{\underline{R}}}$	Ra	$E_{y,m,n}^{(g)+}$	pfEy	$E_{y,m,n}^{(g)-}$	mfEy
$\overleftarrow{\underline{\underline{T}}}$	Tb	$E_{z,m,n}^{(g)+}$	pfEz	$E_{z,m,n}^{(g)-}$	mfEz
$\overleftarrow{\underline{\underline{R}}}$	Rb	$H_{x,m,n}^{(g)+}$	pfHx	$H_{x,m,n}^{(g)-}$	mfHx
$\underline{\underline{C}}_a$	CCa	$H_{y,m,n}^{(g)+}$	pfHy	$H_{y,m,n}^{(g)-}$	mfHy
$\underline{\underline{C}}_b$	CCb	$H_{z,m,n}^{(g)+}$	pfHz	$H_{z,m,n}^{(g)-}$	mfHz
		$k_z^{(g)+}$	pevalue	$k_z^{(g)-}$	mevalue

The Fourier series coefficients of the positive eigenmodes and positive eigenvalues are saved to the variables, *PfEx*, *PfEy*, *PfEz*, *PfHx*, *PfHy*, *pfHz*, and *Peigvalue*, respectively. The Fourier series coefficients of the negative eigenmodes and negative eigenvalues are saved to the variables, *mf_Ex*, *mf_Ey*, *mf_Ez*, *mf_Hx*, *mf_Hy*, *mf_Hz*, and *meigvalue*, respectively. The boundary S-matrix analysis of free space is performed in *Bdr_SMat_infr_outfr.m*. The boundary S-matrix analysis script, *Bdr_SMat_wg.m*, corresponds to Equations (3.52)~(3.55f).

In the case of *FMM_single_block.m*, $\underline{E}_z$ and $\underline{H}_z$ are obtained by Equations (3.17b) and (3.17c):

$$\underline{E}_z = \underline{\underline{\varepsilon}}_z^{-1}(j(\underline{K}_y/k_0)\underline{H}_x - j(\underline{K}_x/k_0)\underline{H}_y)$$ (3.17b)

$$\underline{H}_z = \underline{\underline{\mu}}_z^{-1}(j(\underline{K}_y/k_0)\underline{E}_x - j(\underline{K}_x/k_0)\underline{E}_y)$$ (3.17c)

and are coded in lines 216~217 and 228~229 of *FMM_single_block.m*. In the case of *FMM_single_block_tensor.m*, $\underline{E}_z$ and $\underline{H}_z$ are obtained by Equations (3.18f) and (3.18c):

$$\underline{E}_z = \underline{\underline{\varepsilon}}_{zz}^{-1}[j(\underline{K}_y/k_0)\underline{H}_x - j(\underline{K}_x/k_0)\underline{H}_y - (\underline{\underline{\varepsilon}}_{zx}\underline{E}_x + \underline{\underline{\varepsilon}}_{zy}\underline{E}_y)]$$ (3.18f)

$$\underline{H}_z = \underline{\underline{\mu}}_{zz}^{-1}[j(\underline{K}_y/k_0)\underline{E}_x - j(\underline{K}_x/k_0)\underline{E}_y - (\underline{\underline{\mu}}_{zx}\underline{H}_x + \underline{\underline{\mu}}_{zy}\underline{H}_y)]$$ (3.18c)

coded in lines 284~285 and 295~296.

Step 3: S-matrix method (lines 88~143 *FMM_Lwg_Rwg.m*). The recursive algorithms of the S-matrix method of Equations (3.46a)~(3.46d) and Equations (3.50a)~(3.50d) are coded in lines 95~155 of *FMM_Lwg_Rwg.m*.

Step 4: Half-infinite block interconnection and field visualization. The directional field visualization is selected by the parameter *direct_* (line 61 *FMM_Lwg_Rwg.m*). For the case of left-to-right directional characterization, the parameter *direct_* is set to *direct_=1*; for the case of right-to-left directional characterization, the parameter *direct_* is set to *direct_=2*. As explained in Section 3.2, the analysis is classified into four cases with respect to four possible combinations of finite-sized multiblock and half-infinite blocks: (1) half-infinite homogeneous space–multiblock structure–half-infinite homogeneous space, (2) half-infinite homogeneous space–multiblock structure–half-infinite inhomogeneous space, (3) half-infinite inhomogeneous space–multiblock structure–half-infinite homogeneous space, and (4) half-infinite inhomogeneous space–multiblock structure–half-infinite inhomogeneous space.

Cases A to D are implemented in *Bdr_Smat_case1.m, Bdr_Smat_case2.m, Bdr_Smat_case3.m,* and *Bdr_Smat_case4.m*, respectively. See lines 193~215 in *FMM_Lwg_Rwg.m*.

3.4 Applications

3.4.1 Extraordinary Optical Transmission Phenomenon

According to classical aperture theory, light propagating through an aperture inscribed on an opaque screen with a size on the subwavelength scale undergoes so much diffraction that the intensity of the light decreases with the square of the inverse of the distance from the slit, indicating that light cannot be detected by far-field measurement [1]. Ebbesen et al. discovered, however, that the surface plasmon resonance effect around the subwavelength aperture on a metal screen can significantly enhance the transmission [2]. This phenomenon is referred to as the EOT and has attracted considerable interest from researchers in various fields [3].

On the other hand, recent experimental studies have revealed that there are other properties in the transmission of light through an array of subwavelength holes or slits [4,5]. In addition, the extraordinary optical transmission suppression was introduced [6]. The interference between the incident light and the surface plasmon modes is strongly dependent on the distance of the neighboring slits. If a specific condition is satisfied, then those two components interfere with each other destructively, resulting in significant suppression in the transmission. This phenomenon is called EOT suppression and can be adopted for the optical switch with a high extinction ratio [7].

The MATLAB code implementation for the simulation of the EOT suppression is provided here. Based on the calculated results, which are in good agreement with the published data, the physical interpretation of the phenomenon is also provided.

Let us first consider the geometry of this problem, shown in Figure 3.13. The detail specification is from reference [7]. The operating wavelength in free space λ_0 is 514.5 nm (see line 58 on *main.m*). A thin silver film with the thickness h of 200 nm (see line 5 on *Grating_gen_TriangleGrating2.m*) is on the dielectric substrate with the refractive index of 1.46 (see line 21 on *PRCWA_basic.m*). It is assumed that the silver film is covered by index-matching fluid; i.e., the refractive index of the surrounding material is the same as that of the substrate (see line 25 on *PRCWA_basic.m*). The dielectric function of silver at the operating wavelength is $-9.3 + 0.18i$ (see line 10 on *PRCWA_Gen_diagonal_TriangleGrating2.m*). In the given configuration both geometry and physical quantities are not changed along the out-of-plane direction (see line 35 on *PRCWA_basic.m*). The slit array has the width w of 50 nm (see line 40 on *Grating_gen_TriangleGrating2.m*) and its period varies from 150 nm to 950 nm with a 5 nm step (see line 60 on *main.m*). Figure 3.14 shows the dielectric constant distribution constructed by the Fourier representation of the configuration for period of $\Lambda = 307$ nm (see lines 65 to 88 on *PRCWA_Gen_diagonal_TriangleGrating2.m*).

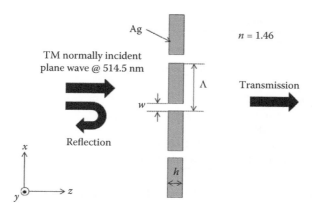

FIGURE 3.13
Schematic diagram for the extraordinary optical transmission suppression.

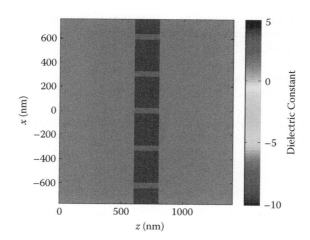

FIGURE 3.14
Schematic diagram for the EOT.

MATLAB Code 3.8: *main.m*

```
58     lambda=514.5*nm;
59
60     Tx_set=(150:5:950);Tx_len=length(Tx_set);
103    alpha_tm=1;
104    beta_tm=1;
105    %Off_diagonal_tensor_SMM;
106    Diagonal_SMM;
185    result_refle(:,Tx_idx)=DEt1;
186    result_trans(:,Tx_idx)=DEt3;
```

MATLAB Code 3.9: *PRCWA_Gen_diagonal_TriangleGrating2.m*

```
8      epra=ni^2;
9      eprb=epra;
10     eprm=-9.3+0.18*i;
65     %% permittivity profile testing
66     xx=5*[-Tx/2:Tx*0.01:Tx/2];
67     %yy=5*[-Ty/2:Ty*0.01:Ty/2];
68     yy=0;
69
70     Gr_str_bg=real(ni)*ones(length(xx),length(yy));
71     Gr_str_gr=zeros(length(xx),length(yy));
72     [ya xa]=meshgrid(yy,xx);
73
74     for k=-2*nx:2*nx
75       for l=-2*ny:2*ny
76
77         Gr_str_gr=Gr_str_gr+eps_xx(k+NBx,l+NBy,1)*exp(j*(k*x
             a*2*pi/Tx+l*ya*2*pi/Ty));
78
79       end;
80     end;
81
82     figure 5);set(gca,'fontsize',16);set(gca,'fontname','ti
       mes new roman');
83     imagesc((0:1399),xx/nm,[repmat(Gr_str_bg,1,600)
       repmat(real(Gr_str_gr),1,200)
84     repmat(Gr_str_bg,1,600)]);set(gca,'ydir','normal');
85     set(gca,'fontname','times new roman');
86     axis equal;axis([0 1400 xx(1)/nm xx(end)/nm]);xlabel('z
       (nm)');ylabel('x (nm)');set(gca,'fontname','times new
87     roman');
88     caxis([-10 5]);colorbar;set(gca,'fontsize',16);set(gca,'
       fontname','times new roman');
```

MATLAB Code 3.10: *PRCWA_Gen_diagonal_TriangleGrating2.m*

```
4    Wx=50*nano;
5    Height=0.2*um;
23   eps=eprb *rect_2D_mesh(m,n,1,Tx,Ty,-wx/2,wx/2,-
     Ty/2,Ty/2) + eprm*( rect_2D_mesh(m,n,1,Tx,Ty,-
     Tx/2,Tx/2,-Ty/2,Ty/2)-rect_2D_mesh(m,n,1,Tx,Ty,-
     wx/2,wx/2,-Ty/2,Ty/2));
24   aps=1/eprb*rect_2D_mesh(m,n,1,Tx,Ty,-wx/2,wx/2,-
     Ty/2,Ty/2) + 1/eprm*( rect_2D_mesh(m,n,1,Tx,Ty,-
     Tx/2,Tx/2,-Ty/2,Ty/2)-rect_2D_mesh(m,n,1,Tx,Ty,-
     wx/2,wx/2,-Ty/2,Ty/2));
25   mu=mur0*rect_2D_mesh(m,n,1,Tx,Ty,-Tx/2,Tx/2,-
     Ty/2,Ty/2);
26   bu=1/mur0*rect_2D_mesh(m,n,1,Tx,Ty,-Tx/2,Tx/2,-
     Ty/2,Ty/2);
```

A transverse magnetic (TM) plane wave is incident normally from the left (see lines 109 to 113 on *Field_visualization_3D_xz_case1_Lfree_Rfree_leftright.m*). After solving the block S-matrix for each block (see lines 103 to 106 on *main.m*), the extended Redheffer star product is invoked to obtain the S-matrix for the overall multisuperblock in the file *Bdr_Smat_case1.m*. The reflection and transmission coefficients are now calculated and saved (see lines 185 and 186 on *main.m*).

MATLAB Code 3.11: *Field_visualization_3D_xz_case1_Lfree_Rfree_leftright*.m

```
109   %% . input field - plane wave
110   %
111   Uy=tm_Uy;
112   Ux=tm_Ux;
113   Uz=tm_Uz;
```

Figure 3.15 shows the transmission coefficient through the slit array as a function of the period of the array (see lines 192 to 205 on *main.m*). Note that the ordinate is in the logarithmic scale. Three vertical lines denote periods of integer multiples of the surface plasmon wave λ_{SPP}. Transmission minima occur when the period of array matches to an integer multiple of λ_{SPP}. This is in good agreement with previously published data [7]. It turned out that the transmission suppression originates from the fact that the surface plasmon wave excited on the interface is π out of phase compared to the incident plane wave, resulting in destructive interference.

MATLAB Code 3.12: *main.m*

```
192   min_val=min(sum(result_trans));
193   max_val=max(sum(result_refle));
194
```

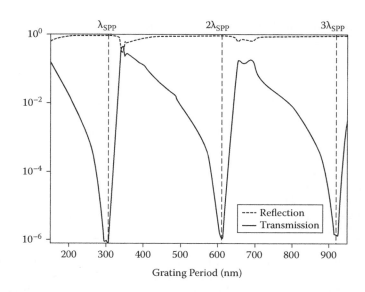

FIGURE 3.15
Transmission coefficient through the slit array.

```
195    figure 11);set(gca,'fontsize',16);set(gca,'fontname','t
       imes new roman');box on;
196    semilogy(Tx_set,sum(result_
       refle),':r','linewidth',2);hold on;
197    semilogy(Tx_set,sum(result_trans),'-b','linewidth',2);
198    axis([Tx_set(1) Tx_set(end) min_val*1.0 max_val*1.1]);s
       et(gca,'fontname','times new roman');
199    xlabel('Grating period (nm)');set(gca,'fontname','times
       new roman');
200    legend('Reflection','Transmission');set(gca,'fontname',
       'times new roman');
201
202    lambda_spp=307;
203    line(1*lambda_spp*ones(1,10),linspace(min_val,max_
       val,10),'linestyle','--','color','k');
204    line(2*lambda_spp*ones(1,10),linspace(min_val,max_
       val,10),'linestyle','--','color','k');
205    line(3*lambda_spp*ones(1,10),linspace(min_val,max_
       val,10),'linestyle','--','color','k');
```

To prove the argument above, Figure 3.16 shows the electromagnetic field distribution on the x-z plane for selected periods of the slit array. The field distribution is calculated in the file *Field_visualization_3D_xz_case1_Lfree_Rfree_leftright.m*. Figure 3.16(a) and (c) corresponds to the field distribution for

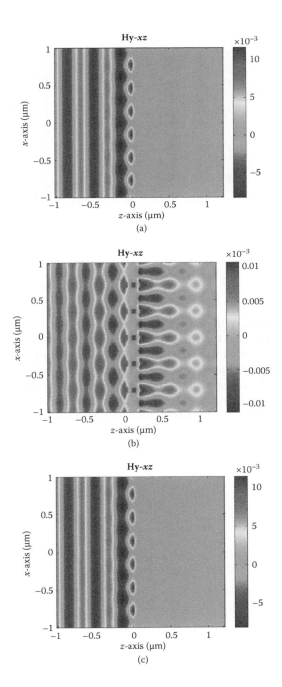

FIGURE 3.16
Field distribution of y-directional component of the magnetic field (H_y) for slit period of (a) 307 nm, (b) 350 nm, (c) 614 nm, and (d) 700 nm.

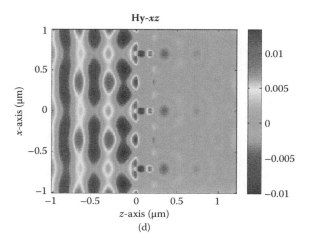

FIGURE 3.16
(Continued)

transmission minima with $\Lambda = 307$ nm and $\Lambda = 614$ nm, respectively. It can clearly be seen that the field distribution on the transmitted region almost vanishes, indicating that the transmission is greatly suppressed. This can be ascribed to destructive interference between the incident wave and the surface plasmon mode [6]. In contrast, if the slit period does not match to an integer multiple of λ_{SPP}, then the transmission can be enhanced. This is shown in Figure 3.16(b) and (d) for transmission maxima with $\Lambda = 350$ nm and $\Lambda = 700$ nm, respectively.

4

A Perfect Matched Layer for Fourier Modal Method

In this chapter, a perfect matched layer (PML) for Fourier modal method (FMM) is described. Although actual space is made up of half-infinite spaces with reflectionless outgoing waves, every simulation method, such as FMM, finite-difference time-domain (FDTD), and finite-element method (FEM), works, in practice, in a finite-sized computation space. Therefore it is inevitable that this computational region would be surrounded by boundaries that perform as if there were half-infinite spaces outside those boundaries. A PML is an artificial layer that permits the accurate calculation of the distribution of an electromagnetic field inside the computational region by satisfying outgoing wave conditions.

Since FMM was first developed to analyze diffraction properties of binary gratings with a fixed period, it has employed periodic boundary conditions and assumes an infinite space along the lateral direction. A PML is thus not always required in FMM, especially when electromagnetic fields with periodic geometries are being examined. However, if we investigate optical properties in isolated geometries such as transmission through a single slit, the generation of the surface plasmon polariton (SPP) waves from a single groove illuminated by a plane wave, or optical problems related to various waveguides, then a PML is necessary, in order to prevent undesired cross talk and interference from neighboring computation cells.

The performance of a PML is specified in terms of isolation between neighboring cells and the preservation of properties of eigenmodes. A PML can be achieved via a variety of methods, including anisotropic media, stretched coordinate transformation, and absorbing boundary layers. In this chapter, we introduce the concept of absorbing boundary layers and a nonlinear transform coordinate PML.

4.1 An Absorbing Boundary Layer for Fourier Modal Method

Figure 4.1 shows a schematic diagram of a subwavelength single slit in a thin metal film without the absorbing boundary layer illuminated by a normally incident plane wave. As can be seen in this figure, the outgoing wave leaks into the neighboring computation cell and undesired interference

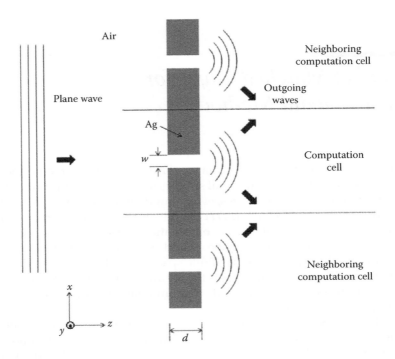

FIGURE 4.1
Schematic of subwavelength single slit without a PML.

occurs if there is no PML between the computation cells. Introducing a properly designed absorbing boundary layer enables each computation cell to be isolated. The goal of the problem is to design such an absorbing boundary layer.

The absorption of the electromagnetic field in a material is described by the imaginary part of the refractive index of that material. Since we are interested in developing an artificial layer with good absorption properties, the first simple guess for such a layer would be to adopt a modified refractive index distribution in the homogeneous surrounding material as follows:

$$
n_{ABL}^{(1)}(x) = \begin{cases} n_s & \left(|x| < \dfrac{\Lambda - w_{ABL}}{2} \right) \\[4mm] n_s + in_i & \left(\dfrac{\Lambda - w_{ABL}}{2} < |x| < \dfrac{\Lambda}{2} \right) \end{cases} \tag{4.1}
$$

where n_s denotes the refractive index of the homogeneous surrounding material at the output region and n_i represents the imaginary part of the absorption layer. However, an abrupt change in the complex refractive index distribution in $n_{ABL}^{(1)}$ may lead to a significant reflection of electromagnetic waves. The modified version can be given as

$$
n_{ABL}^{(2)}(x) = \begin{cases} n_s & \left(|x| < \dfrac{\Lambda - w_{ABL}}{2} \right) \\[4mm] n_s + i \dfrac{2n_i}{\Lambda - w_{ABL}} \left(x - \dfrac{w_{ABL}}{2} \right) & \left(\dfrac{\Lambda - w_{ABL}}{2} < |x| < \dfrac{\Lambda}{2} \right) \end{cases}
\tag{4.2}
$$

This corresponds to a linear variation of the imaginary part of the complex refractive index of the surrounding material. The reflection can be reduced for a wide angle of incidence, depending on the parameters n_i and w_{ABL}.

In order to reduce the reflection more efficiently, it was proposed that both the real and imaginary parts of the refractive index of the absorption boundary layer be modified. The performance of the absorbing boundary can be optimized by invoking an appropriate modification as follows:

$$
n_{ABL}^{(3)}(x) = \begin{cases} n_s & \left(|x| < \dfrac{\Lambda - w_{ABL}}{2} \right) \\[4mm] n_s + \left[p_1 \left(x - \dfrac{w_{ABL}}{2} \right) \right]^{p_3} + i \left[p_2 \left(x - \dfrac{w_{ABL}}{2} \right) \right]^{p_4} & \left(\dfrac{\Lambda - w_{ABL}}{2} < |x| < \dfrac{\Lambda}{2} \right) \end{cases}
\tag{4.3}
$$

According to results from Klaus and his coworkers, the optimum combination was found to be $(p_1, p_2, p_3, p_4) = (0.947, 1.043, 4.552, 7.343)$ [1].

Before applying the above absorption boundary layers to the subwavelength single slit diffraction problem, we tested the reflection from and the transmission through these absorption boundary layers using the transmission matrix method (TMM). Figure 4.2(a) shows the distribution of the real (solid line) and imaginary (dotted line) part of the absorption boundary layer for type 1, i.e., $n_{ABL}^{(1)}$, and the inset shows a schematic diagram. It can be seen that there is an abrupt change in the imaginary part of the refractive index, which functions as an absorber. Because of this abrupt change, it would be expected that the reflection properties would deteriorate. Corresponding reflection and transmission properties are shown in Figure 4.2(b) in a linear scale and in Figure 4.2(c) in a dB level. Here, the wavelength λ_0 is chosen to be 532 nm and the half thickness of the absorption boundary layer $w_{ABL}/2$ is fixed at $5\lambda_0$. The polarization is a transverse magnetic (TM), since we are mainly interested in plasmonic nanostructures.

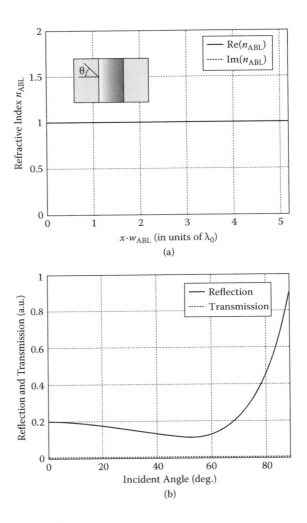

FIGURE 4.2
(a) Refractive index of the absorbing boundary layer of Equation (4.1) and its transmission and reflection spectra in a linear scale (b) and a dB level (c). (d)–(f) and (g)–(i) correspond to those of Equations (4.2) and (4.3), respectively.

4.1.1 MATLAB® Implementation

Figure 4.3(a) shows a schematic view of the real part of the refractive index profile for the absorbing boundary layer represented by $n_{ABL}^{(3)}$. Its distribution is continuous with respect to x. It is difficult to obtain a Fourier representation of $n_{ABL}^{(3)}$ in an analytical way, especially when the absorbing boundary is accompanied with other configurations of interest,

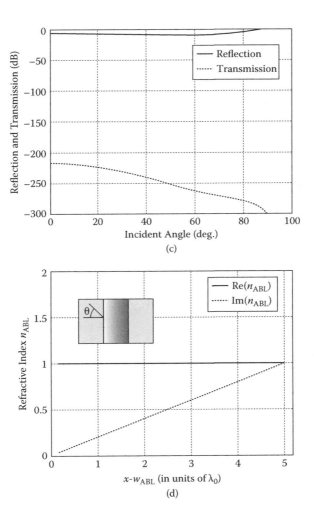

FIGURE 4.2
(Continued)

such as metal slits and dielectric gratings. Therefore we adopt the staircase approximation for $n_{ABL}^{(3)}$ in the transversal direction. A smooth profile of the absorbing boundary layer is split into a narrow single region, as shown in Figure 4.3(b). Each has a finite fixed value of the refractive index, and it is easy to achieve the Fourier representation of a staircase approximation profile. We choose appropriate values for the number of splice N_{ABL} and the total width of the absorbing boundary layer w_{ABL}. Here, $N_{ABL} = 30$ and $w_{ABL} = 5\lambda_0$.

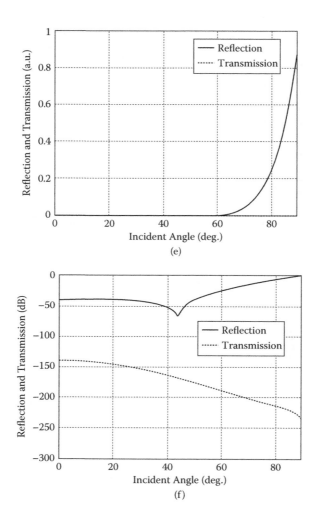

FIGURE 4.2
(Continued)

The Fourier representation of each single square is obtained by using the function *rect_2D*, the code implementation of which is as follows:

MATLAB Code 4.1: *rect_2D.m*

```
1      % rect_2D
2
3      function y=rect_2D(k,l,ep,Tx,Ty,x1,x2,y1,y2)
4
5      dx=abs(x1-x2);
6      dy=abs(y1-y2);
```

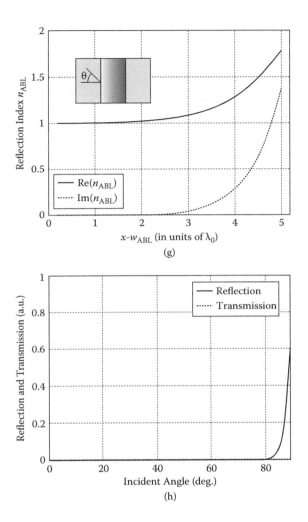

FIGURE 4.2
(Continued)

```
7
8     y=ep*((dx/Tx)*sinc(dx*k/Tx)*exp(-j*pi*k/Tx*(x1+x2)))*
      ((dy/Ty)*sinc(dy*l/Ty)*exp(-j*pi*l/Ty*(y1+y2)));
```

The function *rect_2D* gets (x_1, y_1) and (x_2, y_2) as coordinates of the left bottom and right top corners, respectively, and returns (k,l)th coefficients of the Fourier representation. Note that the Fourier representation of a square is given by the sinc function, which can be found in line 8 in *rect_2D.m*.

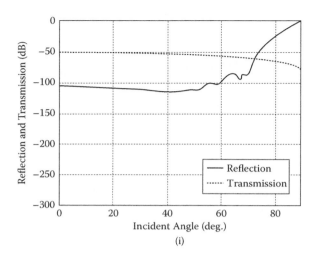

FIGURE 4.2
(Continued)

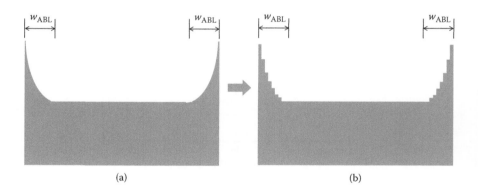

FIGURE 4.3
(a) Refractive index profile $n_{ABL}^{(3)}$ for the absorbing boundary layer, (b) with staircase approximation of (a).

The representative usage of the function *rect_2D* is as follows:

MATLAB Code 4.2: *PRCWA_Gen_PML2D.m*

```
32      for mm=1:num_hx
33          for nn=1:num_hy
34
35              m=mm-NBx;
36              n=nn-NBy;
```

```
37
38                    Epsr_PML(mm,nn)=epra*rect_2D(m,n,1,Tx,Ty,-f_
                      pml*Tx/2,f_pml*Tx/2, -Ty/2,Ty/2);
39                    Apsr_PML(mm,nn)=1/epra*rect_2D(m,n,1,Tx,Ty,-f_
                      pml*Tx/2,f_pml*Tx/2,-Ty/2,Ty/2);
40            end; % for nn
41        end; % for mm
```

As can be seen, double for-loops in the context of *mm* and *nn* are covering the caller of *rect_2D*. The variable *epra* denotes the relative permittivity of air. y_1 and y_2 in this example code are $-T_y/2$ and $T_y/2$, respectively, implying that the geometry is uniform and does not vary with respect to the y-coordinate. For the sake of convenience and ease of understanding, we introduce a simplified geometry with a 1D variation.

The whole Fourier representation of the absorbing boundary layer is given by summation of that from each square region of the absorbing boundary layer. The MATLAB implementation is summarized in the process named *PRCWA_Gen_PML2D.m* as follows:

MATLAB Code 4.3: *PRCWA_Gen_PML2D.m*

```
4     %%% PML setting %%%
5     %-----------------------------------------------------------%
6     pml_width=5*lambda; % PML thickness
7     pml_N=30; % number of staircases along a single direction
8     %-----------------------------------------------------------%

10    n_surr=n0;
11    epra=n_surr^2;
12    % PML basic parameters
13    pml_num=[1:pml_N];
14    p_factor=[0.947 1.043 4.552 7.343];
15    pml_index=n_surr+(p_factor(1)*(pml_num/pml_N)).^p_
      factor(3)...
16        +j*(p_factor(2)*(pml_num/pml_N)).^p_factor(4);
17
18    gambda_x=Tx;
19    pml_epsr=pml_index.^2;
20    pml_fill_factor=(gambda_x-2*pml_width*fliplr(pml_num)/
      pml_N)/gambda_x;
48    for k=1:pml_N-1
49        f_pml1=pml_fill_factor(k);
50        f_pml2=pml_fill_factor(k+1);
55            Epsr_PML(mm,nn)=Epsr_PML(mm,nn)+pml_epsr(k)*...
```

```
56              (rect_2D(m,n,1,Tx,Ty,-f_pml2*Tx/2,f_
                 pml2*Tx/2,-Ty/2,Ty/2) -
                 rect_2D(m,n,1,Tx,Ty,-f_pml1*Tx/2,f_
                 pml1*Tx/2,-Ty/2,Ty/2) );

61    end;

63              f_pml1=pml_fill_factor(pml_N);
64              f_pml2=1;

69              Epsr_PML(mm,nn)=Epsr_PML(mm,nn)+pml_
                 epsr(pml_N)*...
70              (rect_2D(m,n,1,Tx,Ty,-f_pml2*Tx/2,f_pml2*Tx/2,-
                 Ty/2,Ty/2) - rect_2D(m,n,1,Tx,Ty,-f_
                 pml1*Tx/2,f_pml1*Tx/2,-Ty/2,Ty/2));
```

First, the width of the absorbing boundary layer is set in line 6 (*pml_width=5*lambda*) and the number of splits for the absorbing boundary layer is determined in line 7 (*pml_N=30*). The optimum parameters of $n_{ABL}^{(3)}$ are implemented as an array in a variable *p_factor* in line 14. The piecewise-constant refractive index of the absorbing boundary layer is calculated and saved in the variable *pml_index* in line 15. The double for-loops for *mm* and *nn* are not shown in the text for the sake of convenience. We focus on the usage of accumulative calling of the variable *Epsr_PML* in lines 55 to 56. The relative permittivity of the *k*th region of the absorbing boundary layer is denoted by *pml_epsr(k)*. By subtracting the rectangle region with the filling factor of the type 1 from that with the filling factor of the type 2, which are denoted by *f_pml1* and *f_pml2* in lines 49 and 50, respectively, the Fourier coefficients of the *k*th region of the absorbing boundary layer are obtained. In multiplication of *pml_epsr(k)*, as can be seen in line 55, the *k*th region is fully represented in the Fourier space.

Figure 4.4(a) shows the resultant Fourier representation of the absorbing boundary layer added on free space. The relative permittivity distribution can be easily obtained by taking the inverse Fourier series as follows:

MATLAB Code 4.4: *pml_viewer_for_book.m*

```
1    xx=(-Tx/2:Tx*0.001:Tx/2);
2    Gr_str=zeros(1,length(xx));
3
4    for k=-2*nx:2*nx
5         Gr_str=Gr_str+Epsr_PML(k+NBx)*exp(j*(k*xx*2*pi/Tx));
6    end
7    figure 31);set(gca,'fontsize',16);set(gca,'fontname','
       times new roman');
```

```
8    plot(real(Epsr_PML));
9    figure 32);set(gca,'fontsize',16);set(gca,'fontname','
     times new roman');
10   plot(xx/Tx,real(Gr_str),'r','linewidth',2);hold
     on;set(gca,'fontname','times new roman');
11   xlabel('x (in unit of T_x)');set(gca,'fontname','times
     new roman');
12   ylabel('\epsilon');set(gca,'fontname','times new
     roman');
13   plot(xx/Tx,imag(Gr_str),':b','linewidth',2);
```

The resultant relative permittivity distribution in the real space is shown in Figure 4.4(b). The solid line denotes the real value of ε, whereas the dotted line is the imaginary of that. It is seen that the imaginary part responsible for the absorption is increased smoothly.

Let us discuss the application of the absorbing boundary layer to FMM with a simple example. The absorbing boundary layer represented by $n_{ABL}^{(3)}$ is applied to the geometry shown in Figure 4.1. Figure 4.5(a) and (b) shows the simulation results for the case without an absorbing boundary layer and that with an absorbing boundary layer, respectively. We note that there is strong interference in the former case, whereas the latter shows a clear spherical wave pattern originating from perfect isolation between neighboring computation super-cells. In order to examine the propagating properties of the diffractive field with the absorbing boundary layer, we show, in Figure 4.6, the Fourier angular spectrum of the diffractive field. The bell-shaped Fourier angular spectrum indicates that the field from a single slit is diffractive in all directions, including evanescent field components.

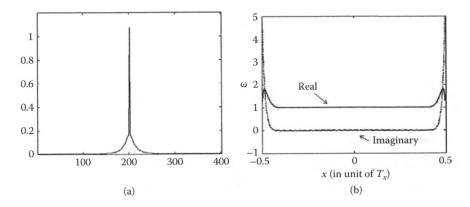

(a)   (b)

FIGURE 4.4
Absorbing boundary layer with $n_{ABL}^{(3)}$ represented in (a) Fourier domain and (b) real space domain.

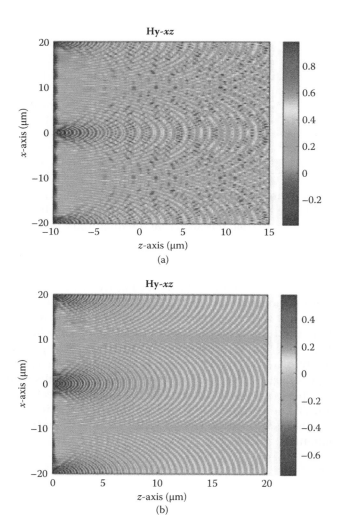

FIGURE 4.5
Diffraction pattern from a single slit illuminated by plane wave for (a) the case without the absorbing boundary layer and (b) that with the absorbing boundary layer.

4.2 Nonlinear Coordinate Transformed Perfect Matched Layer for Fourier Modal Method

Figure 4.7 shows a schematic diagram of an interconnector that transfers the incoming free space plane wave into a guided mode propagating along the metal-insulator-metal (MIM) plasmonic waveguide. The SPP waves are excited at the corners of the interconnector, and they propagate along both sides

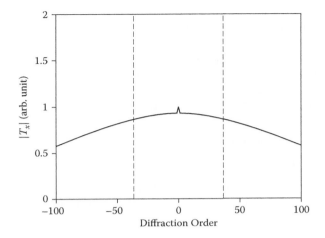

FIGURE 4.6
Fourier angular spectrum of the field distribution shown in Figure 4.5(b).

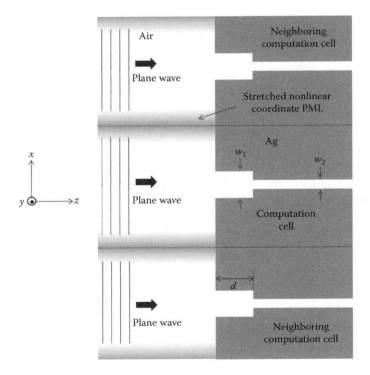

FIGURE 4.7
Schematic of the interconnector that converts an incoming plane wave into the propagating mode in the MIM plasmonic waveguide. It is necessary to isolate the incoming plane wave of neighboring cells from each other.

upward and downward. If there is no PML layer between neighboring computation cells in the incident region, then interference by the SPP waves arising from adjacent cells may result in numerical errors. To prevent this problem, one may be inclined to employ the absorbing boundary layer discussed above.

However, the incoming plane wave cannot be described in the inhomogeneous region with an artificial absorbing boundary layer. Rather, it is recommended to invoke the PML based on the nonlinear coordinate transformation.

4.2.1 Mathematical Model

4.2.1.1 Split-Field PML

We first start with a brief introduction of the split-field PML. After the theory of the split-field PML was first reported by Berenger in 1994 [2], considerable progress was made in this field, and it has been successfully adopted for use in various electromagnetic calculations such as the FDTD and the FEM [3].

For a material with finite conductivity, the electromagnetic fields on a fixed xy-plane in the TM mode with respect to the z-coordinate can be decomposed as follows:

$$\varepsilon \frac{\partial E_x}{\partial t} + \sigma E_x = \frac{\partial H_z}{\partial y} \tag{4.4a}$$

$$\varepsilon \frac{\partial E_y}{\partial t} + \sigma E_y = -\frac{\partial H_z}{\partial x} \tag{4.4b}$$

$$\mu \frac{\partial H_z}{\partial t} + \sigma^* H_z = \frac{\partial E_x}{\partial y} - \frac{\partial E_y}{\partial x} \tag{4.4c}$$

Here, σ and σ^* are the electric and magnetic conductivities, respectively. In time harmonic fields, the time dependence is given by $\exp(-j\omega t)$. Therefore, Equations (4.4a) to (4.4c) can be rewritten as

$$E_x = \frac{-1}{j\omega\varepsilon s}\left(\frac{\partial H_z}{\partial y}\right) \tag{4.5a}$$

$$E_y = \frac{1}{j\omega\varepsilon s}\left(\frac{\partial H_z}{\partial x}\right) \tag{4.5b}$$

$$H_z = \frac{-1}{j\omega\mu s^*}\left(\frac{\partial E_x}{\partial y} - \frac{\partial E_y}{\partial x}\right) \tag{4.5c}$$

where

$$s = 1 - \frac{\sigma}{j\omega\varepsilon} \tag{4.6a}$$

$$s^* = 1 - \frac{\sigma^*}{j\omega\mu} \tag{4.6b}$$

Note that the lossless case is represented by $s = s^* = 1$ in this notation.

Taking partial derivatives with respect to y and x to Equations (4.5a) and (4.5b), respectively, and putting them into Equation (4.5c), we obtain a second-order differential equation of H_z.

$$\left(\frac{1}{s^*}\frac{\partial}{\partial x}\frac{1}{s}\frac{\partial}{\partial x} + \frac{1}{s^*}\frac{\partial}{\partial y}\frac{1}{s}\frac{\partial}{\partial y} - \omega^2\varepsilon\mu \right)H_z = 0 \tag{4.7}$$

H_z can hence be represented as

$$H_z = C\exp\left[-j\left(\sqrt{ss^*}\beta_x x + \sqrt{ss^*}\beta_y y \right) \right] \tag{4.8}$$

where the momentum conservation relation is obtained as

$$\beta_x^2 + \beta_y^2 = \omega^2\varepsilon\mu \tag{4.9}$$

Now let us consider the reflection and transmission properties of a plane wave incident from a region of $x < 0$ with ε_1 and μ_1 to a region of $x > 0$ with ε_2, μ_2, and s, s^*. The wave number in the x and y directions is

$$\beta_{ix}^2 + \beta_{iy}^2 = \omega^2\varepsilon_i\mu_i, \quad (i = 1,2) \tag{4.10}$$

From the boundary condition we obtain the reflection coefficient as follows:

$$\Gamma = \frac{\dfrac{\beta_{1x}}{\varepsilon_1} - \dfrac{\sqrt{ss^*}\beta_{2x}}{\varepsilon_2 s}}{\dfrac{\beta_{1x}}{\varepsilon_1} + \dfrac{\sqrt{ss^*}\beta_{2x}}{\varepsilon_2 s}} \tag{4.11}$$

We focus on the impedance matched case, i.e., $\varepsilon_2 = \varepsilon_1$ and $\mu_2 = \mu_1$. For the boundary condition to be satisfied along the interface of $x = 0$, $\sqrt{ss^*}\beta_{2y}$ should be equal to β_{1y}. Consequently, the reflectionless condition $\Gamma = 0$ can be rewritten as

$$\frac{\sqrt{\omega^2\varepsilon_1\mu_1 - \beta_{1y}^2}}{\varepsilon_1} - \frac{\sqrt{\omega^2\varepsilon_1\mu_1 ss^* - \frac{ss^*}{ss^*}\beta_{1y}^2}}{\varepsilon_1 s} = 0 \tag{4.12}$$

which can be summarized as

$$\beta_{1y}^2\left(1-\frac{1}{s^2}\frac{ss^*}{ss^*}\right)-\omega^2\varepsilon_1\mu_1\left(1-\frac{ss^*}{s^2}\right)=0 \qquad (4.13)$$

Note that, for the sake of convenience of comparison that will be discussed below, we did not cancel out $(ss^*)/(ss^*)$. For Equation (4.13) to be satisfied regardless of the incident angle, $s = s^* = 1$. However, this indicates that there is no absorption. Therefore we can conclude that an incident angle-independent perfect absorption is not possible. The only way for reflectionless absorption is in the normally incident case, i.e., $\beta_{1y} = 0$. In this case, once $s = s^*$ is satisfied, then there is no reflection.

Now we are led to discussion on the split-field PML. The analogue of Equations (4.5) to (4.13) will be derived. By investigating differences, we can easily find the fundamental physics of the split-field PML. The starting point is somewhat complex.

$$\varepsilon\frac{\partial E_x}{\partial t}+\sigma_y E_x=\frac{\partial H_z}{\partial y} \qquad (4.14a)$$

$$\varepsilon\frac{\partial E_y}{\partial t}+\sigma_x E_y=-\frac{\partial H_z}{\partial x} \qquad (4.14b)$$

$$\mu\frac{\partial H_{zy}}{\partial t}+\sigma_y^* H_{zy}=\frac{\partial E_x}{\partial y} \qquad (4.14c)$$

$$\mu\frac{\partial H_{zx}}{\partial t}+\sigma_x^* H_{zx}=-\frac{\partial E_y}{\partial x} \qquad (4.14d)$$

where $H_z = H_{zx} + H_{zy}$. Note that Equation (4.5c) is now split into Equations (4.14c) and (4.14d). The time harmonic representation is given as

$$E_x=\frac{-1}{j\omega\varepsilon s_y}\left(\frac{\partial H_z}{\partial y}\right) \qquad (4.15a)$$

$$E_y=\frac{1}{j\omega\varepsilon s_x}\left(\frac{\partial H_z}{\partial x}\right) \qquad (4.15b)$$

$$H_{zy}=\frac{-1}{j\omega\mu s_y^*}\left(\frac{\partial E_x}{\partial y}\right) \qquad (4.15c)$$

$$H_{zx}=\frac{1}{j\omega\mu s_x^*}\left(\frac{\partial E_y}{\partial x}\right) \qquad (4.15d)$$

where

$$s_x = 1 - \frac{\sigma_x}{j\omega\varepsilon} \tag{4.16a}$$

$$s_y = 1 - \frac{\sigma_y}{j\omega\varepsilon} \tag{4.16b}$$

$$s_x^* = 1 - \frac{\sigma_x^*}{j\omega\mu} \tag{4.17a}$$

$$s_y^* = 1 - \frac{\sigma_y^*}{j\omega\mu} \tag{4.17b}$$

Substitution of Equation (4.15a) into Equation (4.15c) and Equation (4.15b) into Equation (4.15d) leads to

$$\left(\frac{1}{s_x^*} \frac{\partial}{\partial x} \frac{1}{s_x} \frac{\partial}{\partial x} + \frac{1}{s_y^*} \frac{\partial}{\partial y} \frac{1}{s_y} \frac{\partial}{\partial y} - \omega^2 \varepsilon\mu \right) H_z = 0 \tag{4.18}$$

H_z has a form of

$$H_z = C \exp\left[-j\left(\sqrt{s_x s_x^*} \beta_x x + \sqrt{s_y s_y^*} \beta_y y \right) \right] \tag{4.19}$$

with the momentum conservation relation of

$$\beta_x^2 + \beta_y^2 = \omega^2 \varepsilon\mu \tag{4.20}$$

The reflection coefficient is now given by

$$\Gamma = \frac{\dfrac{\beta_{1x}}{\varepsilon_1} - \dfrac{\sqrt{s_x s_x^*} \beta_{2x}}{\varepsilon_2 s_x}}{\dfrac{\beta_{1x}}{\varepsilon_1} + \dfrac{\sqrt{s_x s_x^*} \beta_{2x}}{\varepsilon_2 s_x}} \tag{4.21}$$

The relation between y-directional wave numbers is $\sqrt{s_y s_y^*} \beta_{2y} = \beta_{1y}$. We assume that $\varepsilon_2 = \varepsilon_1$ and $\mu_2 = \mu_1$. The reflectionless condition $\Gamma = 0$ is summarized as

$$\frac{\sqrt{\omega^2 \varepsilon_1 \mu_1 - \beta_{1y}^2}}{\varepsilon_1} - \frac{\sqrt{\omega^2 \varepsilon_1 \mu_1 s_x s_x^* - \frac{s_x s_x^*}{s_y s_y^*} \beta_{1y}^2}}{\varepsilon_1 s_x} = 0 \tag{4.22}$$

which is rewritten as

$$\beta_{1y}^2 \left(1 - \frac{1}{s_x^2} \frac{s_x s_x^*}{s_y s_y^*} \right) - \omega^2 \varepsilon_1 \mu_1 \left(1 - \frac{s_x s_x^*}{s_x^2} \right) = 0 \tag{4.23}$$

The separation of s variables with respect to x and y, which can be seen by comparing the reflectionless condition in the ordinary lossy medium in Equation (4.13) and that in the split-field PML in Equation (4.23), allows a novel way for satisfying the incident angle-independent PML. By setting $s_y s_y^* = 1$ and $s_x = s_x^*$, Equation (4.23) can be satisfied for any β_{1y}.

A question now under consideration is how to implement the aforementioned split-field PML. It will be shown that the uniaxial permittivity and permeability tensors are identical to the split-field PML. Let us assume that a TE$_z$ incident wave represented by $H_z = H_0 \exp(j\beta_{1x}x + j\beta_{1y}y)$ is propagating from an isotropic region 1 $(x < 0)$ having the permittivity ε_1 and permeability μ_1 to a uniaxial anisotropic region 2 $(x > 0)$ with

$$\overline{\overline{\varepsilon}}_2 = \varepsilon_2 \begin{pmatrix} a & 0 & 0 \\ 0 & b & 0 \\ 0 & 0 & b \end{pmatrix} \qquad (4.24)$$

$$\overline{\overline{\mu}}_2 = \mu_2 \begin{pmatrix} c & 0 & 0 \\ 0 & d & 0 \\ 0 & 0 & d \end{pmatrix} \qquad (4.25)$$

The dispersion relation can be derived from the nontrivial field components using Maxwell's curl equations, which reads as follows.

$$k_2^2 - \beta_{2x}^2 b^{-1} d^{-1} - \beta_{2y}^2 a^{-1} d^{-1} = 0 \quad \text{for TE}_z \qquad (4.26)$$

$$k_2^2 - \beta_{2x}^2 b^{-1} d^{-1} - \beta_{2y}^2 b^{-1} c^{-1} = 0 \quad \text{for TM}_z \qquad (4.27)$$

where $k_2^2 = \omega^2 \mu_2 \varepsilon_2$. By using the boundary condition in which the tangential electric and magnetic fields are continuous across the interface between two media $(x = 0)$, we then obtain the reflection coefficient as

$$\Gamma = \frac{\dfrac{\beta_{1x}}{\varepsilon_1} - \dfrac{\beta_{2x}}{\varepsilon_2} b^{-1}}{\dfrac{\beta_{1x}}{\varepsilon_1} + \dfrac{\beta_{2x}}{\varepsilon_2} b^{-1}} \qquad (4.28)$$

Here, since the transverse wavenumber should be continuous, we adopt $\beta_{2y} = \beta_{1y}$. The incident angle-independent reflectionless condition under the impedance matched case ($\varepsilon_1 = \varepsilon_2$ and $\mu_1 = \mu_2$) is hence reduced to

$$\frac{\sqrt{\omega^2 \varepsilon_1 \mu_1 - \beta_{1y}^2}}{\varepsilon_1} - \frac{\sqrt{\omega^2 \varepsilon_1 \mu_1 bd - \beta_{1y}^2 a^{-1} b}}{\varepsilon_1 b} = 0 \qquad (4.29)$$

which is represented as

$$\beta_{1y}^2 \left(1 - \frac{1}{ab} \right) - \omega^2 \varepsilon_1 \mu_1 \left(1 - \frac{d}{b} \right) = 0 \qquad (4.30)$$

Consequently, if $a = b^{-1}$ and $d = b$, then reflection vanishes regardless of the incident angle. Likewise, one can easily derive the reflection-free condition for the TM_z case, which gives $c = d^{-1}$ and $b = d$. To sum up, a uniaxial anisotropic material with a form of

$$\underline{\underline{\varepsilon}}_2 = \varepsilon_1 \underline{\underline{m}}, \qquad \underline{\underline{\mu}}_2 = \mu_1 \underline{\underline{m}}, \qquad (4.31)$$

$$\underline{\underline{m}} = \begin{pmatrix} s_x^{-1} & 0 & 0 \\ 0 & s_x & 0 \\ 0 & 0 & s_x \end{pmatrix} \qquad (4.32)$$

exhibits reflectionless wave propagation from region 1 to region 2 for any incident angle, any polarization, and operating frequency [3]. The PML based on uniaxial material is referred to as UPML.

The extension of the UPML theory in the 2D case to the 3D case can also be easily achieved. Figure 4.8(a) and (b) shows the 2D and 3D computation cells surrounded by the UPML, respectively. With the definition of $\underline{\underline{\varepsilon}} = \varepsilon_0 \underline{\underline{m}}$ and $\underline{\underline{\mu}} = \mu_0 \underline{\underline{m}}$, we have the general tensor formula for regions A_{2D} and B_{2D} as follows:

$$\underline{\underline{m}}_{A_{2D}} = \begin{pmatrix} 1 & 0 & 0 \\ 0 & 1 & 0 \\ 0 & 0 & 1 \end{pmatrix} \qquad (4.33a)$$

$$\underline{\underline{m}}_{B_{2D}} = \begin{pmatrix} s_x^{-1} & 0 & 0 \\ 0 & s_x & 0 \\ 0 & 0 & s_x \end{pmatrix} \qquad (4.33b)$$

Likewise, the permittivity and permeability have a common tensor, as shown below.

$$\underline{\underline{m}}_{A_{3D}} = \begin{pmatrix} 1 & 0 & 0 \\ 0 & 1 & 0 \\ 0 & 0 & 1 \end{pmatrix} \qquad (4.34a)$$

$$\underline{\underline{m}}_{B_{3D}} = \begin{pmatrix} s_x^{-1} & 0 & 0 \\ 0 & s_x & 0 \\ 0 & 0 & s_x \end{pmatrix} \qquad (4.34b)$$

$$\underline{\underline{m}}_{C_{3D}} = \begin{pmatrix} s_y & 0 & 0 \\ 0 & s_y^{-1} & 0 \\ 0 & 0 & s_y \end{pmatrix} \qquad (4.34c)$$

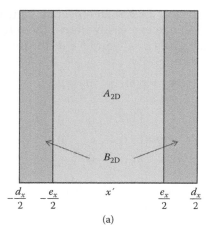

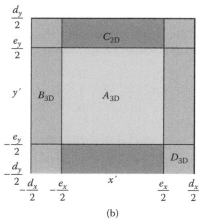

FIGURE 4.8
Classification of domain in (a) 2D and (b) 3D computation cell.

$$
\underline{\underline{m}}_{D_{3D}} =
\begin{pmatrix}
s_x^{-1}s_y & 0 & 0 \\
0 & s_x s_y^{-1} & 0 \\
0 & 0 & s_x s_y
\end{pmatrix}
\tag{4.34d}
$$

4.2.1.2 Stretched Nonlinear Coordinate Transformation

Due to the finite thickness of the PML discussed above, the absorption in that PML is not complete. Therefore, an additional method that allows for the complete isolation is required. Here we invoke the stretched nonlinear coordinate transformation [4]. Figure 4.9(a) shows the virtual space coordinate x' with the range of $|x'| < d/2$. On the other hand, Figure 4.9(b) illustrates

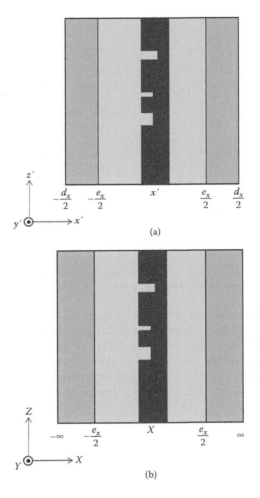

FIGURE 4.9
Schematic diagrams of (a) the coordinate transformed region (x', y', z') and (b) the original frame region (X, Y, Z).

the real space coordinate of X, which ranges from $-\infty$ to ∞. In the range of $|x'| < e/2$, $X = x'$. As x' increases from $e/2$ to $d/2$, the half-infinite space of $X \in (e/2, \infty)$ is spanned.

A mathematical representation of this coordinate transformation $X = F(x')$ is given as

$$
F(x') = \begin{cases} x' & |x'| < e/2 \\[2mm] \dfrac{x'}{|x'|}\left[\dfrac{e}{2} + \dfrac{q}{\pi(1-\gamma)} \tan\left(\pi \dfrac{|x'| - e/2}{q} \right) \right] & e/2 < |x'| < d/2 \end{cases} \quad (4.35)
$$

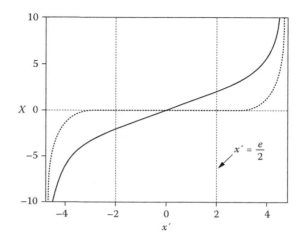

FIGURE 4.10

Mapping of the infinite real space coordinate X from the finite virtual space coordinate x'. The solid and dotted lines correspond to the real and imaginary parts of X.

where $q = d - e$. Here we choose $\gamma = 1/(1 - i)$. At $x' = e/2$, the tangent term vanishes and $F(x') = e/2$. As x' approaches $e/2$, it diverges to infinity, indicating that $X = F(x')$ covers a virtual half-infinite space, as shown in Figure 4.10.

Now let us consider the general coordinate transform theory for Maxwell's equation. For (X, Y, Z) being coordinates in the original frame and (x', y', z') those in a transformed frame, the Jacobian matrix can be obtained as follows:

$$
J = \begin{pmatrix}
\dfrac{\partial x'}{\partial X} & \dfrac{\partial x'}{\partial Y} & \dfrac{\partial x'}{\partial X} \\[2mm]
\dfrac{\partial y'}{\partial X} & \dfrac{\partial y'}{\partial Y} & \dfrac{\partial y'}{\partial Z} \\[2mm]
\dfrac{\partial z'}{\partial X} & \dfrac{\partial z'}{\partial Y} & \dfrac{\partial z'}{\partial Z}
\end{pmatrix}
\tag{4.36}
$$

As in the case of the PML, the mapping functions X, Y, and Z differ in various regions. In the 2D computation cell, there is no variance along the y'-direction.

$$
X = \begin{cases}
x' & \text{in } A_{2D} \\
F(x') & \text{in } B_{2D}
\end{cases}
\tag{4.37}
$$

and $Y = y'$, $Z = z'$ everywhere. Therefore, the Jacobian matrices in the 2D computation cells are given as

$$
J_{A_{2D}} = \begin{pmatrix} 1 & 0 & 0 \\ 0 & 1 & 0 \\ 0 & 0 & 1 \end{pmatrix} \tag{4.38}
$$

$$
J_{B_{2D}} = \begin{pmatrix} \dfrac{\partial x'}{\partial X} & 0 & 0 \\ 0 & 1 & 0 \\ 0 & 0 & 1 \end{pmatrix} \tag{4.39}
$$

From $X = F(x')$, we derive

$$
P(x') \triangleq \partial x'/\partial X = (dF/dx')^{-1} = (1-\gamma)\cos^2\left(\pi\frac{|x'|-e/2}{q}\right) \tag{4.40}
$$

Here, we choose $\gamma = 1/(1-i)$. In coordinate transformation optics, Maxwell's equations are represented by using the Jacobian tensor:

$$
\underline{\underline{\varepsilon}} \rightarrow \underline{\underline{\varepsilon}}' = \frac{J^{-1}\underline{\underline{\varepsilon}}J}{\det(J)} \tag{4.41}
$$

$$
\underline{\underline{\mu}} \rightarrow \underline{\underline{\mu}}' = \frac{J^{-1}\underline{\underline{\mu}}J}{\det(J)} \tag{4.42}
$$

The original permittivity $\underline{\underline{\varepsilon}}$ and permeability $\underline{\underline{\mu}}$ tensors arise from the aforementioned split-field PML or UPML, as follows:

$$
\underline{\underline{\varepsilon}} = \varepsilon_{A_{2D}} \begin{pmatrix} s_x^{-1} & 0 & 0 \\ 0 & s_x & 0 \\ 0 & 0 & s_x \end{pmatrix} \tag{4.43}
$$

$$
\underline{\underline{\mu}} = \mu_{A_{2D}} \begin{pmatrix} s_x^{-1} & 0 & 0 \\ 0 & s_x & 0 \\ 0 & 0 & s_x \end{pmatrix} \tag{4.44}
$$

where $\varepsilon_{A_{2D}}$ and $\mu_{A_{2D}}$ denote the relative permittivity and permeability of region A_{2D}, which are chosen so that the impedance is matched across the interface between regions A_{2D} and B_{2D}. Here the value of s_x is selected as $1 + 0.01j$. This is a general choice and is not optimized. One may evaluate the performance of the nonlinear stretched coordinate transformation PML with various values of s_x to obtain optimal results.

Substitution of Equations (4.43) and (4.44) into Equations (4.41) and (4.42) gives

$$\underline{\underline{\varepsilon}}'(x') = \varepsilon_{A_{2D}} \underline{\underline{m}}(x') \tag{4.45}$$

$$\underline{\underline{\mu}}'(x') = \varepsilon_{A_{2D}} \underline{\underline{m}}(x') \tag{4.46}$$

where

$$\underline{\underline{m}}(x') = \begin{pmatrix} [P(x')]^2 s_x^{-1} & 0 & 0 \\ 0 & [P(x')]^{-2} s_x & 0 \\ 0 & 0 & [P(x')]^{-2} s_x \end{pmatrix} \tag{4.47}$$

4.2.2 MATLAB® Implementation

The MATLAB implementation of the nonlinear stretched coordinate transformation PML is similar to that of the absorption boundary layer and is implemented in MATLAB code *PRCWA_Gen_PML1D_stretched.m* and *PML_stretched_evaluate.m*. The permittivity distribution in region A_{2D} is represented in the variable *rect_region0* in line 71 on *PRCWA_Gen_PML1D_stretched.m*. The term $[P(x')]^2 s_x^{-1}$ is evaluated in the variable named *cos2sx* in line 78 on *PRCWA_Gen_PML1D_stretched.m*.

MATLAB Code 4.5: *PRCWA_Gen_PML1D_stretched.m*

```
59    pml_num=(1:pml_N);
60    pml_x=(Tx/2-pml_width) + pml_width*([pml_num pml_N+1]-1)/
      pml_N;
61    % pml_y=(Ty/2-pml_width) + pml_width*([pml_num pml_N+1]-
      1)/paml_N;
62
63
64    for mm=1:num_hx
65    %         disp([num2str(mm) '/' num2str(num_hx)]);
66        m=mm-NBx;
67        for nn=1:num_hy
68            n=nn-NBy;
69
70            % region 0
71            rect_region0(mm,nn)=rect_2D(m,n,1,Tx,Ty,-
              e_x/2,e_x/2,-Ty/2,Ty/2);
72
73            for k=1:pml_N-0 %
74
```

```
75              % region 1 and 2
76              pml_x1=pml_x(k);
77              pml_x2=pml_x(k+1);
78              cos2sx=(1-gam)*cos(pi * (abs(pml_x1)-
                e_x/2) / q_x)^2/sx;
79              cos2sy=1;
```

After calculating $[P(x')]^2 s_x^{-1}$, we are able to evaluate $\underline{\underline{m}}(x')$. This is implemented in MATLAB code *PML_stretched_evaluate.m*. The variable has the naming rule of *(E,A,M,B)psr_PML_(xx,yy,zz)*, where *E*, *A*, *M*, and *B* denote the electrical relative permittivity, the reciprocal electrical relative permittivity, the magnetic relative permeability, and the reciprocal relative permeability, respectively. *xx*, *yy*, and *zz* denote the (1,1), (2,2), and (3,3) elements of $\underline{\underline{m}}(x')$. The inverse and reciprocal relations of these variables are carefully implemented from line 1 to line 12.

MATLAB Code 4.6: *PML_stretched_evaluate.m*

```
1    Epsr_PML_xx(mm,nn)=Epsr_PML_xx(mm,nn) + 1*(eps_
     surr*cos2sx/cos2sy)*rect_form;
2    Epsr_PML_yy(mm,nn)=Epsr_PML_yy(mm,nn) + 1*(eps_surr/
     cos2sx*cos2sy)*rect_form;
3    Epsr_PML_zz(mm,nn)=Epsr_PML_zz(mm,nn) + 1*(eps_surr/
     cos2sx/cos2sy)*rect_form;
4    Apsr_PML_xx(mm,nn)=Apsr_PML_xx(mm,nn) + 1/(eps_
     surr*cos2sx/cos2sy)*rect_form;
5    Apsr_PML_yy(mm,nn)=Apsr_PML_yy(mm,nn) + 1/(eps_surr/
     cos2sx*cos2sy)*rect_form;
6    Apsr_PML_zz(mm,nn)=Apsr_PML_zz(mm,nn) + 1/(eps_surr/
     cos2sx/cos2sy)*rect_form;
7    Mpsr_PML_xx(mm,nn)=Mpsr_PML_xx(mm,nn) + 1*(muu_
     surr*cos2sx/cos2sy)*rect_form;
8    Mpsr_PML_yy(mm,nn)=Mpsr_PML_yy(mm,nn) + 1*(muu_surr/
     cos2sx*cos2sy)*rect_form;
9    Mpsr_PML_zz(mm,nn)=Mpsr_PML_zz(mm,nn) + 1*(muu_surr/
     cos2sx/cos2sy)*rect_form;
10   Bpsr_PML_xx(mm,nn)=Bpsr_PML_xx(mm,nn) + 1/(muu_
     surr*cos2sx/cos2sy)*rect_form;
11   Bpsr_PML_yy(mm,nn)=Bpsr_PML_yy(mm,nn) + 1/(muu_surr/
     cos2sx*cos2sy)*rect_form;
12   Bpsr_PML_zz(mm,nn)=Bpsr_PML_zz(mm,nn) + 1/(muu_surr/
     cos2sx/cos2sy)*rect_form;
```

Figure 4.11(a) shows the H_y field distribution of an interconnector that transfers the incoming free space plane into a guided mode propagating along the MIM

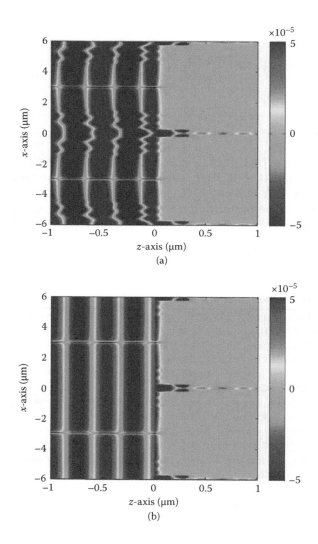

FIGURE 4.11
(a) H_y field distribution of an interconnector of the plane wave and the MIM plasmonic wave-guide. (b) H_y field distribution without the reflected wave in the region $z < 0$.

plasmonic waveguide, which was described in the introduction of this section. Here the period of the computation super-cell is 6 μm. It can be seen that a node line exists along the line $x = 3$ μm and $x = -3$ μm, originating from the presence of the stretched nonlinear coordinate PML. The width and the length of the inter-connector are set at 400 nm and 300 nm, respectively. The width of the MIM plas-monic waveguide is 100 nm. Note that the symmetric mode is excited in both the interconnector and the MIM plasmonic waveguide. It is noteworthy that there is no interference pattern in the PML region ($z < 0$), indicating that the neighboring computation cells have been successfully isolated. In order to examine the field distribution of the incident plane wave, we plot in Figure 4.11(b) only the incoming plane wave, omitting the reflected wave. It can be seen that the wavefront in the region $z < 0$ is perpendicular to the direction of propagation, even in the presence of the PML. This property cannot be accomplished by using absorbing boundary layer-based PML methodology.

4.3 Applications

4.3.1 Plasmonic Beaming

The manipulation of light has always been a critical issue in optics. Extensive research has been done in attempts to achieve a well-collimated beam from a subwavelength metal slit by taking advantage of the plasmonic effect [5–8]. This beaming of light via plasmonics has attracted substantial interest due to its various potential applications, such as free space optical interconnections and plasmonic light-emitting devices. In particular, considerable efforts have been devoted to the generation of the optical beaming diffracted from subwavelength nanostructures [9–12]. One of the possible methods for optical beaming from a subwavelength metal slit is to employ surface dielectric gratings arranged around the subwavelength metal slit. Here, we employ FMM implementation for on- and off-axis optical beamings, based on reference [9].

Since optical beaming from a single subwavelength slit on a metal film requires calculations of the electromagnetic fields both near and far from the slit, a large memory space is necessary for simulation methods such as FEM or FDTD. To the contrary, FMM can be used to solve an electromagnetic field in the spatial Fourier domain. This configuration is a good example showing that, by using the PML, the numerical simulation of an isolated supercell is possible in FMM. The postprocessing step for calculating the angular Fourier spectrum of the beam profile is also provided. Readers will see the advantages of using FMM implementation in this example. This is because the far-field calculation can easily be carried out without any additional calculation overload after the fundamental calculation. As a result, a relatively small memory space is required.

Figure 4.12 shows a schematic diagram of the on-axis optical beaming structure. The transverse magnetic (TM) plane wave is normally incident from the glass substrate under a silver film with thickness $d = 300$ nm (see line 9 in *PRCWA_Gen_SPP_beaming.m*). The refractive index of the glass is assumed to be 1.5 (see line 61 in *SPP_beaming.m*). The operation wavelength is 532 nm (see line 60 in *main.m*), where the relative permittivity of the silver is obtained to be $-10.2 + 0.83i$ from Palik's empirical data (see line 7 in *PRCWA_Gen_SPP_beaming.m*). The fundamental SPP mode inside the slit with width $w = 100$ nm is generated by the incident light and propagates along the slit

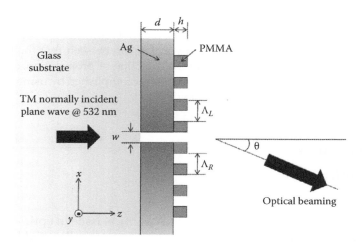

FIGURE 4.12
Schematic diagram of the off-axis optical beaming by the symmetric dielectric surface gratings arranged around the subwavelength metal slit.

(see line 3 in *Grating_gen_SPP_beaming.m*). At the exit region, the fundamental SPP mode excites the single-interface SPP mode, which propagates along the interface between the metal and the free space. This single-interface SPP then experiences diffraction by the polymethyl methacrylate (PMMA) dielectric gratings arranged around the slit. The period of the dielectric grating is designed in such a way that the diffracted light results in a well-collimated optical beam in the free space region. The detailed design process will be discussed below. Here, the relative permittivity of the PMMA dielectric gratings is 2.96 (see lines 44 and 45 in *Grating_gen_SPP_beaming.m*). The height of the dielectric gratings h is 80 nm (see line 10 in *PRCWA_Gen_SPP_beaming.m*). The period of the computation cell and the number of the x-directional and y-directional Fourier harmonics are 20 µm and 100, respectively (see lines 66 to 69 in *main.m*).

MATLAB Code 4.7: *main.m*

```
60   lambda=532*nm;          % operation wavelength
61   ni=1.5;                 % refractive index in region I
62   nf=1;                   % refractive index in region II
63   Tx=20.000001*micro;     % x-size of the computation cell
64   Ty=Tx;                  % y-size of the computation cell
65   nx=100;                 % x direction truncation order
66   ny=0;                   % y direction truncation order
67   theta=0*pi/180;         % incident angle
68   phi=0;                  % azimuthal angle
69   psi=0;                  % polarization angle : TM=0, TE=pi/2
```

MATLAB Code 4.8: *PRCWA_Gen_SPP_beaming.m*

```
6    epra=1;                       % air
7    eprm=(fun_Ag_nk(lambda))^2;   % silver
8
9    film_thick=300*nm;            % metal film thickness
10   grating_thick=80*nm;          % dielectric grating
                                        thickness
```

MATLAB Code 4.9: *PRCWA_Gen_SPP_beaming.m*

```
2
3         Wx=100*nm;

41            KK=8;
42            facR=0.5;
43            facL=0.5;
44            eprgR=1.72^2;
45            eprgL=eprgR;
46            TgR=380*nm;
47            TgL=TgR;
48   %        TgR=305*nm;
49   %        TgL=505*nm;
```

The relationship between the grating period Λ_g and the diffraction angle θ is given by the SPP diffraction equation as follows [4.9]:

$$k_{SPP} + mG_x = k_0 \sin \theta \qquad (4.48)$$

where k_{SPP} is the wavenumber of the single-interface SPP mode. Here, m is an integer and denotes the order of the diffraction. G_x corresponds to the wavenumber of the grating vector and is given by $2\pi/\Lambda_g$. k_0 is the wavenumber in free space. The diffraction angle is defined as the angle of the direction of propagation of the negative first-order diffraction mode ($m = -1$). The plus sign in the diffraction angle indicates that the x-directional wavenumber of the negative first-order diffraction mode is the same as that of the single-interface SPP mode entering into the dielectric grating, whereas the minus sign corresponds to the reverse x-directional wavenumber. The former is also referred to as the diverging mode and the latter the converging mode [9]. Figure 4.13 shows the diffraction angle as a function of the grating period. It is seen that the diffraction angle is negative if the grating period is short. As the grating period increases, the diffraction angle is increased. For a critical value of the grating period, the diffraction angle exhibits zero. In addition, the diffraction angle grows monotonically for a grating period larger than this critical value.

Let us first examine the on-axis optical beaming generated by the dielectric surface gratings. We choose a period of the dielectric grating that results in a zero diffraction angle (the grating A in Figure 4.13). The left and right

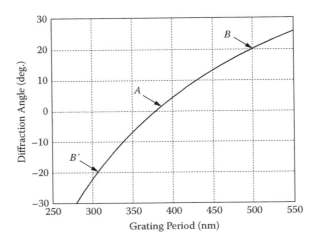

FIGURE 4.13
Relationship between the diffraction angle and the grating period.

PMMA dielectric gratings have the same period of $\Lambda_L = \Lambda_R = 380$ nm (see lines 46 and 47 in *Grating_gen_SPP_beaming.m*). The field intensity distribution is shown in Figure 4.14. The well-collimated optical beam with a deviation angle of zero is observed. To investigate the angular Fourier spectrum of the beam pattern, the field distribution is decomposed to the Rayleigh representation using the relation

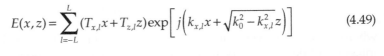

$$E(x,z) = \sum_{l=-L}^{L} (T_{x,l}x + T_{z,l}z) \exp\left[j\left(k_{x,l}x + \sqrt{k_0^2 - k_{x,l}^2}\, z \right)\right] \tag{4.49}$$

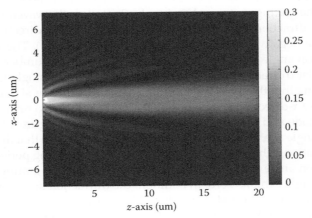

FIGURE 4.14
Field intensity distribution for the on-axis optical beaming.

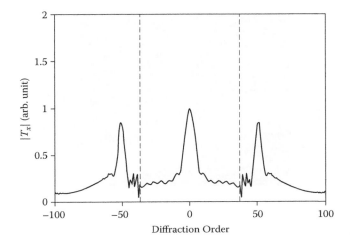

FIGURE 4.15
Angular Fourier spectrum of the on-axis optical beaming.

where l is the diffraction order and $k_{x,l}$ is the x-directional wavenumber of the lth-order spectral component. $T_{x,l}$ and $T_{z,l}$ correspond to the x- and z-components of the transmission coefficient. The resultant angular Fourier spectrum is depicted in Figure 4.15. Two vertical dashed lines denote the boundary of the propagating and evanescent modes; i.e., the spectral components lying inside these lines can propagate in free space, forming an optical beam pattern, whereas those lying outside these lines are evanescent. The transmission coefficient of the propagating mode exhibits the center symmetry. It is also noteworthy that the angular Fourier spectrum distribution shape of the optical beam pattern in Figure 4.15 is similar to that of collimated Gaussian beams. The MATLAB code implementation for the angular Fourier spectrum is shown in lines 178 to 184 in *main.m*.

MATLAB Code 4.10: *main.m*

```
178    figure 2);plot((-nx:nx),abs(pfEx)/
       max(abs(pfEx)),'k');hold on;
179    set(gca,'fontsize',16);set(gca,'fontname','times new
       roman');
180    xlabel('Diffraction order');ylabel('|T_x| (arb.unit)');
181    axis([-nx nx 0 2]);
182    mm=floor(Tx/lambda);
183    line( mm*ones(1,10),linspace(0,2,10),'linestyle','--',
       'color','k');
184    line(-mm*ones(1,10),linspace(0,2,10),'linestyle','--',
       'color','k');
```

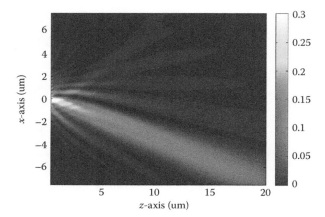

FIGURE 4.16
Field intensity distribution for the off-axis optical beaming.

Now, we are led to a discussion of off-axis optical beaming. Here, off-axis describes the case where the angle of deviation of the optical beam is not zero. The off-axis optical beaming is accomplished by asymmetric dielectric surface gratings arranged around the subwavelength metal slit, which is different from the on-axis optical beaming where the symmetric surface gratings with a diffraction angle of zero were used [9]. In Figure 4.13, we note that the diffraction angle can be either positive or negative depending on the grating period. If the period of B with the diffraction angle of +20° is chosen for the right surface grating (Λ_R) and the period of B′ with the diffraction angle of −20° is selected for the left surface grating (Λ_L), then the overall interference pattern in free space exhibits an off-axis optical beam with a deviation angle of 20°. The MATLAB code implementation requires just a little modification. In the file *Grating_gen_SPP_beaming.m*, we comment out lines 46 and 47, and uncomment lines 48 and 49 for that point. The simulation result is shown in Figure 4.16. We can observe the well-collimated optical beam with the nonzero deviation angle. The angular Fourier spectrum shown in Figure 4.17 shows a shift in the center peak of the transmission coefficient, which comes from the nonzero deviation angle of the optical beam pattern.

4.3.2 Plasmonic Hot Spot and Vortex

One of the hottest issues in optics has been the confinement of light. The isolated bright point of a light is referred to as an optical hot spot. An intense, sharp hot spot has many advantages in various applications, such as optical microscopy, biosensing, and optical data storage. Therefore tremendous efforts have been expended to achieve an intense, sharp hot spot.

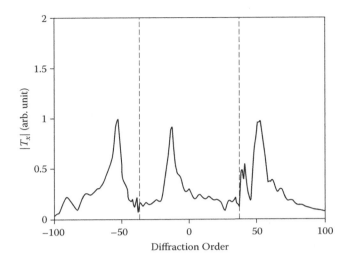

FIGURE 4.17
Angular Fourier spectrum of the off-axis optical beaming.

In particular, a hot spot generated by SPs has attracted considerable interest, mainly due to the fact that the effective wavelength of SP can be shorter than that of light in free space. According to the theory of the diffraction limit, it is theoretically impossible to achieve a hot spot whose size is smaller than half the wavelength. In plasmonics, however, there is no upper limit for the effective refractive index of SPs propagating along the metal-dielectric interface. This leads to the shorter effective wavelength of SP waves, and thus the smaller hot spot. Various methods to achieve plasmonic hot spots have been reported [13–15]. Generation of rotating fields has also played an important role in a range of research fields such as optical trapping, biosensing, and microfluidics. By adopting diffractive slit structures, we can design the plasmonic vortex lens [16].

This section is devoted to their implementation via FMM. In particular, the polarization sensitivity of surface plasmons has interesting and promising features. Once the basic calculation is done, then the effect of the polarization of the incident light can easily be obtained by using FMM. This is one of the strong points of the linear modal analysis on the optical linear structure. Here the generation of a plasmonic hot spot and a surface plasmonic vortex are also presented.

Figure 4.18 shows a side view of a plasmonic lens. A 300 nm thick ($h = 300$ nm) silver film with dielectric constant $\varepsilon_m = -15 + 0.18i$ is evaporated on the glass substrate with refractive index $n_l = 1.5$. The operating wavelength in free space λ_0 is 660 nm. It is assumed that, in the experiment, the slit is formed by milling out the metal film using a focused ion beam. The width of the slit w is 100 nm. The plane wave is normally incident from the bottom

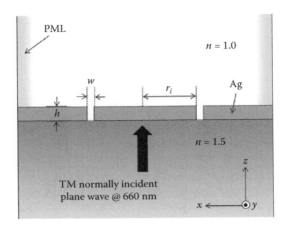

FIGURE 4.18
Side view of the plasmonic lens.

glass substrate and the SP mode in the slit is coupled. A single-interface SP wave propagating along the interface between the silver film and free space is then excited.

Figure 4.19(a) shows the top view of the plasmonic lens. It has an inner radius R_i of 4.0 μm and an outer radius R_o of 4.1μm. The overall pattern is circular. The Fourier representations of such a coaxial circular pattern can be analytically derived using the Bessel function of the first kind. However, as can be seen below, other complex patterns will be dealt with throughout this section. A simple mathematical derivation of the Fourier representation of such a complex structure cannot be readily obtained. Therefore we employ a more general method to obtain the Fourier representation. This is organized in two substeps. The complex slit pattern is first defined in real space. Its Fourier representation is then obtained by calling the function *fft2* provided by the MATLAB. The material constants corresponding to metal and free space are then put into the spatial geometry information. The function *fft2* then converts the spatial geometry information in real space into a Fourier representation. The reciprocal permittivity profile is derived in a similar manner.

To observe the interference of the SP mode along the interface between the metal and dielectric, it is necessary to isolate the computation cell from each other. This can be achieved by adopting the PML along the boundaries between computational super-cells. A detailed theoretical discussion of the PML has already been introduced in the previous section. The readers are referred to Section 4.2.

The x-directional linear polarized light beam is incident from the bottom. In Figure 4.19(b), we depict the intensity distribution inside the plasmonic lens. It is shown that there are two hot spots near the center of the structure. In order to explain the origin of the formation of two hot spots, we show in

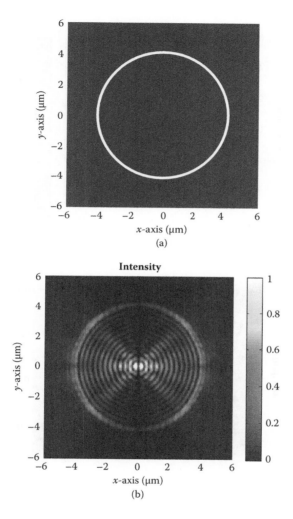

FIGURE 4.19
(a) Top view of the plasmonic lens. (b) Intensity profile on the plasmonic lens illuminated by x-directional linearly polarized light beam. (c)–(e) x-, y-, and z-directional electric field distribution.

Figure 4.19(c) to (e) the x-, y-, and z-directional electric fields, respectively. It is observed that the E_z has two hot spots near the center of the geometry, whereas the E_x field has a single hot spot at the center of the structure. This discrepancy between the E_z and E_x field distributions originates from the phase difference at the slit [16]. The phases of E_z at two antipodal slits are out of phase (π phase difference), which results in the destructive interference at the center of the hot spot. The phases of E_x of two antipodal slits are in phase (zero phase difference), which gives rise to the constructive interference at

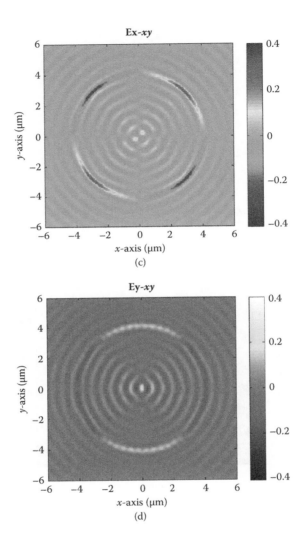

FIGURE 4.19
(Continued)

the center of the hot spot. The intensity profile is given by the linear combination of each field intensity. Here, it is noteworthy that the field amplitude of E_z is much stronger than that of E_x. Therefore, the effect of E_z is dominant in the total intensity.

In order to achieve an isolated single hot spot, the slit pattern can be modified as shown in Figure 4.20(a). Here, the left slit is shifted outward by half the effective surface plasmon wavelength. This causes constructive interference in the E_z fields coming from two antipodal slits, whereas the E_x fields

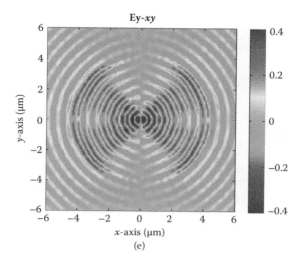

Ey-*xy*

(e)

FIGURE 4.19
(Continued)

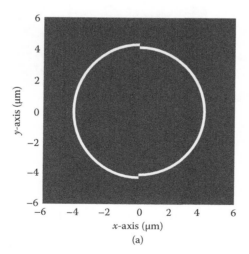

(a)

FIGURE 4.20
(a) Shifted plasmonic lens. The left slit is shifted outward by the half λ_{SPP}. (b) Intensity distribution on the shifted plasmonic lens illuminated by the x-directional linearly polarized light beam. (c) Plasmonic lens. (d) Intensity distribution on the plasmonic lens illuminated by the radially polarized light beam.

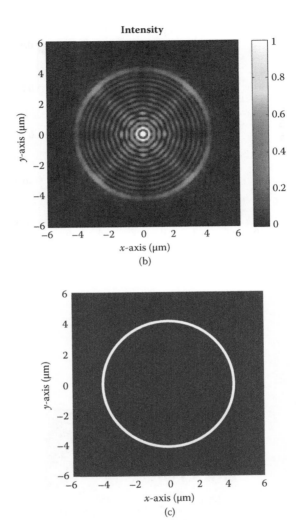

FIGURE 4.20
(Continued)

interfere destructively. Since the amplitude of the E_z field is much larger than that of the E_x field, the proposed geometry exhibits a single hot spot as shown in Figure 4.20(b) [14]. Another way is to illuminate the plasmonic lens with a radially polarized light beam [15]. This is shown in Figure 4.20(c) and (d). In this case, the phases of the E_z field at all slit components are the same. As a result, it is guaranteed that the center of the structure has a single hot spot.

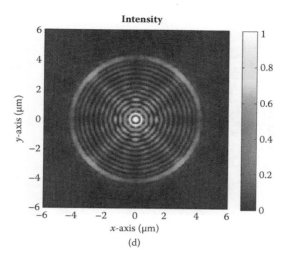

(d)

FIGURE 4.20
(Continued)

Meanwhile, our FMM code can also be adopted for the implementation of the plasmonic vortex lens. By gradually increasing the radius of the slit with the azimuthal angle, as shown in Figure 4.21(a), the phase of the E_z field arriving at the center of the structure experiences retardation. The amount of the phase retardation is proportional to the azimuthal angle, resulting in

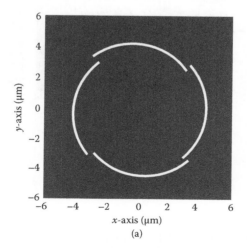

(a)

FIGURE 4.21
(a) Plasmonic vortex lens with geometrical topological charge of 4. Intensity distribution on the plasmonic vortex lens illuminated by (b) the righthanded circularly polarized light beam, (c) the radially polarized light beam, and (d) the lefthanded circularly polarized light beam.

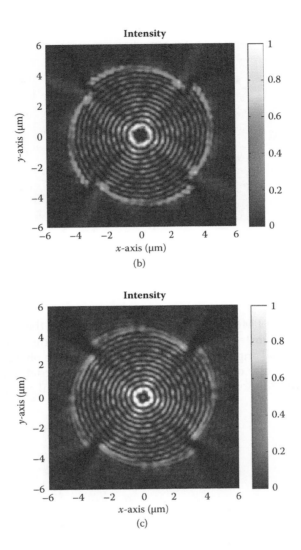

FIGURE 4.21
(Continued)

a rotating field distribution at the center of the structure. In Figure 4.21(b) to (d), we show the intensity field distribution on the plasmonic vortex lens illuminated by the righthanded circularly polarized light beam, the radially polarized light beam, and the lefthanded circularly polarized light beam, respectively [16]. Those field distributions show dark spots at the center

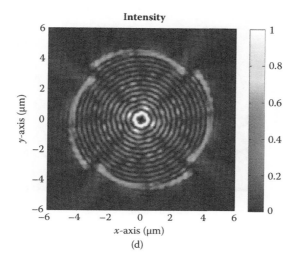

(d)

FIGURE 4.21
(Continued)

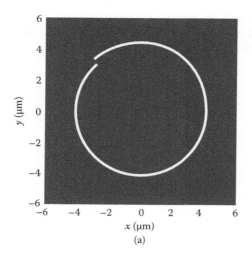

(a)

FIGURE 4.22
(a) Plasmonic vortex lens with geometrical topological charge of 1. (b) Intensity distribution on the plasmonic vortex lens illuminated by the lefthanded circularly polarized light beam.

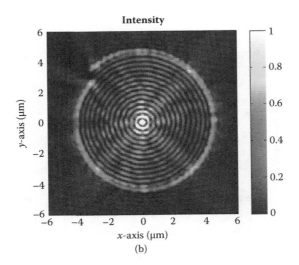

FIGURE 4.22
(Continued)

of the structure. A bright ring is surrounding these dark spots, which is referred to as the primary ring of the vortex. It is noteworthy that the size of the primary ring is dependent on the polarization state of the incident light beam.

The aforementioned plasmonic vortex lens can also be used to generate a single hot spot. Figure 4.22(a) shows the plasmonic vortex lens with the geometrical topolocal charge of 1. If this geometry is illuminated by the lefthanded circularly polarized light beam, then the inner E_z field distribution exhibits the zeroth order of the Bessel function of the first kind.

5

Local Fourier Modal Method

Frameworks for FMM and S-matrix method were established in previous chapters. In this chapter, the mathematical generalization of FMM for local FMM (LFMM) is described. The structure to be treated by LFMM is a super-block structure that can be envisioned as a multiblock comprised of multiblocks. Numerous nanophotonic devices such as photonic crystal waveguides and meta-materials can be approximated by super-block structures. LFMM is formulated to analyze Bloch eigenmodes in super-block structures. In Section 5.1, the framework of LFMM is developed and the S-matrix for a single super-block structure is analyzed within the framework of LFMM. LFMM for a single-super-block with three-dimensional permittivity and permeability profile variation is the cornerstone for the multi-super-block LFMM in Section 5.2. The mathematical extention of LFMM of single-super-blocks to LFMM of multi-super-blocks can be straightforwardly attained by the S-matrix method with a block S-matrix and a boundary S-matrix that are newly derived for super-block structures. In Section 5.3, the MATLAB® standard package of local fourier modal analysis (LFMA) is examined with practical examples of photonic crystal device modeling.

5.1 Local Fourier Modal Analysis of Single-Super-Block Structures

In Figure 5.1(a), a 2D photonic crystal waveguide structure is illustrated, which can be seen as a periodic arrangement of a unit multiblock shown in Figure 5.1(b). When the total structure is a periodic arrangement of a unit multiblock, the structure is referred to as a single-super-block. Thus the 2D photonic crystal waveguide structure is seen as a single-super-block structure. Contrary to the single-block dealt with in the previous chapters, the single-super-block structure has a 3D spatial function form of $\varepsilon(x, y, z)$ and $\mu(x, y, z)$ with both transversal and longitudinal profiles. Maxwell's equations in the single-super-block with both transversal and longitudinal permittivity and permeability variations are given by

$$\nabla \times \mathbf{E} = j\omega\mu_0\mu(x, y, z)\mathbf{H} \tag{5.1a}$$

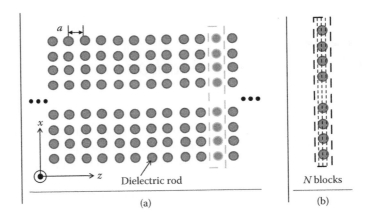

FIGURE 5.1
(a) Photonic crystal waveguide structure; single-super-block composed of periodic arrangement of several unit multiblocks. (b) Unit multiblock modeled by N single blocks.

$$\nabla \times \mathbf{H} = -j\omega\varepsilon_0\varepsilon(x,y,z)\mathbf{E} \qquad (5.1b)$$

$$\nabla \cdot [\varepsilon(x,y,z)\mathbf{E}] = 0 \qquad (5.1c)$$

$$\nabla \cdot [\mu(x,y,z)\mathbf{H}] = 0 \qquad (5.1d)$$

A single-super-block can be represented by the periodic arrangement of
a multiblock structure under the staircase approximation. In this section,
an LFMM framework for an S-matrix analysis of a staircase-approximated
single-super-block is developed based on the elements of FMM. LFMM is
devised to find the set of Bloch eigenmodes. The set of Bloch eigenmodes
analyzed by LFMM is exploited to find the S-matrix of the single-super-
block. One of the Bloch eigenmodes of the exemplary photonic crystal waveguide is presented in Figure 5.2.

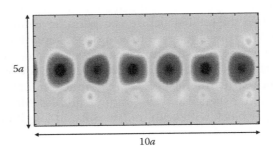

FIGURE 5.2
Bloch eigenmode (electric field) of the photonic crystal waveguide.

The first step of LFMM is to perform an S-matrix analysis of the unit multiblock shown in Figure 5.1(b). The second step is to solve the algebraic eigenvalue equation for extracting the Bloch eigenmode, which is based on the results of the first step. In Equations (5.1a) to (5.1d), the material of single-super-block is supposed to be isotropic for the sake of convenience, but the unfold formulation is valid for general anisotropic media. Thus, LFMM for a single-super-block structure is divided into three logical steps: (1) Bloch eigenmode analysis, (2) S-matrix and coupling coefficient analysis of single-super-block structure, and (3) field visualization.

Step 1: Algebraic Maxwell eigenvalue equation for obtaining Bloch eigenmode. FMM described in Chapter 3 is used to solve the S-matrix and coupling coefficient operator of the unit multiblock of the single-super-block. Let us take a Q-multiblock $M^{(1,Q)}$ with the longitudinal size of Λ_z for the unit multiblock. The S-matrix analysis produces $\vec{\underline{T}}$, $\vec{\underline{R}}$, $\overleftarrow{\underline{R}}$, $\overleftarrow{\underline{T}}$, $\underline{C}_a$, and $\underline{C}_b$ of the unit multiblock and the set of the eigenvalue and eigenvector of the internal field distribution. This work can be done by FMM framework studied through Chapter 3.

Let us assume that all the necessary information is at hand at this stage and, with this, we can then proceed to LFMM, i.e., a Bloch mode analysis of single-super-block structures. From the Bloch theorem, Bloch eigenmodes of Maxwell's equations generally take the form of a pseudoperiodic vector function:

$$\mathbf{E} = (\mathbf{E}_x, \mathbf{E}_y, \mathbf{E}_z) = e^{j(k_{x,0}x + k_{y,0}y + k_z z)}(E_x(x,y,z), E_y(x,y,z), E_z(x,y,z)) \quad (5.2a)$$

$$\mathbf{H} = (\mathbf{H}_x, \mathbf{H}_y, \mathbf{H}_z) = e^{j(k_{x,0}x + k_{y,0}y + k_z z)}(H_x(x,y,z), H_y(x,y,z), H_z(x,y,z)) \quad (5.2b)$$

where the six envelope components of $E_x(x, y, z)$, $E_y(x, y, z)$, $E_z(x, y, z)$, $H_x(x, y, z)$, $H_y(x, y, z)$, and $H_z(x, y, z)$ are periodic functions with the x-directional, y-directional, and z-directional periods of T_x, T_y, and Λ_z, respectively. In Figure 5.3, the concept of the Bloch mode computation is illustrated. The field profile of a Bloch eigenmode at the right boundary takes the form of the field profile at the left boundary multiplied by the eigenvalue $\beta = \exp(jk_{z,0}\Lambda_z)$ as indicated in Figure 5.3, where $\vec{\underline{T}}^{(1,Q)}$, $\vec{\underline{R}}^{(1,Q)}$, $\overleftarrow{\underline{R}}^{(1,Q)}$, and $\overleftarrow{\underline{T}}^{(1,Q)}$ are the components of the total block S-matrix. The coupling coefficient operators, $\underline{C}_a^{(1,Q)}$ and $\underline{C}_b^{(1,Q)}$, contain the information of coupling coefficients of all individual blocks comprising the unit multiblock $M^{(1,Q)}$.

Let the Fourier spectra of the right direction and the left direction propagating portions of a Bloch eigenmode at the left boundary be denoted by $\vec{w}$ and $\overleftarrow{w}$, respectively. Then $\vec{w}$ and $\overleftarrow{w}$ should satisfy

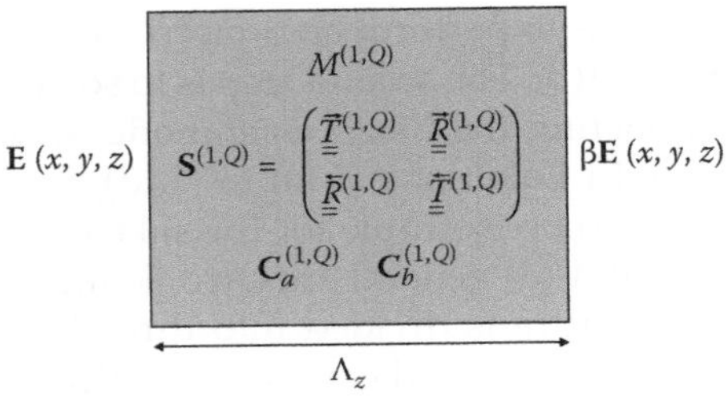

$$\underset{\Lambda_z}{\underbrace{
\begin{array}{c}
M^{(1,Q)} \\[4pt]
\mathbf{E}\,(x,y,z) \quad \mathbf{S}^{(1,Q)} =
\begin{pmatrix}
\vec{\underline{\underline{T}}}^{(1,Q)} & \vec{\underline{\underline{R}}}^{(1,Q)} \\[4pt]
\underset{=}{\vec{R}}^{(1,Q)} & \underset{=}{\vec{T}}^{(1,Q)}
\end{pmatrix} \quad \beta \mathbf{E}\,(x,y,z) \\[6pt]
\mathbf{C}_a^{(1,Q)} \quad \mathbf{C}_b^{(1,Q)}
\end{array}
}}$$

FIGURE 5.3
Schematic diagram of the eigenvalue equation of the Bloch eigenmode with the eigenvalue $\beta = \exp(jk_{z,0}\Lambda_z)$.

the Bloch mode condition:

$$\begin{pmatrix} \beta\vec{w} \\ \underset{=}{\vec{w}} \end{pmatrix} = \begin{pmatrix} \vec{\underline{\underline{T}}}^{(1,Q)} & \vec{\underline{\underline{R}}}^{(1,Q)} \\ \underset{=}{\vec{R}}^{(1,Q)} & \underset{=}{\vec{T}}^{(1,Q)} \end{pmatrix} \begin{pmatrix} \vec{w} \\ \beta\underset{=}{\vec{w}} \end{pmatrix} \tag{5.3a}$$

where $\vec{\underline{\underline{T}}}^{(1,Q)}$, $\underset{=}{\vec{R}}^{(1,Q)}$, $\vec{\underline{\underline{R}}}^{(1,Q)}$, and $\underset{=}{\vec{T}}^{(1,Q)}$ are the S-matrix component of the multiblock $M^{(1,Q)}$. Equation (5.3a) is manipulated to the eigenvalue equation with β as the eigenvalue [1]

$$\begin{pmatrix} \vec{\underline{\underline{T}}}^{(1,Q)} & 0 \\ \underset{=}{\vec{R}}^{(1,Q)} & -\underline{\underline{I}} \end{pmatrix} \begin{pmatrix} \vec{w} \\ \underset{=}{\vec{w}} \end{pmatrix} = \beta \begin{pmatrix} \underline{\underline{I}} & -\vec{\underline{\underline{R}}}^{(1,Q)} \\ 0 & -\underset{=}{\vec{T}}^{(1,Q)} \end{pmatrix} \begin{pmatrix} \vec{w} \\ \underset{=}{\vec{w}} \end{pmatrix} \tag{5.3b}$$

Let the gth eigenvalue and eigenvector of Equation (5.3b) be denoted by $\beta^{(g)}$ and $(\vec{w}^{(g)}, \underset{=}{\vec{w}}^{(g)})$, respectively. The set of internal coupling coefficient operators is $\mathbf{C}_a^{(1,Q)} = \{\underline{C}_{a,(1)}^{(1,Q)}, \underline{C}_{a,(2)}^{(1,Q)}, \ldots, \underline{C}_{a,(Q)}^{(1,Q)}\}$ and $\mathbf{C}_b^{(1,Q)} = \{\underline{C}_{b,(1)}^{(1,Q)}, \underline{C}_{b,(2)}^{(1,Q)}, \ldots, \underline{C}_{b,(Q)}^{(1,Q)}\}$. The internal coupling coefficients of the gth Bloch eigenmode, $\underline{C}_{(q)}^{(g)}$, of the qth block are then determined by

$$\underline{C}_{(q)}^{(g)} = \underline{C}_{a,(q)}^{(1,Q)}\vec{w}^{(g)} + \beta^{(g)}\underline{C}_{b,(q)}^{(1,Q)}\underset{=}{\vec{w}}^{(g)} \quad \text{for} \quad q = 1,2,\ldots Q \tag{5.3c}$$

In practical implementation, for confirming the numerical stability in solving Equation (5.3b), two-step eigenvalue analysis should be adopted. Let the eigenvalues of the positive and negative eigenmodes be denoted by $\beta^+ = \exp(jk_{z,0}^+\Lambda_z)$ and $\beta^- = \exp(jk_{z,0}^-\Lambda_z)$, respectively. The absolute value of the eigenvalue of negative eigenmode β^- can be so large to exceed the precision limitation of practical computers. As illustrated in Figure 5.4, the eigenvalue equations for obtaining positive eigenmodes and negative eigenmodes are taken, respectively, as

$$\begin{pmatrix} \vec{\underline{\underline{T}}}^{(1,Q)} & 0 \\ \underset{=}{\vec{R}}^{(1,Q)} & -\underline{\underline{I}} \end{pmatrix} \begin{pmatrix} \vec{w} \\ \underset{=}{\vec{w}} \end{pmatrix} = \beta^+ \begin{pmatrix} \underline{\underline{I}} & -\vec{\underline{\underline{R}}}^{(1,Q)} \\ 0 & -\underset{=}{\vec{T}}^{(1,Q)} \end{pmatrix} \begin{pmatrix} \vec{w} \\ \underset{=}{\vec{w}} \end{pmatrix} \tag{5.4a}$$

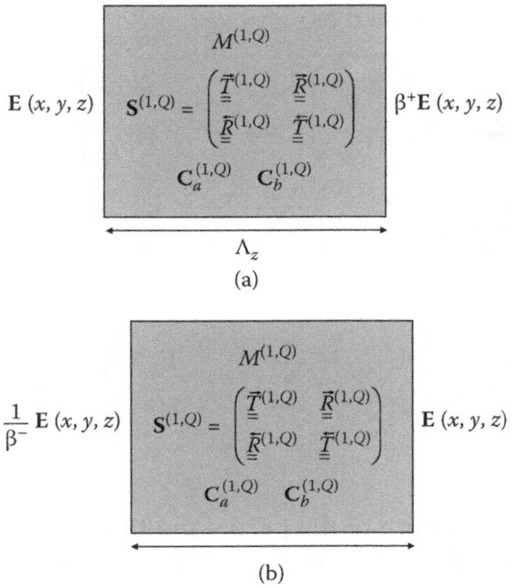

FIGURE 5.4
Schematic diagrams of (a) eigenvalue equation of positive Bloch eigenmodes and (b) eigenvalue equation of negative Bloch eigenmodes.

$$\frac{1}{\beta^-}\left(\begin{array}{cc} \vec{T}^{(1,Q)} & \underline{\underline{0}} \\ \vec{R}^{(1,Q)} & -\underline{\underline{I}} \end{array}\right)\left(\begin{array}{c} \vec{w} \\ \tilde{w} \end{array}\right) = \left(\begin{array}{cc} \underline{\underline{I}} & -\vec{R}^{(1,Q)} \\ \underline{\underline{0}} & -\tilde{T}^{(1,Q)} \end{array}\right)\left(\begin{array}{c} \vec{w} \\ \tilde{w} \end{array}\right) \tag{5.4b}$$

These eigenvalue equations are the fundamental equations of LFMM.

The classification rule of eigenvalues is still applied to LFMM. The eigenmodes with eigenvalues of $jk_{z,0}^{(g)} = a^{(g)} + jb^{(g)}$ or $jk_{z,0}^{(g)} = a^{(g)} - jb^{(g)}$ are negative mode, and the notation $k_{z,0}^{(g)-}$ with the minus superscript is used to indicate the negative mode. The eigenmodes with eigenvalues of $jk_{z,0}^{(g)} = -a^{(g)} + jb^{(g)}$ and $jk_{z,0}^{(g)} = -a^{(g)} - jb^{(g)}$ are positive mode, and the notation $k_{z,0}^{(g)+}$ with the plus superscript is used to indicate the positive mode. The eigenmodes with pure real eigenvalues of $jk_{z,0}^{(g)} = jb^{(g)}$ and $jk_{z,0}^{(g)} = -jb^{(g)}$ with $a^{(g)} = 0$ are classified to the positive mode. The number of the positive modes and that of the negative modes are denoted by M^+ and M^-, respectively. The sum of M^+ and M^- is $M^+ + M^- = 4(2M+1)(2N+1)$.

The coupling coefficient $\underline{C}_{(q)}^{(g)} = \underline{C}_{a,(q)}^{(1,Q)}\tilde{w}^{(g)} + \beta^{(g)}\underline{C}_{b,(q)}^{(1,Q)}\tilde{w}^{(g)}$ of the Bloch eigenmode shows that the Bloch mode is represented by the linear superposition of eigenmodes of each block. The internal electromagnetic field is represented by the superposition of the Bloch eigenmodes with their own specific coupling coefficients. The gth Bloch eigenmode is represented as

$$
\begin{pmatrix} \mathbf{E}^{(g)} \\ \mathbf{H}^{(g)} \end{pmatrix} = \begin{cases} \displaystyle\sum_{i=1}^{M^+} C_{(1),i}^{(g)+} \begin{pmatrix} \mathbf{E}_{(1)}^{(i)+} \\ \mathbf{H}_{(1)}^{(i)+} \end{pmatrix} + \sum_{i=1}^{M^-} C_{(1),i}^{(g)-} \begin{pmatrix} \mathbf{E}_{(1)}^{(i)-} \\ \mathbf{H}_{(1)}^{(i)-} \end{pmatrix} & \text{for } 0 \le z \le l_{1,1} \\[2em] \displaystyle\sum_{i=1}^{M^+} C_{(2),i}^{(g)+} \begin{pmatrix} \mathbf{E}_{(2)}^{(i)+} \\ \mathbf{H}_{(2)}^{(i)+} \end{pmatrix} + \sum_{i=1}^{M^-} C_{(2),i}^{(g)-} \begin{pmatrix} \mathbf{E}_{(2)}^{(i)-} \\ \mathbf{H}_{(2)}^{(i)-} \end{pmatrix} & \text{for } l_{1,1} \le z \le l_{1,2} \\[1em] \qquad\qquad\qquad\qquad\vdots \\[1em] \displaystyle\sum_{i=1}^{M^+} C_{(Q),i}^{(g)+} \begin{pmatrix} \mathbf{E}_{(N)}^{(i)+} \\ \mathbf{H}_{(N)}^{(i)+} \end{pmatrix} + \sum_{i=1}^{M^-} C_{(Q),i}^{(g)-} \begin{pmatrix} \mathbf{E}_{(N)}^{(i)-} \\ \mathbf{H}_{(N)}^{(i)-} \end{pmatrix} & \text{for } l_{1,Q-1} \le z \le l_{1,Q} \end{cases}
$$

$$(5.5a)$$

where $l_{1,q}$ is the length of the partial multiblock from the first block to the qth block. $\mathbf{E}_{(q)}^{(i)+}$, $\mathbf{E}_{(q)}^{(i)-}$, $\mathbf{H}_{(q)}^{(i)+}$, and $\mathbf{H}_{(q)}^{(i)-}$ are given, respectively, by

$$
\mathbf{E}_{(q)}^{(i)+}(x,y,z) = \sum_{m=-M}^{M} \sum_{n=-N}^{N} \left(E_{(q),x,m,n}^{(i)+}, E_{(q),y,m,n}^{(i)+}, E_{(q),z,m,n}^{(i)+} \right) e^{j\left(k_{x,m}x + k_{y,n}y + k_{(q),z}^{(i)+}(z-l_{1,q-1})\right)}
$$

$$(5.5b)$$

$$
\mathbf{E}_{(q)}^{(i)-}(x,y,z) = \sum_{m=-M}^{M} \sum_{n=-N}^{N} \left(E_{(q),x,m,n}^{(i)-}, E_{(q),y,m,n}^{(i)-}, E_{(q),z,m,n}^{(i)-} \right) e^{j\left(k_{x,m}x + k_{y,n}y + k_{(q),z}^{(i)-}(z-l_{1,q})\right)}
$$

$$(5.5c)$$

$$
\mathbf{H}_{(q)}^{(i)+}(x,y,z) = \sum_{m=-M}^{M} \sum_{n=-N}^{N} \left(H_{(q),x,m,n}^{(i)+}, H_{(q),y,m,n}^{(i)+}, H_{(q),z,m,n}^{(i)+} \right) e^{j\left(k_{x,m}x + k_{y,n}y + k_{(q),z}^{(i)+}(z-l_{1,q-1})\right)}
$$

$$(5.5d)$$

$$
\mathbf{H}_{(q)}^{(i)-}(x,y,z) = \sum_{m=-M}^{M} \sum_{n=-N}^{N} \left(H_{(q),x,m,n}^{(i)-}, H_{(q),y,m,n}^{(i)-}, H_{(q),z,m,n}^{(i)-} \right) e^{j\left(k_{x,m}x + k_{y,n}y + k_{(q),z}^{(i)-}(z-l_q)\right)}
$$

$$(5.5e)$$

Now, the obtained Bloch eigenmode assumes the conventional form of FMM, that is, a separate 2D pseudo-Fourier representation allowed to each block with its own coupling coefficients, as seen in Equation (5.5a). The part of the gth Bloch eigenmode in the qth block is represented by

$$
\begin{pmatrix} \mathbf{E}_{(q)}^{(g)} \\ \mathbf{H}_{(q)}^{(g)} \end{pmatrix} = \sum_{m=-M}^{M} \sum_{n=-N}^{N} \begin{pmatrix} E_{(q),x,mng}^{(g)}(z), E_{(q),y,mng}^{(g)}(z), E_{(q),z,mng}^{(g)}(z) \\ H_{(q),x,mng}^{(g)}(z), H_{(q),y,mng}^{(g)}(z), H_{(q),z,mng}^{(g)}(z) \end{pmatrix} e^{j(k_{x,m}x + k_{y,n}y)} \quad (5.6a)
$$

where the Fourier coefficients of the electric field and the magnetic field are the function of z-variable. Equation (5.6a) is manipulated as

$$
\begin{pmatrix} \mathbf{E}_{(q)}^{(g)} \\ \mathbf{H}_{(q)}^{(g)} \end{pmatrix} = \sum_{m=-M}^{M} \sum_{n=-N}^{N} \begin{pmatrix} E_{(q),x,mng}^{(g)}(z), E_{(q),y,mng}^{(g)}(z), E_{(q),z,mng}^{(g)}(z) \\ H_{(q),x,mng}^{(g)}(z), H_{(q),y,mng}^{(g)}(z), H_{(q),z,mng}^{(g)}(z) \end{pmatrix} e^{j(k_{x,m}x + k_{y,n}y)}
$$

$$
= \sum_{i=1}^{M^+} C_{(q),i}^{(g)+} \begin{pmatrix} \mathbf{E}_{(q)}^{(i)+} \\ \mathbf{H}_{(q)}^{(i)+} \end{pmatrix} + \sum_{i=1}^{M^-} C_{(q),i}^{(g)-} \begin{pmatrix} \mathbf{E}_{(q)}^{(i)-} \\ \mathbf{H}_{(q)}^{(i)-} \end{pmatrix}
$$

$$
= \sum_{m=-M}^{M} \sum_{n=-N}^{N} \left[\sum_{i=1}^{M^+} C_{(q),i}^{(g)+} \begin{pmatrix} \left(E_{(q),x,m,n}^{(i)+}, E_{(q),y,m,n}^{(i)+}, E_{(q),z,m,n}^{(i)+} \right) e^{jk_{(q),z}^{(i)+}(z - l_{1,q-1})} \\ \left(H_{(q),x,m,n}^{(i)+}, H_{(q),y,m,n}^{(i)+}, H_{(q),z,m,n}^{(i)+} \right) e^{jk_{(q),z}^{(i)+}(z - l_{1,q-1})} \end{pmatrix} \right.
$$

$$
\left. + \sum_{i=1}^{M^-} C_{(q),i}^{(g)-} \begin{pmatrix} \left(E_{(q),x,m,n}^{(i)-}, E_{(q),y,m,n}^{(i)-}, E_{(q),z,m,n}^{(i)-} \right) e^{jk_{(q),z}^{(i)-}(z - l_{1,q})} \\ \left(H_{(q),x,m,n}^{(i)-}, H_{(q),y,m,n}^{(i)-}, H_{(q),z,m,n}^{(i)-} \right) e^{jk_{(q),z}^{(i)-}(z - l_q)} \end{pmatrix} \right] e^{j(k_{x,m}x + k_{y,n}y)}
$$

$$
\text{for} \quad 1 \le q \le Q \tag{5.6b}
$$

The pseudo-Fourier representation of the Bloch eigenmode is obtained by approximating the z-dependent functions in Equations (5.5a) and (5.5b) to symmetrically truncated Fourier series as follows. Let d_q denote the thickness of the qth block and then $l_{n,n+m}$ is given by $l_{n,n+m} = d_n + d_{n+1} + \cdots d_{n+m}$. Using the simple discrete Fourier transform (DFT), we can find the equivalent Fourier expansion of these z-variable-dependent Fourier coefficients as

$$
\begin{cases}
\exp\left(-jk_{z,0}^{(g)}z \right) E_{(1),x,mng}^{(1,Q)}(z) & \text{for} \quad 0 \le z \le l_{1,1} \\
\exp\left(-jk_{z,0}^{(g)}z \right) E_{(2),x,mng}^{(1,Q)}(z) & \text{for} \quad l_{1,1} \le z \le l_{1,2} \\
\quad \vdots \\
\exp\left(-jk_{z,0}^{(g)}z \right) E_{(Q),x,mng}^{(1,Q)}(z) & \text{for} \quad l_{1,Q-1} \le z \le l_{1,Q}
\end{cases}
\approx \sum_{p=-H}^{H} E_{x,m,n,p}^{(g)} \exp(jG_{z,p}z)
$$

$$
\tag{5.7}
$$

where $G_{z,p}$ is the z-direction reciprocal vector defined by $G_{z,p} = \frac{2\pi}{\Lambda_z}p$ and $k_{z,0}^{(g)}$ is the propagation constant of the Bloch mode ($\beta = \exp(jk_{z,0}^{(g)}\Lambda_z)$). By the same manner, we can find the equivalent Fourier series representation of other Fourier coefficients, $E_{(q),y,mng}^{(g)}(z)$, $E_{(q),z,mng}^{(g)}(z)$, $H_{(q),x,mng}^{(g)}(z)$, $H_{(q),y,mng}^{(g)}(z)$, and $H_{(q),z,mng}^{(g)}(z)$. Therefore, with the Fourier series approximation of the z-dependent parts, the gth Bloch eigenmode takes the form of the pseudo-Fourier representation as

$$
\begin{pmatrix} \mathbf{E}^{(g)} \\ \mathbf{H}^{(g)} \end{pmatrix} = \begin{pmatrix} \mathbf{E}_x^{(g)}, \mathbf{E}_y^{(g)}, \mathbf{E}_z^{(g)} \\ \mathbf{H}_x^{(g)}, \mathbf{H}_y^{(g)}, \mathbf{H}_z^{(g)} \end{pmatrix}
$$

$$
= \sum_{m=-M}^{M} \sum_{n=-N}^{N} \sum_{p=-H}^{H} \begin{pmatrix} E_{x,m,n,p}^{(g)}, E_{y,m,n,p}^{(g)}, E_{z,m,n,p}^{(g)} \\ H_{x,m,n,p}^{(g)}, H_{y,m,n,p}^{(g)}, H_{z,m,n,p}^{(g)} \end{pmatrix} e^{j\left(k_{x,m}x+k_{y,n}y+k_{z,p}^{(g)}z\right)} \quad (5.8)
$$

where $k_{z,p}^{(g)}$ is defined by $k_{z,p}^{(g)} = k_{z,0}^{(g)} + 2\pi p/\Lambda_z$. With these conventions, the gth positive eigenmode $(\mathbf{E}_{(g)}^+, \mathbf{H}_{(g)}^+)$ and the gth negative eigenmode $(\mathbf{E}_{(g)}^-, \mathbf{H}_{(g)}^-)$ are represented as, respectively,

$$
\begin{pmatrix} \mathbf{E}^{(g)+} \\ \mathbf{H}^{(g)+} \end{pmatrix} = \begin{pmatrix} \mathbf{E}_x^{(g)+}, \mathbf{E}_y^{(g)+}, \mathbf{E}_z^{(g)+} \\ \mathbf{H}_x^{(g)+}, \mathbf{H}_y^{(g)+}, \mathbf{H}_z^{(g)+} \end{pmatrix}
$$

$$
= \sum_{m=-M}^{M} \sum_{n=-N}^{N} \sum_{p=-H}^{H} \begin{pmatrix} E_{x,m,n,p}^{(g)+}, E_{y,m,n,p}^{(g)+}, E_{z,m,n,p}^{(g)+} \\ H_{x,m,n,p}^{(g)+}, H_{y,m,n,p}^{(g)+}, H_{z,m,n,p}^{(g)+} \end{pmatrix} e^{\left[j\left(k_{x,m}x+k_{y,n}y+k_{z,p}^{(g)+}z\right)\right]}
$$

$$
(5.9a)
$$

$$
\begin{pmatrix} \mathbf{E}^{(g)-} \\ \mathbf{H}^{(g)-} \end{pmatrix} = \begin{pmatrix} \mathbf{E}_x^{(g)-}, \mathbf{E}_y^{(g)-}, \mathbf{E}_z^{(g)-} \\ \mathbf{H}_x^{(g)-}, \mathbf{H}_y^{(g)-}, \mathbf{H}_z^{(g)-} \end{pmatrix}
$$

$$
= \sum_{m=-M}^{M} \sum_{n=-N}^{N} \sum_{p=-H}^{H} \begin{pmatrix} E_{x,m,n,p}^{(g)-}, E_{y,m,n,p}^{(g)-}, E_{z,m,n,p}^{(g)-} \\ H_{x,m,n,p}^{(g)-}, H_{y,m,n,p}^{(g)-}, H_{z,m,n,p}^{(g)-} \end{pmatrix} e^{j\left(k_{x,m}x+k_{y,n}y+k_{z,p}^{(g)-}z\right)}
$$

$$
(5.9b)
$$

The pseudo-Fourier representation of the Bloch eigenmodes is an essential element in the modeling and analysis of nanophotonic networks, which are introduced in the next chapter. Hereafter, the Bloch eigenmodes represented by the pseudo-Fourier series are adopted as the mathematical basis of LFMM.

As an example of the described LFMA, a Bloch eigenmode analysis of a 2D photonic crystal waveguide structure [2–5] is presented. Figure 5.5(a) shows the 2D photonic crystal waveguide with the period of the circular rod of a, the diameter of the rod

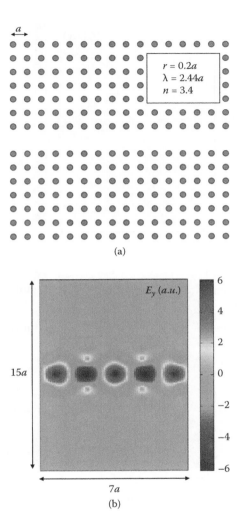

FIGURE 5.5
(a) Two-dimensional photonic crystal waveguide and the guided Bloch eigenmode:
(b) *y*-polarization electric field distribution, (c) *x*-polarization magnetic field distribution, and
(d) *z*-polarization magnetic field distribution. (From H. Kim and B. Lee, *J. Opt. Soc. Am. B*, 25(4),
2008. With permission.)

0.4*a*, wavelength 2.44*a*, and refractive index 3.4, respectively. This
waveguide structure is known to be a single-mode waveguide. This
point can be confirmed by analyzing the dispersion relation of the
photonic crystal waveguide. In Figure 5.5(b) to (d), the *y*-polariza-
tion electric field, *x*-polarization magnetic field, and *z*-polarization

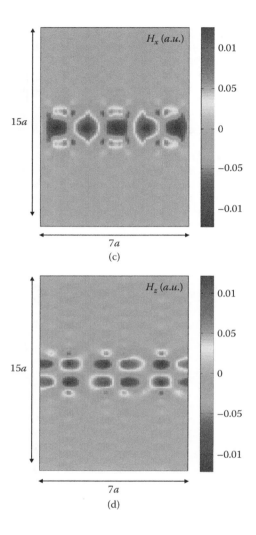

FIGURE 5.5
(Continued)

magnetic field distributions of the dominant Bloch eigenmode are presented, respectively.

Step 2: S-matrix and coupling coefficient operator calculation. In the first step, the Bloch eigenmodes in a single-super-block have been analyzed in general fashion. The next task is to find the S-matrix of the single-super-block structure. The internal electromagnetic field is represented by the super-position of the Bloch eigenmodes with their own specific coupling coefficients as

$$
\begin{pmatrix} \mathbf{E} \\ \mathbf{H} \end{pmatrix} = \sum_{g=1}^{M^+} C_{a,g}^+ \begin{pmatrix} \mathbf{E}^{(g)+} \\ \mathbf{H}^{(g)+} \end{pmatrix} + \sum_{g=1}^{M^-} C_{a,g}^- \begin{pmatrix} \mathbf{E}^{(g)-} \\ \mathbf{H}^{(g)-} \end{pmatrix}
\tag{5.10a}
$$

where M^+ and M^- should be $2(2M + 1)(2N + 1)$ simultaneously. Here, the representations of the positive Bloch eigenmode and the negative Bloch eigenmode are slightly modified, from Equation (5.9), by the Bloch phase terms of $e^{j(k_x x + k_y y + k_z^{(g)+}(z-z_-))}$ and $e^{j(k_x x + k_y y + k_z^{(g)-}(z-z_+))}$, respectively, as

$$
\begin{pmatrix} \mathbf{E}^{(g)+} \\ \mathbf{H}^{(g)+} \end{pmatrix} = \begin{pmatrix} \mathbf{E}_x^{(g)+}, \mathbf{E}_y^{(g)+}, \mathbf{E}_z^{(g)+} \\ \mathbf{H}_x^{(g)+}, \mathbf{H}_y^{(g)+}, \mathbf{H}_z^{(g)+} \end{pmatrix}
$$

$$
= \sum_{m=-M}^{M} \sum_{n=-N}^{N} \sum_{p=-H}^{H} \begin{pmatrix} E_{x,m,n,p}^{(g)+}, E_{y,m,n,p}^{(g)+}, E_{z,m,n,p}^{(g)+} \\ H_{x,m,n,p}^{(g)+}, H_{y,m,n,p}^{(g)+}, H_{z,m,n,p}^{(g)+} \end{pmatrix} e^{\left[j\left(k_{x,m} x + k_{y,n} y + k_{z,p}^{(g)+}(z-z_-) \right) \right]}
\tag{5.10b}
$$

$$
\begin{pmatrix} \mathbf{E}^{(g)-} \\ \mathbf{H}^{(g)-} \end{pmatrix} = \begin{pmatrix} \mathbf{E}_x^{(g)-}, \mathbf{E}_y^{(g)-}, \mathbf{E}_z^{(g)-} \\ \mathbf{H}_x^{(g)-}, \mathbf{H}_y^{(g)-}, \mathbf{H}_z^{(g)-} \end{pmatrix}
$$

$$
= \sum_{m=-M}^{M} \sum_{n=-N}^{N} \sum_{p=-H}^{H} \begin{pmatrix} E_{x,m,n,p}^{(g)-}, E_{y,m,n,p}^{(g)-}, E_{z,m,n,p}^{(g)-} \\ H_{x,m,n,p}^{(g)-}, H_{y,m,n,p}^{(g)-}, H_{z,m,n,p}^{(g)-} \end{pmatrix} e^{j\left(k_{x,m} x + k_{y,n} y + k_{z,p}^{(g)-}(z-z_+) \right)}
\tag{5.10c}
$$

For the left-to-right characterization, the representation of the incident optical field, the reflected optical field in the lefthand free space, and the transmitted optical field in the righthand free space are given, respectively, as

$$
\begin{pmatrix} \vec{\mathbf{E}}_i \\ \vec{\mathbf{H}}_i \end{pmatrix} = \begin{pmatrix} \vec{\mathbf{E}}_{i,x}, \vec{\mathbf{E}}_{i,y}, \vec{\mathbf{E}}_{i,z} \\ \vec{\mathbf{H}}_{i,x}, \vec{\mathbf{H}}_{i,y}, \vec{\mathbf{H}}_{i,z} \end{pmatrix}
$$

$$
= \sum_{m=-M}^{M} \sum_{n=-N}^{N} \begin{pmatrix} \vec{E}_{i,x,m,n}, \vec{E}_{i,y,m,n}, \vec{E}_{i,z,m,n} \\ \vec{H}_{i,x,m,n}, \vec{H}_{i,y,m,n}, \vec{H}_{i,z,m,n} \end{pmatrix} e^{j(k_{x,m,n} x + k_{ym,n} y + k_{z,m,n}(z-z_-))}
\tag{5.11a}
$$

$$\begin{pmatrix} \vec{\mathbf{E}}_r \\ \vec{\mathbf{H}}_r \end{pmatrix} = \begin{pmatrix} \vec{\mathbf{E}}_{r,x}, \vec{\mathbf{E}}_{r,y}, \vec{\mathbf{E}}_{r,z} \\ \vec{\mathbf{H}}_{r,x}, \vec{\mathbf{H}}_{r,y}, \vec{\mathbf{H}}_{r,z} \end{pmatrix}$$

$$= \sum_{m=-M}^{M} \sum_{n=-N}^{N} \begin{pmatrix} \vec{\mathbf{E}}_{r,x,m,n}, \vec{\mathbf{E}}_{r,y,m,n}, \vec{\mathbf{E}}_{r,z,m,n} \\ \vec{\mathbf{H}}_{r,x,m,n}, \vec{\mathbf{H}}_{r,y,m,n}, \vec{\mathbf{H}}_{r,z,m,n} \end{pmatrix} e^{j(k_{x,m,n}x + k_{ym,n}y - k_{z,m,n}(z-z_-))}$$

$$\text{(5.11b)}$$

$$\begin{pmatrix} \vec{\mathbf{E}}_t \\ \vec{\mathbf{H}}_t \end{pmatrix} = \begin{pmatrix} \vec{\mathbf{E}}_{t,x}, \vec{\mathbf{E}}_{t,y}, \vec{\mathbf{E}}_{t,z} \\ \vec{\mathbf{H}}_{t,x}, \vec{\mathbf{H}}_{t,y}, \vec{\mathbf{H}}_{t,z} \end{pmatrix}$$

$$= \sum_{m=-M}^{M} \sum_{n=-N}^{N} \begin{pmatrix} \vec{\mathbf{E}}_{t,x,m,n}, \vec{\mathbf{E}}_{t,y,m,n}, \vec{\mathbf{E}}_{t,z,m,n} \\ \vec{\mathbf{H}}_{t,x,m,n}, \vec{\mathbf{H}}_{t,y,m,n}, \vec{\mathbf{H}}_{t,z,m,n} \end{pmatrix} e^{j(k_{x,m,n}x + k_{y,m,n}y + k_{z,m,n}(z-z_+))}$$

$$\text{(5.11c)}$$

The tangential components of the incidence, reflection, and transmission magnetic fields are solved in Maxwell's equations by the electric field components as follows:

$$\vec{\mathbf{H}}_{i,y} = \frac{1}{j\omega\mu_0}\left(\frac{\partial \vec{\mathbf{E}}_{i,x}}{\partial z} - \frac{\partial \vec{\mathbf{E}}_{i,z}}{\partial x} \right) \Leftrightarrow \vec{H}_{i,y,m,n} = \frac{1}{\omega\mu_0}(k_{z,m,n}\vec{E}_{i,x,m,n} - k_{x,m,n}\vec{E}_{i,z,m,n}),$$

$$\text{(5.12a)}$$

$$\vec{\mathbf{H}}_{i,x} = \frac{1}{j\omega\mu_0}\left(\frac{\partial \vec{\mathbf{E}}_{i,z}}{\partial y} - \frac{\partial \vec{\mathbf{E}}_{i,y}}{\partial z} \right) \Leftrightarrow \vec{H}_{i,x,m,n} = \frac{1}{\omega\mu_0}(k_{y,m,n}\vec{E}_{i,z,m,n} - k_{z,m,n}\vec{E}_{i,y,m,n}),$$

$$\text{(5.12b)}$$

$$\vec{\mathbf{H}}_{r,y} = \frac{1}{j\omega\mu_0}\left(\frac{\partial \vec{\mathbf{E}}_{r,x}}{\partial z} - \frac{\partial \vec{\mathbf{E}}_{r,z}}{\partial x} \right) \Leftrightarrow \vec{H}_{r,y,m,n} = \frac{1}{\omega\mu_0}(-k_{z,m,n}\vec{E}_{r,x,m,n} - k_{x,m,n}\vec{E}_{r,z,m,n}),$$

$$\text{(5.12c)}$$

$$\vec{\mathbf{H}}_{r,x} = \frac{1}{j\omega\mu_0}\left(\frac{\partial \vec{\mathbf{E}}_{r,z}}{\partial y} - \frac{\partial \vec{\mathbf{E}}_{r,y}}{\partial z} \right) \Leftrightarrow \vec{H}_{r,x,m,n} = \frac{1}{\omega\mu_0}(k_{y,m,n}\vec{E}_{r,z,m,n} + k_{z,m,n}\vec{E}_{r,y,m,n}),$$

$$\text{(5.12d)}$$

$$\vec{\mathbf{H}}_{t,y} = \frac{1}{j\omega\mu_0}\left(\frac{\partial\vec{\mathbf{E}}_{t,x}}{\partial z} - \frac{\partial\vec{\mathbf{E}}_{t,z}}{\partial x}\right) \Leftrightarrow \vec{H}_{t,y} = \frac{1}{\omega\mu_0}(k_{z,m,n}\vec{E}_{t,x,m,n} - k_{x,m,n}\vec{E}_{t,z,m,n}),$$

(5.12e)

$$\vec{\mathbf{H}}_{t,x} = \frac{1}{j\omega\mu_0}\left(\frac{\partial\vec{\mathbf{E}}_{t,z}}{\partial y} - \frac{\partial\vec{\mathbf{E}}_{t,y}}{\partial z}\right) \Leftrightarrow \vec{H}_{t,x} = \frac{1}{\omega\mu_0}(k_{y,m,n}\vec{E}_{t,z,m,n} - k_{z,m,n}\vec{E}_{t,y,m,n}).$$

(5.12f)

The tangential H-field coefficients, $\vec{H}_{i,y,m,n}$, $\vec{H}_{i,x,m,n}$, are represented by the three E-field coefficients, $\vec{E}_{i,x,m,n}$, $\vec{E}_{i,y,m,n}$, and $\vec{E}_{i,z,m,n}$. The pairs of $\vec{H}_{r,y,m,n}$ and $\vec{H}_{r,x,m,n}$, and $\vec{H}_{t,y,m,n}$ and $\vec{H}_{t,x,m,n}$, are represented by the corresponding three electric field coefficients. However, considering the following plane wave condition in free space,

$$k_{x,m,n}\vec{E}_{i,x,m,n} + k_{y,m,n}\vec{E}_{i,y,m,n} + k_{z,m,n}\vec{E}_{i,z,m,n} = 0 \Leftrightarrow \vec{E}_{i,z,m,n}$$

$$= -\frac{k_{x,m,n}\vec{E}_{i,x,m,n} + k_{y,m,n}\vec{E}_{i,y,m,n}}{k_{z,m,n}}$$

(5.13a)

$$k_{x,m,n}\vec{E}_{r,x,m,n} + k_{y,m,n}\vec{E}_{r,y,m,n} - k_{z,m,n}\vec{E}_{r,z,m,n} = 0 \Leftrightarrow \vec{E}_{r,z,m,n}$$

$$= \frac{k_{x,m,n}\vec{E}_{r,x,m,n} + k_{y,m,n}\vec{E}_{r,y,m,n}}{k_{z,m,n}}$$

(5.13b)

$$k_{x,m,n}\vec{E}_{t,x,m,n} + k_{y,m,n}\vec{E}_{t,y,m,n} + k_{z,m,n}\vec{E}_{t,z,m,n} = 0 \Leftrightarrow \vec{E}_{t,z,m,n}$$

$$= -\frac{k_{x,m,n}\vec{E}_{t,x,m,n} + k_{y,m,n}\vec{E}_{t,y,m,n}}{k_{z,m,n}}$$

(5.13c)

the z-directional components, $\vec{E}_{i,z,m,n}$, $\vec{E}_{r,z,m,n}$, and $\vec{E}_{t,z,m,n}$, can be eliminated by substituting these terms by the tangential electric field coefficients. According to the transverse field continuation condition with respect to the harmonic term, $\exp(j(k_{x,m,n}x + k_{y,m,n}y))$, the boundary condition at the left boundary of $z = z_-$ can be expressed by the matrix equations for $-M \le m \le M$ and $-N \le n \le N$,

$$
\begin{pmatrix}
I & 0 & I & 0 \\
0 & I & 0 & I \\
\dfrac{1}{\omega\mu_0}\dfrac{k_{x,m}k_{y,n}}{k_{z,m,n}} & \dfrac{1}{\omega\mu_0}\dfrac{\left(k_{z,m,n}^2+k_{x,m}^2\right)}{k_{z,m,n}} & -\dfrac{1}{\omega\mu_0}\dfrac{k_{x,m}k_{y,n}}{k_{z,m,n}} & -\dfrac{1}{\omega\mu_0}\dfrac{\left(k_{z,m,n}^2+k_{x,m}^2\right)}{k_{z,m,n}} \\
-\dfrac{1}{\omega\mu_0}\dfrac{\left(k_{y,n}^2+k_{z,m,n}^2\right)}{k_{z,m,n}} & -\dfrac{1}{\omega\mu_0}\dfrac{k_{y,n}k_{x,m}}{k_{z,m,n}} & \dfrac{1}{\omega\mu_0}\dfrac{\left(k_{y,n}^2+k_{z,m,n}^2\right)}{k_{z,m,n}} & \dfrac{1}{\omega\mu_0}\dfrac{k_{y,n}k_{x,m}}{k_{z,m,n}}
\end{pmatrix}
$$

$$
\times
\begin{pmatrix}
\vec{E}_{i,y,m,n} \\
\vec{E}_{i,x,m,n} \\
\vec{E}_{r,y,m,n} \\
\vec{E}_{r,x,m,n}
\end{pmatrix}
$$

$$
=
\begin{pmatrix}
\sum\limits_{p=-H}^{H} E_{y,m,n,p}^{(1)+} & \cdots & \sum\limits_{p=-H}^{H} E_{y,m,n,p}^{(M^+)+} & \sum\limits_{p=-H}^{H} E_{y,m,n,p}^{(1)-}e^{jk_{z,p}^{(1)-}(z_--z_+)} & \cdots & \sum\limits_{p=-H}^{H} E_{y,m,n,p}^{(M^-)-}e^{jk_{z,p}^{(M^-)-}(z_--z_+)} \\
\sum\limits_{p=-H}^{H} E_{x,m,n,p}^{(1)+} & \cdots & \sum\limits_{p=-H}^{H} E_{x,m,n,p}^{(M^+)+} & \sum\limits_{p=-H}^{H} E_{x,m,n,p}^{(1)-}e^{jk_{z,p}^{(1)-}(z_--z_+)} & \cdots & \sum\limits_{p=-H}^{H} E_{x,m,n,p}^{(M^-)-}e^{jk_{z,p}^{(M^-)-}(z_--z_+)} \\
\sum\limits_{p=-H}^{H} H_{y,m,n,p}^{(1)+} & \cdots & \sum\limits_{p=-H}^{H} H_{y,m,n,p}^{(M^+)+} & \sum\limits_{p=-H}^{H} H_{y,m,n,p}^{(1)-}e^{jk_{z,p}^{(1)-}(z_--z_+)} & \cdots & \sum\limits_{p=-H}^{H} H_{y,m,n,p}^{(M^-)-}e^{jk_{z,p}^{(M^-)-}(z_--z_+)} \\
\sum\limits_{p=-H}^{H} H_{x,m,n,p}^{(1)+} & \cdots & \sum\limits_{p=-H}^{H} H_{x,m,n,p}^{(M^+)+} & \sum\limits_{p=-H}^{H} H_{x,m,n,p}^{(1)-}e^{jk_{z,p}^{(1)-}(z_--z_+)} & \cdots & \sum\limits_{p=-H}^{H} H_{x,m,n,p}^{(M^-)-}e^{jk_{z,p}^{(M^-)-}(z_--z_+)}
\end{pmatrix}
$$

$$
\times
\begin{pmatrix}
C_{a,1}^+ \\
\vdots \\
C_{a,M^+}^+ \\
C_{a,1}^- \\
\vdots \\
C_{a,M^-}^-
\end{pmatrix}.
$$

$$(5.14a)$$

The boundary condition at the right boundary is expressed by the matrix equation

$$
\left(
\begin{array}{ccccccc}
\sum\limits_{p=-H}^{H} E_{y,m,n,p}^{(1)+} e^{jk_{z,p}^{(1)+}(z_+-z_-)} & \cdots & \sum\limits_{p=-H}^{H} E_{y,m,n,p}^{(M^+)+} e^{jk_{z,p}^{(M^+)+}(z_+-z_-)} & \sum\limits_{p=-H}^{H} E_{y,m,n,p}^{(1)-} & \cdots & \sum\limits_{p=-H}^{H} E_{y,m,n,p}^{(M^-)-} \\
\sum\limits_{p=-H}^{H} E_{x,m,n,p}^{(1)+} e^{jk_{z,p}^{(1)+}(z_+-z_-)} & \cdots & \sum\limits_{p=-H}^{H} E_{x,m,n,p}^{(M^+)+} e^{jk_{z,p}^{(M^+)+}(z_+-z_-)} & \sum\limits_{p=-H}^{H} E_{x,m,n,p}^{(1)-} & \cdots & \sum\limits_{p=-H}^{H} E_{x,m,n,p}^{(M^-)-} \\
\sum\limits_{p=-H}^{H} H_{y,m,n,p}^{(1)+} e^{jk_{z,p}^{(1)+}(z_+-z_-)} & \cdots & \sum\limits_{p=-H}^{H} H_{y,m,n,p}^{(M^+)+} e^{jk_{z,p}^{(M^+)+}(z_+-z_-)} & \sum\limits_{p=-H}^{H} H_{y,m,n,p}^{(1)-} & \cdots & \sum\limits_{p=-H}^{H} H_{y,m,n,p}^{(M^-)-} \\
\sum\limits_{p=-H}^{H} H_{x,m,n,p}^{(1)+} e^{jk_{z,p}^{(1)+}(z_+-z_-)} & \cdots & \sum\limits_{p=-H}^{H} H_{x,m,n,p}^{(M^+)+} e^{jk_{z,p}^{(M^+)+}(z_+-z_-)} & \sum\limits_{p=-H}^{H} H_{x,m,n,p}^{(1)-} & \cdots & \sum\limits_{p=-H}^{H} H_{x,m,n,p}^{(M^-)-}
\end{array}
\right)
$$

$$
\times
\begin{pmatrix}
C_{a,1}^+ \\
\vdots \\
C_{a,M^+}^+ \\
C_{a,1}^- \\
\vdots \\
C_{a,M^-}^-
\end{pmatrix}
$$

$$
=
\left(
\begin{array}{cccc}
I & 0 & I & 0 \\
0 & I & 0 & I \\
\dfrac{1}{\omega\mu_0}\dfrac{k_{x,m}k_{y,n}}{k_{z,m,n}} & \dfrac{1}{\omega\mu_0}\dfrac{\left(k_{z,m,n}^2+k_{x,m}^2\right)}{k_{z,m,n}} & -\dfrac{1}{\omega\mu_0}\dfrac{k_{x,m}k_{y,n}}{k_{z,m,n}} & -\dfrac{1}{\omega\mu_0}\dfrac{\left(k_{z,m,n}^2+k_{x,m}^2\right)}{k_{z,m,n}} \\
-\dfrac{1}{\omega\mu_0}\dfrac{\left(k_{y,n}^2+k_{z,m,n}^2\right)}{k_{z,m,n}} & -\dfrac{1}{\omega\mu_0}\dfrac{k_{y,n}k_{x,m}}{k_{z,m,n}} & \dfrac{1}{\omega\mu_0}\dfrac{\left(k_{y,n}^2+k_{z,m,n}^2\right)}{k_{z,m,n}} & \dfrac{1}{\omega\mu_0}\dfrac{k_{y,n}k_{x,m}}{k_{z,m,n}}
\end{array}
\right)
$$

$$
\times
\begin{pmatrix}
\vec{E}_{t,y,m,n} \\
\vec{E}_{t,x,m,n} \\
0 \\
0
\end{pmatrix}.
$$

(5.14b)

The boundary condition matching Equations (5.14a) and (5.14b) is represented by the following matrix operator equations:

$$
\begin{pmatrix}
\underline{\underline{W}}_h & \underline{\underline{W}}_h \\
\underline{\underline{V}}_h & -\underline{\underline{V}}_h
\end{pmatrix}
\begin{pmatrix}
\vec{E}_i \\
\vec{E}_r
\end{pmatrix}
=
\begin{pmatrix}
\underline{\underline{W}}^+(0) & \underline{\underline{W}}^-(z_--z_+) \\
\underline{\underline{V}}^+(0) & \underline{\underline{V}}^-(z_--z_+)
\end{pmatrix}
\begin{pmatrix}
\underline{C}_a^+ \\
\underline{C}_a^-
\end{pmatrix}
$$

(5.15a)

$$
\begin{pmatrix} \underline{\underline{W}}^+(z_+ - z_-) & \underline{\underline{W}}^-(0) \\ \underline{\underline{V}}^+(z_+ - z_-) & \underline{\underline{V}}^-(0) \end{pmatrix} \begin{pmatrix} \underline{C}_a^+ \\ \underline{C}_a^- \end{pmatrix} = \begin{pmatrix} \underline{\underline{W}}_h & \underline{\underline{W}}_h \\ \underline{\underline{V}}_h & -\underline{\underline{V}}_h \end{pmatrix} \begin{pmatrix} \vec{E}_t \\ 0 \end{pmatrix} \tag{5.15b}
$$

where $\underline{\underline{W}}_h$ and $\underline{\underline{V}}_h$ are $[2(2M+1)(2N+1)] \times [2(2M+1)(2N+1)]$ matrices given, respectively, by

$$
\underline{\underline{W}}_h = \begin{pmatrix} \underline{\underline{I}} & 0 \\ 0 & \underline{\underline{I}} \end{pmatrix} \tag{5.15c}
$$

$$
\underline{\underline{V}}_h = \begin{pmatrix} \left[\dfrac{1}{\omega\mu_0} \dfrac{k_{x,m}k_{y,n}}{k_{z,m,n}} \right] & \left[\dfrac{1}{\omega\mu_0} \dfrac{\left(k_{z,m,n}^2 + k_{x,m}^2 \right)}{k_{z,m,n}} \right] \\[6mm] \left[-\dfrac{1}{\omega\mu_0} \dfrac{\left(k_{y,n}^2 + k_{z,m,n}^2 \right)}{k_{z,m,n}} \right] & \left[-\dfrac{1}{\omega\mu_0} \dfrac{k_{y,n}k_{x,m}}{k_{z,m,n}} \right] \end{pmatrix} \tag{5.15d}
$$

$\underline{\underline{W}}^+(z)$ and $\underline{\underline{V}}^+(z)$ are $[2(2M+1)(2N+1)] \times M^+$ matrices indicating the part of the positive modes in Equations (5.15a) and (5.15b), given, respectively, by

$$
\underline{\underline{W}}^+(z) = \begin{pmatrix} \left[\displaystyle\sum_{p=-H}^{H} E_{y,m,n,p}^{(1)+} e^{jk_{z,p}^{(1)+}z} \right] & \cdots & \left[\displaystyle\sum_{p=-H}^{H} E_{y,m,n,p}^{(M^+)+} e^{jk_{z,p}^{(M^+)+}z} \right] \\[8mm] \left[\displaystyle\sum_{p=-H}^{H} E_{x,m,n,p}^{(1)+} e^{jk_{z,p}^{(1)+}z} \right] & \cdots & \left[\displaystyle\sum_{p=-H}^{H} E_{x,m,n,p}^{(M^+)+} e^{jk_{z,p}^{(M^+)+}z} \right] \end{pmatrix} \tag{5.16a}
$$

$$
\underline{\underline{V}}^+(z) = \begin{pmatrix} \left[\displaystyle\sum_{p=-H}^{H} H_{y,m,n,p}^{(1)+} e^{jk_{z,p}^{(1)+}z} \right] & \cdots & \left[\displaystyle\sum_{p=-H}^{H} H_{y,m,n,p}^{(M^+)+} e^{jk_{z,p}^{(M^+)+}z} \right] \\[8mm] \left[\displaystyle\sum_{p=-H}^{H} H_{x,m,n,p}^{(1)+} e^{jk_{z,p}^{(1)+}z} \right] & \cdots & \left[\displaystyle\sum_{p=-H}^{H} H_{x,m,n,p}^{(M^+)+} e^{jk_{z,p}^{(M^+)+}z} \right] \end{pmatrix} \tag{5.16b}
$$

$\underline{\underline{W}}^-(z)$, and $\underline{\underline{V}}^-(z)$ are $[2(2M+1)(2N+1)] \times M^-$ matrices indicating the part of the negative modes in Equations (5.15a) and (5.15b), given, respectively, by

$$
\underline{\underline{W}}^{-}(z) = \left(
\begin{array}{ccc}
\left[\displaystyle\sum_{p=-H}^{H} E_{y,m,n,p}^{(1)-} e^{jk_{z,p}^{(1)-}z} \right] & \cdots & \left[\displaystyle\sum_{p=-H}^{H} E_{y,m,n,p}^{(M-)-} e^{jk_{z,p}^{(M-)-}z} \right] \\[2em]
\left[\displaystyle\sum_{p=-H}^{H} E_{x,m,n,p}^{(1)-} e^{jk_{z,p}^{(1)-}z} \right] & \cdots & \left[\displaystyle\sum_{p=-H}^{H} E_{x,m,n,p}^{(M-)-} e^{jk_{z,p}^{(M-)-}z} \right]
\end{array}
\right)
\tag{5.16c}
$$

$$
\underline{\underline{V}}^{-}(z) = \left(
\begin{array}{ccc}
\left[\displaystyle\sum_{p=-H}^{H} H_{y,m,n,p}^{(1)-} e^{jk_{z,p}^{(1)-}z} \right] & \cdots & \left[\displaystyle\sum_{p=-H}^{H} H_{y,m,n,p}^{(M-)-} e^{jk_{z,p}^{(M-)-}z} \right] \\[2em]
\left[\displaystyle\sum_{p=-H}^{H} H_{x,m,n,p}^{(1)-} e^{jk_{z,p}^{(1)-}z} \right] & \cdots & \left[\displaystyle\sum_{p=-H}^{H} H_{x,m,n,p}^{(M-)-} e^{jk_{z,p}^{(M-)-}z} \right]
\end{array}
\right)
\tag{5.16d}
$$

Set $\vec{\underline{U}}$ to be the input operator, an $[2(2M + 1)(2N + 1)] \times [2(2M + 1)(2N + 1)]$ identity matrix, as in the previous chapter. Then $\vec{\underline{R}}$ and $\vec{\underline{T}}$ are the reflection coefficient matrix operator and transmission coefficient matrix operator, respectively. The coupling coefficient matrix operators are denoted by $\underline{\underline{C}}_a^+$ and $\underline{\underline{C}}_a^-$. With these linear operators, we can write Equations (5.14a) and (5.14b) by the operator mathematics form

$$
\begin{pmatrix} \underline{\underline{W}}_h & \underline{\underline{W}}_h \\ \underline{\underline{V}}_h & -\underline{\underline{V}}_h \end{pmatrix}
\begin{pmatrix} \vec{\underline{U}} \\ \vec{\underline{R}} \end{pmatrix}
=
\begin{pmatrix} \underline{\underline{W}}^+(0) & \underline{\underline{W}}^-(z_- - z_+) \\ \underline{\underline{V}}^+(0) & \underline{\underline{V}}^-(z_- - z_+) \end{pmatrix}
\begin{pmatrix} \underline{\underline{C}}_a^+ \\ \underline{\underline{C}}_a^- \end{pmatrix}
\tag{5.17a}
$$

$$
\begin{pmatrix} \underline{\underline{W}}^+(z_+ - z_-) & \underline{\underline{W}}^-(0) \\ \underline{\underline{V}}^+(z_+ - z_-) & \underline{\underline{V}}^-(0) \end{pmatrix}
\begin{pmatrix} \underline{\underline{C}}_a^+ \\ \underline{\underline{C}}_a^- \end{pmatrix}
=
\begin{pmatrix} \underline{\underline{W}}_h & \underline{\underline{W}}_h \\ \underline{\underline{V}}_h & -\underline{\underline{V}}_h \end{pmatrix}
\begin{pmatrix} \vec{\underline{T}} \\ \underline{\underline{0}} \end{pmatrix}
\tag{5.17b}
$$

The solution is obtained as

$$
\begin{pmatrix} \underline{\underline{C}}_a^+ \\ \underline{\underline{C}}_a^- \end{pmatrix}
=
\begin{pmatrix}
\underline{\underline{W}}_h^{-1}\underline{\underline{W}}^+(0) + \underline{\underline{V}}_h^{-1}\underline{\underline{V}}^+(0) & \underline{\underline{W}}_h^{-1}\underline{\underline{W}}^-(z_- - z_+) + \underline{\underline{V}}_h^{-1}\underline{\underline{V}}^-(z_- - z_+) \\
\underline{\underline{W}}_h^{-1}\underline{\underline{W}}^+(z_+ - z_-) - \underline{\underline{V}}_h^{-1}\underline{\underline{V}}^+(z_+ - z_-) & \underline{\underline{W}}_h^{-1}\underline{\underline{W}}^-(0) - \underline{\underline{V}}_h^{-1}\underline{\underline{V}}^-(0)
\end{pmatrix}^{-1}
$$

$$
\times \begin{pmatrix} 2\vec{\underline{U}} \\ \underline{\underline{0}} \end{pmatrix}
\tag{5.18a}
$$

The reflection and transmission operators $\overset{\leftrightarrow}{\underline{\underline{R}}}$ and $\overset{\leftrightarrow}{\underline{\underline{T}}}$ are obtained, respectively, by

$$\underline{\underline{\tilde{R}}} = \underline{\underline{W}}_h^{-1}\left[\underline{\underline{W}}^+(0)\underline{\underline{C}}_a^+ + \underline{\underline{W}}^-(z_- - z_+)\underline{\underline{C}}_a^- - \underline{\underline{W}}_h\underline{\underline{U}}\right] \tag{5.18b}$$

$$\underline{\underline{\tilde{T}}} = \underline{\underline{W}}_h^{-1}\left[\underline{\underline{W}}^+(z_+ - z_-)\underline{\underline{C}}_a^+ + \underline{\underline{W}}^-(0)\underline{\underline{C}}_a^-\right] \tag{5.18c}$$

The obtained matrix operators, $\underline{\underline{\tilde{R}}}$, $\underline{\underline{\tilde{T}}}$, $\underline{\underline{C}}_a^+$, and $\underline{\underline{C}}_a^-$, provide complete information on the left-to-right directional characteristics of the single-super-block. The coupling coefficient matrix operator $\underline{\underline{C}}_a$ is defined by

$$\underline{\underline{C}}_a = \begin{pmatrix} \underline{\underline{C}}_a^+ \\ \underline{\underline{C}}_a^- \end{pmatrix}.$$

In the same way, the right-to-left characterization is performed. In the case of the right-to-left characterization, the excitation field of the right boundary, $\vec{U}$, the reflection field $\vec{R}$, and the transmission field $\vec{T}$ are given, respectively, by

$$\begin{pmatrix} \vec{E}_i \\ \vec{H}_i \end{pmatrix} = \begin{pmatrix} \vec{E}_{i,x}, \vec{E}_{i,y}, \vec{E}_{i,z} \\ \vec{H}_{i,x}, \vec{H}_{i,y}, \vec{H}_{i,z} \end{pmatrix}$$

$$= \sum_{m=-M}^{M} \sum_{n=-N}^{N} \begin{pmatrix} \vec{E}_{i,x,m,n}, \vec{E}_{i,y,m,n}, \vec{E}_{i,z,m,n} \\ \vec{H}_{i,x,m,n}, \vec{H}_{i,y,m,n}, \vec{H}_{i,z,m,n} \end{pmatrix} e^{j(k_{x,m,n}x + k_{ym,n}y - k_{z,m,n}(z-z_+))} \tag{5.19a}$$

$$\begin{pmatrix} \vec{E}_r \\ \vec{H}_r \end{pmatrix} = \begin{pmatrix} \vec{E}_{r,x}, \vec{E}_{r,y}, \vec{E}_{r,z} \\ \vec{H}_{r,x}, \vec{H}_{r,y}, \vec{H}_{r,z} \end{pmatrix}$$

$$= \sum_{m=-M}^{M} \sum_{n=-N}^{N} \begin{pmatrix} \vec{E}_{r,x,m,n}, \vec{E}_{r,y,m,n}, \vec{E}_{r,z,m,n} \\ \vec{H}_{r,x,m,n}, \vec{H}_{r,y,m,n}, \vec{H}_{r,z,m,n} \end{pmatrix} e^{j(k_{x,m,n}x + k_{ym,n}y + k_{z,m,n}(z-z_+))} \tag{5.19b}$$

$$\begin{pmatrix} \vec{E}_t \\ \vec{H}_t \end{pmatrix} = \begin{pmatrix} \vec{E}_{t,x}, \vec{E}_{t,y}, \vec{E}_{t,z} \\ \vec{H}_{t,x}, \vec{H}_{t,y}, \vec{H}_{t,z} \end{pmatrix}$$

$$= \sum_{m=-M}^{M} \sum_{n=-N}^{N} \begin{pmatrix} \vec{E}_{t,x,m,n}, \vec{E}_{t,y,m,n}, \vec{E}_{t,z,m,n} \\ \vec{H}_{t,x,m,n}, \vec{H}_{t,y,m,n}, \vec{H}_{t,z,m,n} \end{pmatrix} e^{j(k_{x,m,n}x + k_{y,m,n}y - k_{z,m,n}(z-z_-))} \tag{5.19c}$$

The boundary conditions at the left and right boundaries are described, respectively, as

$$
\begin{pmatrix} \underline{\underline{W}}_h & \underline{\underline{W}}_h \\ \underline{\underline{V}}_h & -\underline{\underline{V}}_h \end{pmatrix} \begin{pmatrix} \underline{0} \\ \vec{\underline{T}} \end{pmatrix} = \begin{pmatrix} \underline{\underline{W}}^+(0) & \underline{\underline{W}}^-(z_- - z_+) \\ \underline{\underline{V}}^+(0) & \underline{\underline{V}}^-(z_- - z_+) \end{pmatrix} \begin{pmatrix} \underline{C}_b^+ \\ \underline{C}_b^- \end{pmatrix}
\tag{5.20a}
$$

$$
\begin{pmatrix} \underline{\underline{W}}^+(z_+ - z_-) & \underline{\underline{W}}^-(0) \\ \underline{\underline{V}}^+(z_+ - z_-) & \underline{\underline{V}}^-(0) \end{pmatrix} \begin{pmatrix} \underline{C}_b^+ \\ \underline{C}_b^- \end{pmatrix} = \begin{pmatrix} \underline{\underline{W}}_h & \underline{\underline{W}}_h \\ \underline{\underline{V}}_h & -\underline{\underline{V}}_h \end{pmatrix} \begin{pmatrix} \vec{\underline{R}} \\ \vec{\underline{U}} \end{pmatrix}
\tag{5.20b}
$$

The coupling coefficient matrix operators $\underline{C}_b^+$ and $\underline{C}_b^-$ are obtained by

$$
\begin{pmatrix} \underline{C}_b^+ \\ \underline{C}_b^- \end{pmatrix} = \begin{pmatrix} \underline{\underline{W}}_h^{-1}\underline{\underline{W}}^+(0) + \underline{\underline{V}}_h^{-1}\underline{\underline{V}}^+(0) & \underline{\underline{W}}_h^{-1}\underline{\underline{W}}^-(z_- - z_+) + \underline{\underline{V}}_h^{-1}\underline{\underline{V}}^-(z_- - z_+) \\ \underline{\underline{W}}_h^{-1}\underline{\underline{W}}^+(z_+ - z_-) - \underline{\underline{V}}_h^{-1}\underline{\underline{V}}^+(z_+ - z_-) & \underline{\underline{W}}_h^{-1}\underline{\underline{W}}^-(0) - \underline{\underline{V}}_h^{-1}\underline{\underline{V}}^-(0) \end{pmatrix}^{-1}
$$

$$
\times \begin{pmatrix} \underline{0} \\ 2\vec{\underline{U}} \end{pmatrix}
\tag{5.21a}
$$

The reflection and transmission coefficient matrix operators $\vec{\underline{R}}$ and $\vec{\underline{T}}$ are obtained, respectively, as

$$
\vec{\underline{R}} = \underline{\underline{W}}_h^{-1}\left[\underline{\underline{W}}^+(z_+ - z_-)\underline{C}_b^+ + \underline{\underline{W}}^-(0)\underline{C}_b^- - \underline{\underline{W}}_h\vec{\underline{U}} \right]
\tag{5.21b}
$$

$$
\vec{\underline{T}} = \underline{\underline{W}}_h^{-1}\left[\underline{\underline{W}}^+(0)\underline{C}_b^+ + \underline{\underline{W}}^-(z_- - z_+)\underline{C}_b^- \right]
\tag{5.21c}
$$

The obtained matrix operators, $\vec{\underline{R}}$, $\vec{\underline{T}}$, $\underline{C}_b^+$, and $\underline{C}_b^-$ provide complete information on the left-to-right directional characteristics of the single block. The coupling coefficient matrix operator $\underline{\underline{C}}_b$ is defined by

$$
\underline{\underline{C}}_b = \begin{pmatrix} \underline{C}_b^+ \\ \underline{C}_b^- \end{pmatrix}.
$$

5.2 Local Fourier Modal Analysis of Collinear Multi-Super-Block Structures

This section describes the collinear interconnection between multi-super-block structures, as determined by the S-matrix method. A multi-super-block is comprised of optical structures that are composed of collinearly cascaded optical single-super-blocks. The algorithm for obtaining the electromagnetic field distribution in the multi-super-block structure is the same S-matrix method as is used in FMM. It should be noted that the only difference between FMM and LFMM is the eigenmode representation. FMM has the 2D pseudo-Fourier representation as the mathematical basis, while LFMM has the 3D pseudo-Fourier representation for the mathematical basis. In this section, the general analysis scheme with MATLAB implementation of a multi-super-block structure based on LFMA and the S-matrix method is described.

The reader is referred to the schematic diagram of the multiblock structure shown in Figure 5.6. Here, each single block comprising the multiblock structure is a single-super-block structure characterized by LFMA. In Figure 5.6(b), the finite-sized triangular photonic crystal can be considered the multi-super-block model. In the similar manner with FMM, the S-matrices and coupling coefficient matrix operators of single-super-blocks separately characterized by LFMA are combined by the recursion formulas established in the previous chapters. Thus we can apply the S-matrix method in a straightforward manner, without modification of LFMA for analyzing multi-super-block structures.

With LFMA, the reflection and transmission characteristics of the finite-sized 2D photonic crystal waveguide structure shown in Figure 5.6 are analyzed. Figure 5.7(a) illustrates that a plane wave is normally incident from the left free space on the left endface of the photonic crystal waveguide and a diffraction field distribution is generated in the right free space region. The y-polarization electric field, x-polarization magnetic electric field, and z-polarization magnetic field distributions are shown in Figure 5.7(b) to (d), respectively.

With LFMA, the reflection and transmission characteristics of the guided Bloch eigenmode of the half-infinite two-dimensional photonic crystal waveguide structure previously presented are analyzed. As shown in Figure 5.8(a), the guided Bloch eigenmode is incident on the endface of the photonic crystal waveguide. The backward propagating guided Bloch mode is reflected and the diffraction field distribution is generated at the interface of the endface of the photonic crystal waveguide and free space. The y-polarization electric field, x-polarization magnetic field, and z-polarization magnetic field distributions are shown in Figure 5.8(b) to (d), respectively. In Figure 5.9, the excitation of the guided Bloch eigenmode by a normally incident plane wave from the free space region to the interface

(a)

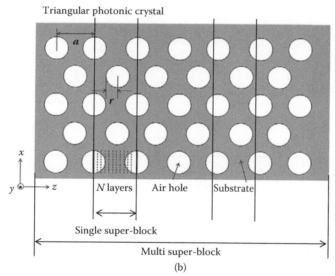

(b)

FIGURE 5.6
(a) Multi-super-block and (b) finite-sized triangular photonic crystal as a multi-super-block.

is presented. The y-polarization electric field, x-polarization magnetic field, and z-polarization magnetic field distributions are shown in Figure 5.9(b) to (d), respectively.

5.2.1 Field Visualization

LFMM has almost the same mathematical structure as FMM except the basis form of the Bloch eigenmodes. Hence, the field visualization of

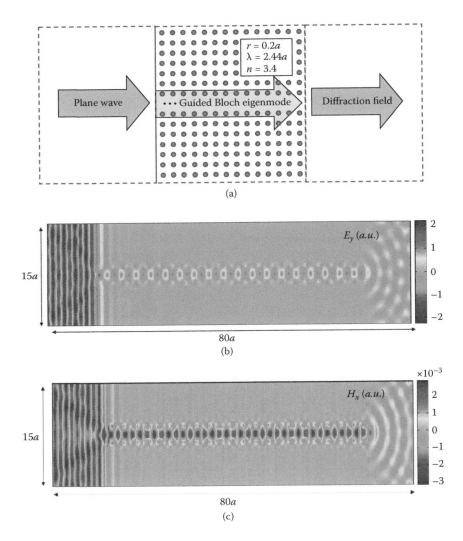

(a)

(b)

(c)

FIGURE 5.7
Transmission and reflection of two-dimensional finite-sized photonic crystal waveguide by a normally incident plane wave: (a) simulation schematic, (b) y-polarization electric field distribution, (c) x-polarization magnetic field distribution, and (d) z-polarization magnetic field distribution. (From H. Kim and B. Lee, *J. Opt. Soc. Am. B*, 25(4), 2008. With permission.)

LFMM holds similar forms to that of FMM. The analysis can be classified into four cases with respect to four possible combinations of finite-sized multiblock and half-infinite blocks: (A) half-infinite homogeneous space–multiblock structure–half-infinite homogeneous space, (B) half-infinite homogeneous space–multiblock structure–half-infinite inhomogeneous space,

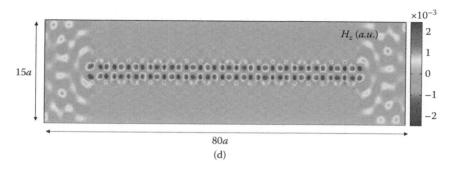

(d)

FIGURE 5.7
(Continued)

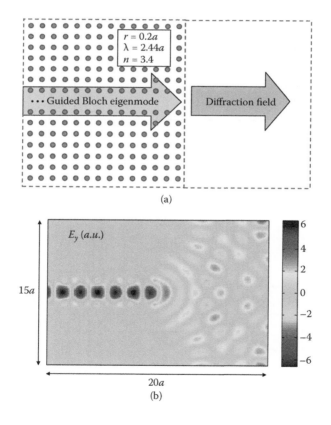

FIGURE 5.8
Diffraction of the guided Bloch eigenmode at the right endface of the two-dimensional half-infinite photonic crystal structure: (a) simulation schematic, (b) y-polarization electric field distribution, (c) x-polarization magnetic field distribution, and (d) z-polarization magnetic field distribution. (From H. Kim and B. Lee, *J. Opt. Soc. Am. B*, 25(4), 2008. With permission.)

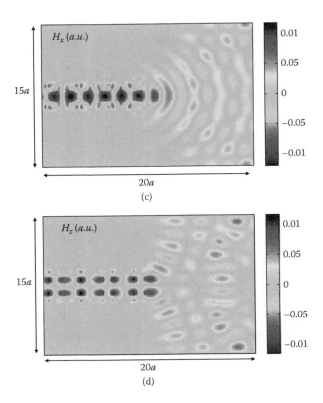

(c)

(d)

FIGURE 5.8
(Continued)

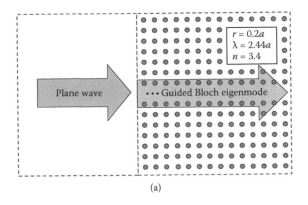

(a)

FIGURE 5.9
Excitation of the guided Bloch eigenmode at the left endface of the two-dimensional half-infinite photonic crystal structure: (a) simulation schematic, (b) *y*-polarization electric field distribution, (c) *x*-polarization magnetic field distribution, and (d) *z*-polarization magnetic field distribution. (From H. Kim and B. Lee, *J. Opt. Soc. Am. B*, 25(4), 2008. With permission.)

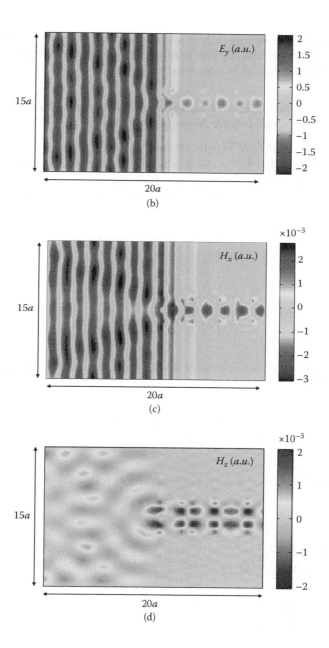

FIGURE 5.9
(Continued)

(C) half-infinite inhomogeneous space–multiblock structure–half-infinite homogeneous space, and (D) half-infinite inhomogeneous space–multiblock structure–half-infinite inhomogeneous space.

5.2.1.1 Case A: Half-Infinite Homogeneous Space–Multiblock Structure–Half-Infinite Homogeneous Space

5.2.1.1.1 Left-to-Right Field Visualization

For the incident field, $\vec{E}_i$, in the left half-infinite block, the reflection field coefficient inside the left half-infinite block and the transmission field coefficient inside the right half-infinite block are given, respectively, as

$$
\begin{pmatrix} \tilde{E}_{r,y} \\ \tilde{E}_{r,x} \end{pmatrix} = \tilde{\underline{\underline{R}}}^{(0,N+1)} \begin{pmatrix} \tilde{E}_{i,y} \\ \tilde{E}_{i,x} \end{pmatrix}
\tag{5.22a}
$$

$$
\begin{pmatrix} \tilde{E}_{t,y} \\ \tilde{E}_{t,x} \end{pmatrix} = \tilde{\underline{\underline{T}}}^{(0,N+1)} \begin{pmatrix} \tilde{E}_{i,y} \\ \tilde{E}_{i,x} \end{pmatrix}
\tag{5.22b}
$$

And the coupling coefficients of the internal optical field distribution are represented by

$$
\left\{ \underline{C}_{a,(1)}^{(1,N)}, \underline{C}_{a,(2)}^{(1,N)}, \ldots, \underline{C}_{a,(N)}^{(1,N)} \right\} = \left\{ \underline{\underline{C}}_{a,(1)}^{(1,N)} \begin{pmatrix} \tilde{E}_{i,y} \\ \tilde{E}_{i,x} \end{pmatrix}, \underline{\underline{C}}_{a,(2)}^{(1,N)} \begin{pmatrix} \tilde{E}_{i,y} \\ \tilde{E}_{i,x} \end{pmatrix}, \ldots, \underline{\underline{C}}_{a,(N)}^{(1,N)} \begin{pmatrix} \tilde{E}_{i,y} \\ \tilde{E}_{i,x} \end{pmatrix} \right\}
\tag{5.22c}
$$

The total field distributions are represented as follows:

1. For $z < z_-$

$$
\begin{pmatrix} \mathbf{E} \\ \mathbf{H} \end{pmatrix} = \begin{pmatrix} \vec{E}_i \\ \vec{H}_i \end{pmatrix} + \begin{pmatrix} \vec{E}_r \\ \vec{H}_r \end{pmatrix}
$$

$$
= \sum_{m=-M}^{M} \sum_{n=-N}^{N} \begin{pmatrix} \tilde{E}_{i,x,m,n}, \tilde{E}_{i,y,m,n}, \tilde{E}_{i,z,m,n} \\ \tilde{H}_{i,x,m,n}, \tilde{H}_{i,y,m,n}, \tilde{H}_{i,z,m,n} \end{pmatrix} e^{j(k_{x,m,n}x + k_{ym,n}y + k_{z,m,n}(z-z_-))}
$$

$$
+ \sum_{m=-M}^{M} \sum_{n=-N}^{N} \begin{pmatrix} \tilde{E}_{r,x,m,n}, \tilde{E}_{r,y,m,n}, \tilde{E}_{r,z,m,n} \\ \tilde{H}_{r,x,m,n}, \tilde{H}_{r,y,m,n}, \tilde{H}_{r,z,m,n} \end{pmatrix} e^{j(k_{x,m,n}x + k_{ym,n}y - k_{z,m,n}(z-z_-))}
$$

$$
\tag{5.23}
$$

2. For $z_- \leq z < z_+$

$$
\begin{pmatrix} \mathbf{E} \\ \mathbf{H} \end{pmatrix} = \begin{cases}
\displaystyle\sum_{g=1}^{M^+} C_{a,(1),g}^+ \begin{pmatrix} \mathbf{E}_{(1)}^{(g)+} \\ \mathbf{H}_{(1)}^{(g)+} \end{pmatrix} + \sum_{g=1}^{M^-} C_{a,(1),g}^- \begin{pmatrix} \mathbf{E}_{(1)}^{(g)-} \\ \mathbf{H}_{(1)}^{(g)-} \end{pmatrix} & \text{for } 0 \leq z \leq l_{1,1} \\[3em]
\displaystyle\sum_{g=1}^{M^+} C_{a,(2),g}^+ \begin{pmatrix} \mathbf{E}_{(2)}^{(g)+} \\ \mathbf{H}_{(2)}^{(g)+} \end{pmatrix} + \sum_{g=1}^{M^-} C_{a,(2),g}^- \begin{pmatrix} \mathbf{E}_{(2)}^{(g)-} \\ \mathbf{H}_{(2)}^{(g)-} \end{pmatrix} & \text{for } l_{1,1} \leq z \leq l_{1,2} \\[2em]
\qquad\qquad\qquad\qquad \vdots \\[1em]
\displaystyle\sum_{g=1}^{M^+} C_{a,(N),g}^+ \begin{pmatrix} \mathbf{E}_{(N)}^{(g)+} \\ \mathbf{H}_{(N)}^{(g)+} \end{pmatrix} + \sum_{g=1}^{M^-} C_{a,(N),g}^- \begin{pmatrix} \mathbf{E}_{(N)}^{(g)-} \\ \mathbf{H}_{(N)}^{(g)-} \end{pmatrix} & \text{for } l_{1,N-1} \leq z \leq l_{1,N}
\end{cases}
$$

(5.24a)

where $\mathbf{E}^{(g)+}$, $\mathbf{E}^{(g)-}$, $\mathbf{H}^{(g)+}$, and $\mathbf{H}^{(g)-}$ are given, respectively, by

$$
\mathbf{E}^{(g)+}(x,y,z) = \sum_{m=-M}^{M} \sum_{n=-N}^{N} \sum_{p=-H}^{H} \left(E_{x,m,n,p}^{(g)+}, E_{y,m,n,p}^{(g)+}, E_{z,m,n,p}^{(g)+} \right) e^{j\left(k_{x,m}x + k_{y,n}y + k_{z,p}^{(g)+}(z-z_-)\right)}
$$

(5.24b)

$$
\mathbf{E}^{(g)-}(x,y,z) = \sum_{m=-M}^{M} \sum_{n=-N}^{N} \sum_{p=-H}^{H} \left(E_{x,m,n,p}^{(g)-}, E_{y,m,n,p}^{(g)-}, E_{z,m,n,p}^{(g)-} \right) e^{j\left(k_{x,m}x + k_{y,n}y + k_{z,p}^{(g)-}(z-z_-)\right)}
$$

(5.24c)

$$
\mathbf{H}^{(g)+}(x,y,z) = \sum_{m=-M}^{M} \sum_{n=-N}^{N} \sum_{p=-H}^{H} \left(H_{x,m,n,p}^{(g)+}, H_{y,m,n,p}^{(g)+}, H_{z,m,n,p}^{(g)+} \right) e^{j\left(k_{x,m}x + k_{y,n}y + k_{z,p}^{(g)+}(z-z_+)\right)}
$$

(5.24d)

$$
\mathbf{H}^{(g)-}(x,y,z) = \sum_{m=-M}^{M} \sum_{n=-N}^{N} \sum_{p=-H}^{H} \left(H_{x,m,n,p}^{(g)-}, H_{y,m,n,p}^{(g)-}, H_{z,m,n,p}^{(g)-} \right) e^{j\left(k_{x,m}x + k_{y,n}y + k_{z,p}^{(g)-}(z-z_-)\right)}
$$

(5.24e)

3. For $z_+ \leq z$

$$
\begin{pmatrix} \vec{\mathbf{E}}_t \\ \vec{\mathbf{H}}_t \end{pmatrix} = \begin{pmatrix} \vec{\mathbf{E}}_{t,x}, \vec{\mathbf{E}}_{t,y}, \vec{\mathbf{E}}_{t,z} \\ \vec{\mathbf{H}}_{t,x}, \vec{\mathbf{H}}_{t,y}, \vec{\mathbf{H}}_{t,z} \end{pmatrix}
$$

$$
= \sum_{m=-M}^{M} \sum_{n=-N}^{N} \begin{pmatrix} \vec{E}_{t,x,m,n}, \vec{E}_{t,y,m,n}, \vec{E}_{t,z,m,n} \\ \vec{H}_{t,x,m,n}, \vec{H}_{t,y,m,n}, \vec{H}_{t,z,m,n} \end{pmatrix} e^{j(k_{x,m,n}x + k_{y,m,n}y + k_{z,m,n}(z-z_+))}
$$

(5.25)

The incident electromagnetic power at the boundary B_0 is equal to the sum of the reflection power normal to the boundary B_0 and the transmission power normal to the boundary B_N.

$$\sum_{m=-M}^{M}\sum_{n=-N}^{N}\left(\left|\bar{E}_{r,x,m,n}\right|^2+\left|\bar{E}_{r,y,m,n}\right|^2+\left|\bar{E}_{r,z,m,n}\right|^2\right)\mathrm{Re}\left(k_{z,m,n}^{(0)}/k_{z,0,0}^{(0)}\right)$$

$$+\sum_{m=-M}^{M}\sum_{n=-N}^{N}\left(\left|\bar{E}_{t,x,m,n}\right|^2+\left|\bar{E}_{t,y,m,n}\right|^2+\left|\bar{E}_{t,z,m,n}\right|^2\right)\mathrm{Re}\left(k_{z,m,n}^{(N+1)}/k_{z,0,0}^{(0)}\right)=1 \qquad (5.26)$$

5.2.1.1.2 Right-to-Left Field Visualization

When the incident field, $\bar{\mathbf{E}}_i$, in the left half-infinite block impinges on the multiblock structure $M^{(1,N)}$ at the boundary B_N, the reflection field coefficient inside the left half-infinite block and the transmission field coefficient inside the right half-infinite block are given, respectively, as

$$\begin{pmatrix} \vec{E}_{r,y} \\ \vec{E}_{r,x} \end{pmatrix} = \vec{\underline{R}}^{(0,N+1)}\begin{pmatrix} \underline{E}_{i,y} \\ \underline{E}_{i,x} \end{pmatrix}, \qquad (5.27a)$$

$$\begin{pmatrix} \vec{E}_{t,y} \\ \vec{E}_{t,x} \end{pmatrix} = \vec{\underline{T}}^{(0,N+1)}\begin{pmatrix} \underline{E}_{i,y} \\ \underline{E}_{i,x} \end{pmatrix}. \qquad (5.27b)$$

And the coupling coefficients of the internal optical field distribution are represented by

$$\left\{\underline{C}_{b,(1)}^{(1,N)},\underline{C}_{b,(2)}^{(1,N)},\ldots,\underline{C}_{b,(N)}^{(1,N)}\right\}=\left\{\underline{\underline{C}}_{b,(1)}^{(1,N)}\begin{pmatrix}\underline{E}_{i,y}\\\underline{E}_{i,x}\end{pmatrix},\underline{\underline{C}}_{b,(2)}^{(1,N)}\begin{pmatrix}\underline{E}_{i,y}\\\underline{E}_{i,x}\end{pmatrix},\ldots,\underline{\underline{C}}_{b,(N)}^{(1,N)}\begin{pmatrix}\underline{E}_{i,y}\\\underline{E}_{i,x}\end{pmatrix}\right\}. \qquad (5.27c)$$

The total field distributions are represented as follows:

1. For $z < z_-$

$$\begin{pmatrix} \bar{\mathbf{E}}_t \\ \bar{\mathbf{H}}_t \end{pmatrix} = \begin{pmatrix} \bar{\mathbf{E}}_{t,x},\bar{\mathbf{E}}_{t,y},\bar{\mathbf{E}}_{t,z} \\ \bar{\mathbf{H}}_{t,x},\bar{\mathbf{H}}_{t,y},\bar{\mathbf{H}}_{t,z}, \end{pmatrix}$$

$$= \sum_{m=-M}^{M}\sum_{n=-N}^{N}\begin{pmatrix} \bar{E}_{t,x,m,n},\bar{E}_{t,y,m,n},\bar{E}_{t,z,m,n} \\ \bar{H}_{t,x,m,n},\bar{H}_{t,y,m,n},\bar{H}_{t,z,m,n} \end{pmatrix}e^{j(k_{x,m,n}x+k_{y,m,n}y-k_{z,m,n}(z-z_-))} \qquad (5.28)$$

2. For $z \leq z < z_+$

$$
\begin{pmatrix} \mathbf{E} \\ \mathbf{H} \end{pmatrix} = \begin{cases}
\displaystyle\sum_{g=1}^{M^+} C^+_{b,(1),g} \begin{pmatrix} \mathbf{E}^{(g)+}_{(1)} \\ \mathbf{H}^{(g)+}_{(1)} \end{pmatrix} + \sum_{g=1}^{M^-} C^-_{b,(1),g} \begin{pmatrix} \mathbf{E}^{(g)-}_{(1)} \\ \mathbf{H}^{(g)-}_{(1)} \end{pmatrix} & \text{for } 0 \leq z \leq l_{1,1} \\[3em]
\displaystyle\sum_{g=1}^{M^+} C^+_{b,(2),g} \begin{pmatrix} \mathbf{E}^{(g)+}_{(2)} \\ \mathbf{H}^{(g)+}_{(2)} \end{pmatrix} + \sum_{g=1}^{M^-} C^-_{b,(2),g} \begin{pmatrix} \mathbf{E}^{(g)-}_{(2)} \\ \mathbf{H}^{(g)-}_{(2)} \end{pmatrix} & \text{for } l_{1,1} \leq z \leq l_{1,2} \\[3em]
\qquad\qquad\qquad\qquad \vdots \\[1em]
\displaystyle\sum_{g=1}^{M^+} C^+_{b,(N),g} \begin{pmatrix} \mathbf{E}^{(g)+}_{(N)} \\ \mathbf{H}^{(g)+}_{(N)} \end{pmatrix} + \sum_{g=1}^{M^-} C^-_{b,(N),g} \begin{pmatrix} \mathbf{E}^{(g)-}_{(N)} \\ \mathbf{H}^{(g)-}_{(N)} \end{pmatrix} & \text{for } l_{1,N-1} \leq z \leq l_{1,N}
\end{cases}
$$

$$(5.29)$$

where $\mathbf{E}^{(g)+}$, $\mathbf{E}^{(g)-}$, $\mathbf{H}^{(g)+}$, and $\mathbf{H}^{(g)-}$ are given in Equations (5.24b)~(5.24e).

3. For $z_+ \leq z$

$$
\begin{pmatrix} \mathbf{E} \\ \mathbf{H} \end{pmatrix} = \begin{pmatrix} \vec{\mathbf{E}}_i \\ \vec{\mathbf{H}}_i \end{pmatrix} + \begin{pmatrix} \vec{\mathbf{E}}_r \\ \vec{\mathbf{H}}_r \end{pmatrix}
$$

$$
= \sum_{m=-M}^{M} \sum_{n=-N}^{N} \begin{pmatrix} \vec{E}_{i,x,m,n}, \vec{E}_{i,y,m,n}, \vec{E}_{i,z,m,n} \\ \vec{H}_{i,x,m,n}, \vec{H}_{i,y,m,n}, \vec{H}_{i,z,m,n} \end{pmatrix} e^{j(k_{x,m,n}x + k_{ym,n}y - k_{z,m,n}(z-z_+))}
$$

$$
+ \sum_{m=-M}^{M} \sum_{n=-N}^{N} \begin{pmatrix} \vec{E}_{r,x,m,n}, \vec{E}_{r,y,m,n}, \vec{E}_{r,z,m,n} \\ \vec{H}_{r,x,m,n}, \vec{H}_{r,y,m,n}, \vec{H}_{r,z,m,n} \end{pmatrix} e^{j(k_{x,m,n}x + k_{ym,n}y + k_{z,m,n}(z-z_+))} \quad (5.30)
$$

The incident electromagnetic power at the boundary B_N is normalized by 1, and is equal to the sum of the reflection power normal to the boundary B_N and the transmission power normal to the boundary B_0.

$$
\sum_{m=-M}^{M} \sum_{n=-N}^{N} \left(\left| \vec{E}_{r,x,m,n} \right|^2 + \left| \vec{E}_{r,y,m,n} \right|^2 + \left| \vec{E}_{r,z,m,n} \right|^2 \right) \mathrm{Re}\left(k^{(0)}_{z,m,n} / k^{(N+1)}_{z,0,0} \right)
$$

$$
+ \sum_{m=-M}^{M} \sum_{n=-N}^{N} \left(\left| \vec{E}_{t,x,m,n} \right|^2 + \left| \vec{E}_{t,y,m,n} \right|^2 + \left| \vec{E}_{t,z,m,n} \right|^2 \right) \mathrm{Re}\left(k^{(N+1)}_{z,m,n} / k^{(N+1)}_{z,0,0} \right) = 1 \quad (5.31)
$$

5.2.1.2 Case B: Semi-Infinite Homogeneous Space–Multiblock Structure–Semi-Infinite Inhomogeneous Space

5.2.1.2.1 Left-to-Right Field Visualization

$$
\begin{pmatrix} \tilde{\vec{E}}_{r,y} \\ \tilde{\vec{E}}_{r,x} \end{pmatrix} = \tilde{\underline{\underline{R}}}^{(0,N+1)} \begin{pmatrix} \tilde{\vec{E}}_{i,y} \\ \tilde{\vec{E}}_{i,x} \end{pmatrix} \tag{5.32a}
$$

$$
\vec{t} = \tilde{\underline{\underline{T}}}^{(0,N+1)} \begin{pmatrix} \tilde{\vec{E}}_{i,y} \\ \tilde{\vec{E}}_{i,x} \end{pmatrix} \tag{5.32b}
$$

$$
\left\{ \underline{C}_{a,(1)}^{(1,N)}, \underline{C}_{a,(2)}^{(1,N)}, \cdots, \underline{C}_{a,(N)}^{(1,N)} \right\} = \left\{ \underline{C}_{a,(1)}^{(1,N)} \begin{pmatrix} \tilde{\vec{E}}_{i,y} \\ \tilde{\vec{E}}_{i,x} \end{pmatrix}, \underline{C}_{a,(2)}^{(1,N)} \begin{pmatrix} \tilde{\vec{E}}_{i,y} \\ \tilde{\vec{E}}_{i,x} \end{pmatrix}, \cdots, \underline{C}_{a,(N)}^{(1,N)} \begin{pmatrix} \tilde{\vec{E}}_{i,y} \\ \tilde{\vec{E}}_{i,x} \end{pmatrix} \right\} \tag{5.32c}
$$

The total field distributions are represented as follows:

1. For $z < z_-$

$$
\begin{pmatrix} \mathbf{E} \\ \mathbf{H} \end{pmatrix} = \begin{pmatrix} \tilde{\vec{E}}_i \\ \tilde{\vec{H}}_i \end{pmatrix} + \begin{pmatrix} \tilde{\vec{E}}_r \\ \tilde{\vec{H}}_r \end{pmatrix}
$$

$$
= \sum_{m=-M}^{M} \sum_{n=-N}^{N} \begin{pmatrix} \tilde{\vec{E}}_{i,x,m,n}, \tilde{\vec{E}}_{i,y,m,n}, \tilde{\vec{E}}_{i,z,m,n} \\ \tilde{\vec{H}}_{i,x,m,n}, \tilde{\vec{H}}_{i,y,m,n}, \tilde{\vec{H}}_{i,z,m,n} \end{pmatrix} e^{j(k_{x,m,n}x + k_{ym,n}y + k_{z,m,n}(z-z_-))}
$$

$$
+ \sum_{m=-M}^{M} \sum_{n=-N}^{N} \begin{pmatrix} \tilde{\vec{E}}_{r,x,m,n}, \tilde{\vec{E}}_{r,y,m,n}, \tilde{\vec{E}}_{r,z,m,n} \\ \tilde{\vec{H}}_{r,x,m,n}, \tilde{\vec{H}}_{r,y,m,n}, \tilde{\vec{H}}_{r,z,m,n} \end{pmatrix} e^{j(k_{x,m,n}x + k_{ym,n}y - k_{z,m,n}(z-z_-))} \tag{5.33}
$$

2. For $z_- \le z < z_+$

$$
\begin{pmatrix} \mathbf{E} \\ \mathbf{H} \end{pmatrix} = \begin{cases} \displaystyle\sum_{g=1}^{M^+} C_{a,(1),g}^+ \begin{pmatrix} \mathbf{E}_{(1)}^{(g)+} \\ \mathbf{H}_{(1)}^{(g)+} \end{pmatrix} + \sum_{g=1}^{M^-} C_{a,(1),g}^- \begin{pmatrix} \mathbf{E}_{(1)}^{(g)-} \\ \mathbf{H}_{(1)}^{(g)-} \end{pmatrix} & \text{for } 0 \le z \le l_{1,1} \\[3ex] \displaystyle\sum_{g=1}^{M^+} C_{a,(2),g}^+ \begin{pmatrix} \mathbf{E}_{(2)}^{(g)+} \\ \mathbf{H}_{(2)}^{(g)+} \end{pmatrix} + \sum_{g=1}^{M^-} C_{a,(2),g}^- \begin{pmatrix} \mathbf{E}_{(2)}^{(g)-} \\ \mathbf{H}_{(2)}^{(g)-} \end{pmatrix} & \text{for } l_{1,1} \le z \le l_{1,2} \\[1ex] \vdots \\[1ex] \displaystyle\sum_{g=1}^{M^+} C_{a,(N),g}^+ \begin{pmatrix} \mathbf{E}_{(N)}^{(g)+} \\ \mathbf{H}_{(N)}^{(g)+} \end{pmatrix} + \sum_{g=1}^{M^-} C_{a,(N),g}^- \begin{pmatrix} \mathbf{E}_{(N)}^{(g)-} \\ \mathbf{H}_{(N)}^{(g)-} \end{pmatrix} & \text{for } l_{1,N-1} \le z \le l_{1,N} \end{cases} \tag{5.34}
$$

3. For $z_+ \leq z$

$$
\begin{pmatrix} \vec{E}_t \\ \vec{H}_t \end{pmatrix} = \sum_{g=1}^{M^+} \vec{t}_g \begin{pmatrix} \mathbf{E}_{(N+1)}^{(g)+} \\ \mathbf{H}_{(N+1)}^{(g)+} \end{pmatrix}
\tag{5.35}
$$

5.2.1.2.2 Right-to-Left Field Visualization

$$
\vec{r} = \vec{\underline{\underline{R}}}^{(0,N+1)} \vec{\tilde{u}}
\tag{5.36a}
$$

$$
\begin{pmatrix} \tilde{E}_{t,y} \\ \tilde{E}_{t,x} \end{pmatrix} = \vec{\underline{\underline{T}}}^{(0,N+1)} \vec{\tilde{u}}
\tag{5.36b}
$$

$$
\left\{ \underline{C}_{b,(1)}^{(1,N)}, \underline{C}_{b,(2)}^{(1,N)}, \ldots, \underline{C}_{b,(N)}^{(1,N)} \right\} = \left\{ \underline{\underline{C}}_{b,(1)}^{(1,N)} \vec{\tilde{u}}, \underline{\underline{C}}_{b,(2)}^{(1,N)} \vec{\tilde{u}}, \cdots, \underline{\underline{C}}_{b,(N)}^{(1,N)} \vec{\tilde{u}} \right\}
\tag{5.36c}
$$

The total field distributions are represented as follows:

1. For $z < z_-$

$$
\begin{pmatrix} \tilde{\mathbf{E}}_t \\ \tilde{\mathbf{H}}_t \end{pmatrix} = \begin{pmatrix} \tilde{E}_{t,x}, \tilde{E}_{t,y}, \tilde{E}_{t,z} \\ \tilde{H}_{t,x}, \tilde{H}_{t,y}, \tilde{H}_{t,z} \end{pmatrix}
$$

$$
= \sum_{m=-M}^{M} \sum_{n=-N}^{N} \begin{pmatrix} \tilde{E}_{t,x,m,n}, \tilde{E}_{t,y,m,n}, \tilde{E}_{t,z,m,n} \\ \tilde{H}_{t,x,m,n}, \tilde{H}_{t,y,m,n}, \tilde{H}_{t,z,m,n} \end{pmatrix} e^{j(k_{x,m,n}x + k_{y,m,n}y - k_{z,m,n}(z-z_-))},
\tag{5.37}
$$

2. For $z_- \leq z < z_+$

$$
\begin{pmatrix} \mathbf{E} \\ \mathbf{H} \end{pmatrix} = \begin{cases}
\displaystyle\sum_{g=1}^{M^+} C_{b,(1),g}^+ \begin{pmatrix} \mathbf{E}_{(1)}^{(g)+} \\ \mathbf{H}_{(1)}^{(g)+} \end{pmatrix} + \sum_{g=1}^{M^-} C_{b,(1),g}^- \begin{pmatrix} \mathbf{E}_{(1)}^{(g)-} \\ \mathbf{H}_{(1)}^{(g)-} \end{pmatrix} & \text{for } 0 \leq z \leq l_{1,1} \\[3ex]
\displaystyle\sum_{g=1}^{M^+} C_{b,(2),g}^+ \begin{pmatrix} \mathbf{E}_{(2)}^{(g)+} \\ \mathbf{H}_{(2)}^{(g)+} \end{pmatrix} + \sum_{g=1}^{M^-} C_{b,(2),g}^- \begin{pmatrix} \mathbf{E}_{(2)}^{(g)-} \\ \mathbf{H}_{(2)}^{(g)-} \end{pmatrix} & \text{for } l_{1,1} \leq z \leq l_{1,2} \\[3ex]
\qquad\qquad\qquad \vdots & \\[2ex]
\displaystyle\sum_{g=1}^{M^+} C_{b,(N),g}^+ \begin{pmatrix} \mathbf{E}_{(N)}^{(g)+} \\ \mathbf{H}_{(N)}^{(g)+} \end{pmatrix} + \sum_{g=1}^{M^-} C_{b,(N),g}^- \begin{pmatrix} \mathbf{E}_{(N)}^{(g)-} \\ \mathbf{H}_{(N)}^{(g)-} \end{pmatrix} & \text{for } l_{1,N-1} \leq z \leq l_{1,N}
\end{cases}
\tag{5.38}
$$

3. For $z_+ \leq z$

$$
\begin{pmatrix} \mathbf{E} \\ \mathbf{H} \end{pmatrix} = \sum_{g=1}^{M^-} \vec{u}_g \begin{pmatrix} \mathbf{E}_{(N+1)}^{(g)-} \\ \mathbf{H}_{(N+1)}^{(g)-} \end{pmatrix} + \sum_{g=1}^{M^+} \vec{r}_g \begin{pmatrix} \mathbf{E}_{(N+1)}^{(g)+} \\ \mathbf{H}_{(N+1)}^{(g)+} \end{pmatrix}
\tag{5.39}
$$

5.2.1.3 Case C: Semi-Infinite Inhomogeneous Waveguide–Multiblock Structure–Semi-Infinite Homogeneous Space

5.2.1.3.1 Left-to-Right Field Visualization

$$\vec{r} = \underline{\underline{\tilde{R}}}^{(0,N+1)}\vec{u} \tag{5.40a}$$

$$\begin{pmatrix} \vec{E}_{t,y} \\ \vec{E}_{t,x} \end{pmatrix} = \underline{\underline{\vec{T}}}^{(0,N+1)}\vec{u} \tag{5.40b}$$

$$\left\{ \underline{C}_{a,(1)}^{(1,N)}, \underline{C}_{a,(2)}^{(1,N)}, \ldots, \underline{C}_{a,(N)}^{(1,N)} \right\} = \left\{ \underline{\underline{C}}_{a,(1)}^{(1,N)}\vec{u}, \underline{\underline{C}}_{a,(2)}^{(1,N)}\vec{u}, \ldots, \underline{\underline{C}}_{a,(N)}^{(1,N)}\vec{u} \right\} \tag{5.40c}$$

1. For $z < z_-$

$$\begin{pmatrix} \mathbf{E} \\ \mathbf{H} \end{pmatrix} = \sum_{g=1}^{M^+} \vec{u}_g \begin{pmatrix} \mathbf{E}_{(0)}^{(g)+} \\ \mathbf{H}_{(0)}^{(g)+} \end{pmatrix} + \sum_{g=1}^{M^-} \tilde{r}_g \begin{pmatrix} \mathbf{E}_{(0)}^{(g)-} \\ \mathbf{H}_{(0)}^{(g)-} \end{pmatrix} \tag{5.41}$$

2. For $z_- \leq z < z_+$

$$\begin{pmatrix} \mathbf{E} \\ \mathbf{H} \end{pmatrix} = \begin{cases} \displaystyle\sum_{g=1}^{M^+} C_{a,(1),g}^+ \begin{pmatrix} \mathbf{E}_{(1)}^{(g)+} \\ \mathbf{H}_{(1)}^{(g)+} \end{pmatrix} + \sum_{g=1}^{M^-} C_{a,(1),g}^- \begin{pmatrix} \mathbf{E}_{(1)}^{(g)-} \\ \mathbf{H}_{(1)}^{(g)-} \end{pmatrix} & \text{for } 0 \leq z \leq l_{1,1} \\[2em] \displaystyle\sum_{g=1}^{M^+} C_{a,(2),g}^+ \begin{pmatrix} \mathbf{E}_{(2)}^{(g)+} \\ \mathbf{H}_{(2)}^{(g)+} \end{pmatrix} + \sum_{g=1}^{M^-} C_{a,(2),g}^- \begin{pmatrix} \mathbf{E}_{(2)}^{(g)-} \\ \mathbf{H}_{(2)}^{(g)-} \end{pmatrix} & \text{for } l_{1,1} \leq z \leq l_{1,2} \\[1em] \qquad\qquad\qquad\qquad \vdots \\[1em] \displaystyle\sum_{g=1}^{M^+} C_{a,(N),g}^+ \begin{pmatrix} \mathbf{E}_{(N)}^{(g)+} \\ \mathbf{H}_{(N)}^{(g)+} \end{pmatrix} + \sum_{g=1}^{M^-} C_{a,(N),g}^- \begin{pmatrix} \mathbf{E}_{(N)}^{(g)-} \\ \mathbf{H}_{(N)}^{(g)-} \end{pmatrix} & \text{for } l_{1,N-1} \leq z \leq l_{1,N} \end{cases}$$

$$\tag{5.42}$$

3. For $z_+ \leq z$

$$\begin{pmatrix} \mathbf{E} \\ \mathbf{H} \end{pmatrix} = \sum_{g=1}^{M^+} \vec{t}_g \begin{pmatrix} \mathbf{E}_{(N+1)}^{(g)+} \\ \mathbf{H}_{(N+1)}^{(g)+} \end{pmatrix} \tag{5.43}$$

5.2.1.3.2 Right-to-Left Field Visualization

$$\begin{pmatrix} \vec{E}_{r,y} \\ \vec{E}_{r,x} \end{pmatrix} = \underline{\underline{\vec{R}}}^{(0,N+1)} \begin{pmatrix} \overleftarrow{E}_{i,y} \\ \overleftarrow{E}_{i,x} \end{pmatrix} \tag{5.44a}$$

$$\underline{t} = \underline{\underline{T}}^{(0,N+1)} \begin{pmatrix} \tilde{E}_{i,y} \\ \tilde{E}_{i,x} \end{pmatrix} \tag{5.44b}$$

And the coupling coefficients of the internal optical field distribution are represented by

$$\left\{ \underline{C}_{b,(1)}^{(1,N)}, \underline{C}_{b,(2)}^{(1,N)}, \ldots, \underline{C}_{b,(N)}^{(1,N)} \right\} = \left\{ \underline{\underline{C}}_{b,(1)}^{(1,N)} \begin{pmatrix} \tilde{E}_{i,y} \\ \tilde{E}_{i,x} \end{pmatrix}, \underline{\underline{C}}_{b,(2)}^{(1,N)} \begin{pmatrix} \tilde{E}_{i,y} \\ \tilde{E}_{i,x} \end{pmatrix}, \ldots, \underline{\underline{C}}_{b,(N)}^{(1,N)} \begin{pmatrix} \tilde{E}_{i,y} \\ \tilde{E}_{i,x} \end{pmatrix} \right\} \tag{5.44c}$$

The total field distributions are represented as follows:

1. For $z < z_-$

$$\begin{pmatrix} \mathbf{E} \\ \mathbf{H} \end{pmatrix} = \sum_{g=1}^{M^+} \tilde{t}_g \begin{pmatrix} \mathbf{E}_{(0)}^{(g)-} \\ \mathbf{H}_{(0)}^{(g)-} \end{pmatrix} \tag{5.45}$$

2. For $z_- \le z < z_+$

$$\begin{pmatrix} \mathbf{E} \\ \mathbf{H} \end{pmatrix} = \begin{cases} \displaystyle\sum_{g=1}^{M^+} C_{b,(1),g}^+ \begin{pmatrix} \mathbf{E}_{(1)}^{(g)+} \\ \mathbf{H}_{(1)}^{(g)+} \end{pmatrix} + \sum_{g=1}^{M^-} C_{b,(1),g}^- \begin{pmatrix} \mathbf{E}_{(1)}^{(g)-} \\ \mathbf{H}_{(1)}^{(g)-} \end{pmatrix} & \text{for } 0 \le z \le l_{1,1} \\[3ex] \displaystyle\sum_{g=1}^{M^+} C_{b,(2),g}^+ \begin{pmatrix} \mathbf{E}_{(2)}^{(g)+} \\ \mathbf{H}_{(2)}^{(g)+} \end{pmatrix} + \sum_{g=1}^{M^-} C_{b,(2),g}^- \begin{pmatrix} \mathbf{E}_{(2)}^{(g)-} \\ \mathbf{H}_{(2)}^{(g)-} \end{pmatrix} & \text{for } l_{1,1} \le z \le l_{1,2} \\[2ex] \qquad\qquad\qquad\qquad \vdots \\[1ex] \displaystyle\sum_{g=1}^{M^+} C_{b,(N),g}^+ \begin{pmatrix} \mathbf{E}_{(N)}^{(g)+} \\ \mathbf{H}_{(N)}^{(g)+} \end{pmatrix} + \sum_{g=1}^{M^-} C_{b,(N),g}^- \begin{pmatrix} \mathbf{E}_{(N)}^{(g)-} \\ \mathbf{H}_{(N)}^{(g)-} \end{pmatrix} & \text{for } l_{1,N-1} \le z \le l_{1,N} \end{cases} \tag{5.46}$$

where $\mathbf{E}^{(g)+}$, $\mathbf{E}^{(g)-}$, $\mathbf{H}^{(g)+}$, and $\mathbf{H}^{(g)-}$ are given in Equation (3.34).

3. For $z_+ \le z$

$$\begin{pmatrix} \mathbf{E} \\ \mathbf{H} \end{pmatrix} = \begin{pmatrix} \vec{E}_i \\ \vec{H}_i \end{pmatrix} + \begin{pmatrix} \vec{E}_r \\ \vec{H}_r \end{pmatrix}$$

$$= \sum_{m=-M}^{M} \sum_{n=-N}^{N} \begin{pmatrix} \tilde{E}_{i,x,m,n}, \tilde{E}_{i,y,m,n}, \tilde{E}_{i,z,m,n} \\ \tilde{H}_{i,x,m,n}, \tilde{H}_{i,y,m,n}, \tilde{H}_{i,z,m,n} \end{pmatrix} e^{j(k_{x,m,n}x + k_{ym,n}y - k_{z,m,n}(z-z_+))}$$

$$+ \sum_{m=-M}^{M} \sum_{n=-N}^{N} \begin{pmatrix} \vec{E}_{r,x,m,n}, \vec{E}_{r,y,m,n}, \vec{E}_{r,z,m,n} \\ \vec{H}_{r,x,m,n}, \vec{H}_{r,y,m,n}, \vec{H}_{r,z,m,n} \end{pmatrix} e^{j(k_{x,m,n}x + k_{ym,n}y + k_{z,m,n}(z-z_+))} \tag{5.47}$$

5.2.1.4 *Case D: Semi-Infinite Inhomogeneous Waveguide–Multiblock Structure–Semi-Infinite Inhomogeneous Space*

5.2.1.4.1 *Left-to-Right Field Visualization*

$$\vec{r} = \underline{\underline{R}}^{(0,N+1)}\vec{u} \tag{5.48a}$$

$$\begin{pmatrix} \vec{E}_{t,y} \\ \vec{E}_{t,x} \end{pmatrix} = \underline{\underline{T}}^{(0,N+1)}\vec{u} \tag{5.48b}$$

$$\left\{ \underline{C}_{a,(1)}^{(1,N)}, \underline{C}_{a,(2)}^{(1,N)}, \dots, \underline{C}_{a,(N)}^{(1,N)} \right\} = \left\{ \underline{C}_{a,(1)}^{(1,N)}\vec{u}, \underline{C}_{a,(2)}^{(1,N)}\vec{u}, \dots, \underline{C}_{a,(N)}^{(1,N)}\vec{u} \right\} \tag{5.48c}$$

The total field distributions are represented as follows:

1. For $z < z_-$

$$\begin{pmatrix} \mathbf{E} \\ \mathbf{H} \end{pmatrix} = \sum_{g=1}^{M^+} \vec{u}_g \begin{pmatrix} \mathbf{E}_{(0)}^{(g)+} \\ \mathbf{H}_{(0)}^{(g)+} \end{pmatrix} + \sum_{g=1}^{M^-} \vec{r}_g \begin{pmatrix} \mathbf{E}_{(0)}^{(g)-} \\ \mathbf{H}_{(0)}^{(g)-} \end{pmatrix} \tag{5.49}$$

2. For $z_- \le z < z_+$

$$\begin{pmatrix} \mathbf{E} \\ \mathbf{H} \end{pmatrix} = \begin{cases} \displaystyle\sum_{g=1}^{M^+} C_{a,(1),g}^+ \begin{pmatrix} \mathbf{E}_{(1)}^{(g)+} \\ \mathbf{H}_{(1)}^{(g)+} \end{pmatrix} + \sum_{g=1}^{M^-} C_{a,(1),g}^- \begin{pmatrix} \mathbf{E}_{(1)}^{(g)-} \\ \mathbf{H}_{(1)}^{(g)-} \end{pmatrix} & \text{for } 0 \le z \le l_{1,1} \\[20pt] \displaystyle\sum_{g=1}^{M^+} C_{a,(2),g}^+ \begin{pmatrix} \mathbf{E}_{(2)}^{(g)+} \\ \mathbf{H}_{(2)}^{(g)+} \end{pmatrix} + \sum_{g=1}^{M^-} C_{a,(2),g}^- \begin{pmatrix} \mathbf{E}_{(2)}^{(g)-} \\ \mathbf{H}_{(2)}^{(g)-} \end{pmatrix} & \text{for } l_{1,1} \le z \le l_{1,2} \\[10pt] \qquad\qquad\qquad\qquad \vdots \\[10pt] \displaystyle\sum_{g=1}^{M^+} C_{a,(N),g}^+ \begin{pmatrix} \mathbf{E}_{(N)}^{(g)+} \\ \mathbf{H}_{(N)}^{(g)+} \end{pmatrix} + \sum_{g=1}^{M^-} C_{a,(N),g}^- \begin{pmatrix} \mathbf{E}_{(N)}^{(g)-} \\ \mathbf{H}_{(N)}^{(g)-} \end{pmatrix} & \text{for } l_{1,N-1} \le z \le l_{1,N} \end{cases} \tag{5.50a}$$

where $\mathbf{E}^{(g)+}$, $\mathbf{E}^{(g)-}$, $\mathbf{H}^{(g)+}$, and $\mathbf{H}^{(g)-}$ are given, respectively, by

$$\mathbf{E}^{(g)+}(x,y,z) = \sum_{m=-M}^{M} \sum_{n=-N}^{N} \left(E_{x,m,n}^{(g)+}, E_{y,m,n}^{(g)+}, E_{z,m,n}^{(g)+} \right) e^{j\left(k_{x,m}x + k_{y,n}y + k_z^{(g)+}(z-z_-)\right)} \tag{5.50b}$$

$$\mathbf{E}^{(g)-}(x,y,z) = \sum_{m=-M}^{M} \sum_{n=-N}^{N} \left(E_{x,m,n}^{(g)-}, E_{y,m,n}^{(g)-}, E_{z,m,n}^{(g)-} \right) e^{j\left(k_{x,m}x + k_{y,n}y + k_z^{(g)-}(z-z_+)\right)} \tag{5.50c}$$

$$\mathbf{H}^{(g)+}(x,y,z) = \sum_{m=-M}^{M} \sum_{n=-N}^{N} \left(H_{x,m,n}^{(g)+}, H_{y,m,n}^{(g)+}, H_{z,m,n}^{(g)+} \right) e^{j\left(k_{x,m}x + k_{y,n}y + k_z^{(g)+}(z-z_-)\right)} \tag{5.50d}$$

$$\mathbf{H}^{(g)-}(x,y,z) = \sum_{m=-M}^{M} \sum_{n=-N}^{N} \left(H_{x,m,n}^{(g)-}, H_{y,m,n}^{(g)-}, H_{z,m,n}^{(g)-} \right) e^{j\left(k_{x,m}x + k_{y,n}y + k_z^{(g)-}(z-z_+)\right)} \quad (5.50e)$$

3. For $z_+ \le z$

$$\begin{pmatrix} \mathbf{E} \\ \mathbf{H} \end{pmatrix} = \sum_{g=1}^{M^+} \vec{t}_g \begin{pmatrix} \mathbf{E}_{(N+1)}^{(g)+} \\ \mathbf{H}_{(N+1)}^{(g)+} \end{pmatrix} \quad (5.51a)$$

where $\mathbf{E}^{(g)+}$ and $\mathbf{H}^{(g)+}$ are given, respectively, by

$$\mathbf{E}^{(g)+}(x,y,z) = \sum_{m=-M}^{M} \sum_{n=-N}^{N} \left(E_{x,m,n}^{(g)+}, E_{y,m,n}^{(g)+}, E_{z,m,n}^{(g)+} \right) e^{j\left(k_{x,m}x + k_{y,n}y + k_z^{(g)+}(z-z_-)\right)} \quad (5.51b)$$

$$\mathbf{H}^{(g)+}(x,y,z) = \sum_{m=-M}^{M} \sum_{n=-N}^{N} \left(H_{x,m,n}^{(g)+}, H_{y,m,n}^{(g)+}, H_{z,m,n}^{(g)+} \right) e^{j\left(k_{x,m}x + k_{y,n}y + k_z^{(g)+}(z-z_-)\right)} \quad (5.51c)$$

5.2.1.4.2 Right-to-Left Field Visualization

$$\vec{r} = \underline{\vec{R}}^{(0,N+1)} \vec{u} \quad (5.52a)$$

$$\vec{t} = \underline{\vec{T}}^{(0,N+1)} \vec{u} \quad (5.52b)$$

$$\left\{ \underline{C}_{b,(1)}^{(1,N)}, \underline{C}_{b,(2)}^{(1,N)}, \ldots, \underline{C}_{b,(N)}^{(1,N)} \right\} = \left\{ \underline{C}_{b,(1)}^{(1,N)} \vec{u}, \underline{C}_{b,(2)}^{(1,N)} \vec{u}, \ldots, \underline{C}_{b,(N)}^{(1,N)} \vec{u} \right\} \quad (5.52c)$$

The total field distributions are represented as follows:

1. For $z < z_-$

$$\begin{pmatrix} \mathbf{E} \\ \mathbf{H} \end{pmatrix} = \sum_{g=1}^{M^+} \vec{t}_g \begin{pmatrix} \mathbf{E}_{(0)}^{(g)-} \\ \mathbf{H}_{(0)}^{(g)-} \end{pmatrix} \quad (5.53)$$

2. For $z_- \le z < z_+$

$$\begin{pmatrix} \mathbf{E} \\ \mathbf{H} \end{pmatrix} = \begin{cases} \displaystyle\sum_{g=1}^{M^+} C_{b,(1),g}^{+} \begin{pmatrix} \mathbf{E}_{(1)}^{(g)+} \\ \mathbf{H}_{(1)}^{(g)+} \end{pmatrix} + \sum_{g=1}^{M^-} C_{b,(1),g}^{-} \begin{pmatrix} \mathbf{E}_{(1)}^{(g)-} \\ \mathbf{H}_{(1)}^{(g)-} \end{pmatrix} & \text{for } 0 \le z \le l_{1,1} \\[3ex] \displaystyle\sum_{g=1}^{M^+} C_{b,(2),g}^{+} \begin{pmatrix} \mathbf{E}_{(2)}^{(g)+} \\ \mathbf{H}_{(2)}^{(g)+} \end{pmatrix} + \sum_{g=1}^{M^-} C_{b,(2),g}^{-} \begin{pmatrix} \mathbf{E}_{(2)}^{(g)-} \\ \mathbf{H}_{(2)}^{(g)-} \end{pmatrix} & \text{for } l_{1,1} \le z \le l_{1,2} \\[2ex] \quad\quad\quad\quad\quad\quad \vdots \\[2ex] \displaystyle\sum_{g=1}^{M^+} C_{b,(N),g}^{+} \begin{pmatrix} \mathbf{E}_{(N)}^{(g)+} \\ \mathbf{H}_{(N)}^{(g)+} \end{pmatrix} + \sum_{g=1}^{M^-} C_{b,(N),g}^{-} \begin{pmatrix} \mathbf{E}_{(N)}^{(g)-} \\ \mathbf{H}_{(N)}^{(g)-} \end{pmatrix} & \text{for } l_{1,N-1} \le z \le l_{1,N} \end{cases} \quad (5.54)$$

3. For $z_+ \leq z$

$$\begin{pmatrix} \mathbf{E} \\ \mathbf{H} \end{pmatrix} = \sum_{g=1}^{M^-} \bar{u}_g \begin{pmatrix} \mathbf{E}_{(N+1)}^{(g)-} \\ \mathbf{H}_{(N+1)}^{(g)-} \end{pmatrix} + \sum_{g=1}^{M^+} \vec{r}_g \begin{pmatrix} \mathbf{E}_{(N+1)}^{(g)+} \\ \mathbf{H}_{(N+1)}^{(g)+} \end{pmatrix} \tag{5.55}$$

5.3 MATLAB® Implementation

A numerical scheme for LFMM is described with working MATLAB codes in this section. The MATLAB implementation of the above-described mathematical model is provided with MATLAB code *LFMM_two_port_Amode.m*. An example of the super-block is the photonic crystal waveguide shown in Figure 5.7. The whole code of *LFMM_two_port_Amode.m* is presented as follows.

MATLAB Code 5.1: *LFMM_two_port_Amode.m*

```
1     %% LFMM two port Amode
2
3     %case 1. free space - finite PC waveguide - free space
4     %case 2. free space - half-infinite PC waveguide
5     %case 3. half-infinite PC waveguide - free space
6
7
8     %% STEP 1
9     clear all;
10    close all;
11    clc;
12
13    addpath( [pwd '\PRCWA_COM']);
14    addpath( [pwd '\FIELD_VISUAL']);
15    addpath( [pwd '\STRUCTURE']);
16
17
18    % length unit
19
20    global nm; global nano;      % nano
21    global um; global micro;     % micro
22    global mm;                   % mili
23    global lambda;               % wavelength
24
25    global k0;                            % wavenumber
26    global c0; global w0;
27    global eps0; global mu0;
```

```
28
29    % zero-thickness buffer refractive index, permittivity,
         permeability
30    global n0; global epr0; global mur0;
31    % refractive index, permittivity, permeability in free
         space I
32    global ni; global epri; global muri;
33    % refractive index, permittivity, permeability in free
         space III
34    global nf; global eprf; global murf;
35
36    % x-directional supercell period, y-directional supercell
         period
37    global aTx; global aTy; global aTz; global nx; global
         ny; global nz;
38    global bTx; global bTy; global bTz; global Tx; global
         Ty; global Tz;
39    global NBx; global NBy; global NBz; global num_hx;
         global num_hy; global k0;
40    global kx_vc; global ky_vc; global kz_vc;
41
42    % input output free space
43    global kix; global kiy; global kiz; global kfz;
44    global kx_ref; global ky_ref; global kz_ref;
45    global kx_tra; global ky_tra; global kz_tra;
46
47
48    PRCWA_basic;      % 3D structure
49    PRCWA_Gen_K;      % zero-thickness buffer
50
51    % structure modeling
52    % The example structure is the photonic crystal
         waveguide structure
53    % approximated by the stepwise multi-blocks. The
         structure modeling is
54    % implemented in the MATLAB code line 59, PRCWA_Gen_
         PCWG.m
55    % In PRCWA_Gen_PCWG.m, the Fourier series of the target
         structure is done
56    % by Grating_gen_Y_branch.m as Grating_Gen_PCWG.m. Open
         PRCWA_Gen_PCWG.m
57    % and see the annotation
58
59    PRCWA_Gen_PCWG; % photonic crystal waveguide
60
61    %% STEP2 Block S-matrix computation of single super-
         block structures
```

```
62    % Bloch eigenmode analysis is performed by
      AmodeBlochAnalysis.m. Let us see
63    % its inside in code
64
65    %%% A mode Bloch Mode Analysis
66    % Fourier coefficients of A Bloch eigenmodes
67
68    ApEFx=zeros(NBx,NBy,NBz,2*L);
69    ApEFy=zeros(NBx,NBy,NBz,2*L);
70    ApEFz=zeros(NBx,NBy,NBz,2*L);
71
72    ApHFx=zeros(NBx,NBy,NBz,2*L);
73    ApHFy=zeros(NBx,NBy,NBz,2*L);
74    ApHFz=zeros(NBx,NBy,NBz,2*L);
75
76    AnEFx=zeros(NBx,NBy,NBz,2*L);
77    AnEFy=zeros(NBx,NBy,NBz,2*L);
78    AnEFz=zeros(NBx,NBy,NBz,2*L);
79
80    AnHFx=zeros(NBx,NBy,NBz,2*L);
81    AnHFy=zeros(NBx,NBy,NBz,2*L);
82    AnHFz=zeros(NBx,NBy,NBz,2*L);
83
84    % Positive & Negative eigenvalue of a modes
85    ap_evalue=zeros(1,2*L);
86    am_evalue=zeros(1,2*L);
87
88    AmodeBlochAnalysis;
89
90    ApEFx=pEFx;
91    ApEFy=pEFy;
92    ApEFz=pEFz;
93
94    ApHFx=pHFx;
95    ApHFy=pHFy;
96    ApHFz=pHFz;
97
98    AnEFx=nEFx;
99    AnEFy=nEFy;
100   AnEFz=nEFz;
101
102   AnHFx=nHFx;
103   AnHFy=nHFy;
104   AnHFz=nHFz;
105
106   ap_evalue=pe_value;
```

```
107    am_evalue=me_value;
108    ap_evector=pe_vector;
109    am_evector=me_vector;
110
111    AModePower_Flow;
112
113
114    %% STEP 3 S-matrix computation
115
116    % Equations (5.1.7a)~(5.1.7d) can be calculated
       efficiently with the use of
117    % bultin DFT function of MATLAB. The field calculation
       of Bloch eigenmodes
118    % is implemented in Amode_visualization_fft.m
119
120    % A mode [2-port Block]
121    % Block S-matrix of Amode
122    AR11=zeros(2*L,2*L); %      port1->port1
123    AT12=zeros(2*L,2*L); %      port1->port2
124    AT21=zeros(2*L,2*L); %      port2->port1
125    AR22=zeros(2*L,2*L); %      port2->port2
126    ACa=zeros(4*L,2*L); %       excitation of port1
127    ACb=zeros(4*L,2*L); %       excitation of port2
128
129    % Left grating S-matrix
130    ALR11=zeros(2*L,2*L); % port1->port1
131    ALT12=zeros(2*L,2*L); % port1->port2
132    ALT21=zeros(2*L,2*L); % port2->port1
133    ALR22=zeros(2*L,2*L); % port2->port2
134
135
136    % Right grating S-matrix
137    ARR11=zeros(2*L,2*L); % port1->port1
138    ART12=zeros(2*L,2*L); % port1->port2
139    ART21=zeros(2*L,2*L); % port2->port1
140    ARR22=zeros(2*L,2*L); % port2->port2
141
142    zm=0;
143    zp=aTz;
144    zc=0;
145
146    SMM_2port_Amode;
147
148    % A mode eigenmode calculation
149    Amode_visualization_fft;
150
```

```
151    %% STEP4 Half-infinite block interconnection & Field
       visualization
152
153    LFMM_Smat_Amode_case1;      % case1
154    %LFMM_Smat_Amode_case2;     % case2
155    %LFMM_Smat_Amode_case3;     % case3
156
157
158    %% Data save
159    save 2portDATA_Amode_all;
160    save 2portDATA_Amode aPG_Ex_xz aPG_Ey_xz aPG_Ez_xz...
161         aPG_Hx_xz aPG_Hy_xz aPG_Hz_xz...
162         aNG_Ex_xz aNG_Ey_xz aNG_Ez_xz...
163         aNG_Hx_xz aNG_Hy_xz aNG_Hz_xz...
164         ALR11 ALT12 ALR22 ALT21...
            ARR11 ART12 ARR22 ART21...
            AR11 AT12 AT21 AR22 ACa ACb...
            ap_evalue am_evalue ap_evector am_evector...
            aPw aNw aTx aTz SAR11 SAT12 SAT21 SAR22 SACa SACb;
```

The main constituents of the MATLAB functions are listed in the following table. In particular, the readers have to get the understanding of the following MATLAB functions:

1. *FMM_single_block_tensor.m*
2. *FMM_single_block.m*
3. *AmodeBlochAnalysis.m*
4. *SMM_2port_Amode.m*
5. *Amode_visualization_fft.m*
6. *LFMM_Smat_Amode_case1.m*
7. *LFMM_Smat_Amode_case2.m*
8. *LFMM_Smat_Amode_case3.m*

These MATLAB functions are the analysis functions of super-block and boundary S-matrices of the target structure.

File Name	Description
LFMM_two_port_Amode.m	Main routine of LFMM and S-matrix method with field visualization
PRCWA_COM/Directory	*Common Library*
FMM_single_block.m	S-matrix calculation of single block with diagonal anisotropic material

File Name		Description
	sWp_gen.m	Transversal boundary electric field distribution at z=z+
	sWm_gen.m	Transversal boundary electric field distribution at z=z–
	sVp_gen.m	Transversal boundary magnetic field distribution at z=z+
	sVm_gen.m	Transversal boundary magnetic field distribution at z=z–
FMM_single_block_tensor.m		S-matrix calculation of single block with general off-diagonal tensor anisotropic material
	Wp_gen.m	Transversal boundary electric field distribution at z=z+
	Wm_gen.m	Transversal boundary electric field distribution at z=z–
	Vp_gen.m	Transversal boundary magnetic field distribution at z=z+
	Vm_gen.m	Transversal boundary magnetic field distribution at z=z–
PRCWA_basic.m		Basic setting of Fourier harmonic orders, supercell period, buffer layer material
PRCWA_Gen_K.m		Setting wavevector grid of Fourier harmonics
AmodeBlochAnalysis.m		Bloch eigenmode analysis
SMM_2port_Amode		N-super-block interconnection with S-matrix method
fun_*.m		Permittivity functions of Ag, Au, Si, PMMA for wavelength
Odd_*.m		FFT shift functions of odd number sampling signal for Fourier transform
LFMM_Smat_Amode_case1.m		Free space–finite photonic crystal waveguide–free space
LFMM_Smat_Amode_case2.m		Free space–half-infinite photonic crystal waveguide
LFMM_Smat_Amode_case3.m		Half-infinite photonic crystal waveguide–free space
FIELD_VISUAL/Directory		*Field Visualization*
Amode_visualization_fft		Bloch eigenmode field distribution calculation
LFMM_Amode_field_visual_xz_*.m		Cross-section of six field components at x-z plane
STRUCTURE/Directory		*Example Structures*
PRCWA_Gen_PCWG.m		Setting the Toeplitz matrix of permittivity and permeability tensor
Grating_gen_PCWG.m		Fourier series coefficients of photonic crystal waveguide

The code structure is divided into the following four steps:

Step 1: Wavevector setting and structure modeling (lines 8~59). In *PRCWA_basic.m*, the permittivity and permeability values of the left and right half-infinite series are determined. The set of the reciprocal vector components is prepared in *PRCWA_Gen_K.m*. The structure modeling is implemented in *PRCWA_Gen_PCWG.m*. In *PRCWA_Gen_PCWG.m*, the Fourier series of the target structure is obtained in *Grating_Gen_PCWG.m*.

Step 2: Block S-matrix computation of single super-block structures (lines 61~111). Bloch eigenmode analysis is performed by *AmodeBlochAnalysis.m*.

Step 2.1: S-matrix and coupling coefficient operator calculation of single-super-block structure. The single-super-block structure is modeled by the multiblock structure. The S-matrices of elementary blocks are computed in *Diagonal_tensor_SMM.m* and interconnected by the recursive interconnection algorithm in Step 2.1 in *AmodeBlochAnalysis.m*.

Step 2.2: Algebraic Maxwell eigenvalue equation for Bloch eigenmode analysis. The algebraic Maxwell eigenvalue equations of Equations (5.4a) and (5.4b) are computed in the part of the positive Bloch mode lines 99~162 and the part of the negative Bloch mode lines 164~222, respectively, in *AmodeBlochAnalysis.m*. The classification rule for the eigenvalues is applied in lines 106~162 for positive Bloch mode and in lines 168~222 for negative Bloch mode.

MATLAB Code 5.2: *AmodeBlochAnalysis.m*

```
99    % positive Bloch mode
101   SBSa=[TTa zeros(2*L,2*L); RRa -I];
102   SBSb=[I -RRb; zeros(2*L,2*L) -TTb];
103
104   [W,D]=eig(SBSa,SBSb);
105
106   for k=1:4*L
107   e_value(k)=log(D(k,k))/d;
108   end;
109
110   pnl=0;
111   mnl=0;
112   for k=1:4*L
113         if abs(real(e_value(k))) >= abs(imag(e_value(k)))
114               pnl=pnl+1;pvalue(pnl)=e_
                  value(k);pvec(:,pnl)=W(:,k);
115         else
116               mnl=mnl+1;mvalue(mnl)=e_
                  value(k);mvec(:,mnl)=W(:,k);
117         end;
118   end;
```

```
119
120    if pnl ~= 0
121
122    [pevv pevn]=sort(abs(real(pvalue)));
123
124    for k=1:pnl
125    e_value(k)=pvalue(pevn(k));
126    e_vector(:,k)=pvec(:,pevn(k));
127    end;
128    end;
129
130    [mevv mevn]=sort(abs(imag(mvalue)));
131    for k=1:mnl
132        e_value(k+pnl)=mvalue(mevn(k));
133        e_vector(:,k+pnl)=mvec(:,mevn(k));
134
135    end;
136
137    pcnt=0;
138
139    for k=1:4*L
140        if real(e_value(k)) <=0
141            pcnt=pcnt+1; % positive mode
142        else
143        ;
144        end;
145    end;
146
147    pe_value=zeros(1,pcnt);    % positive mode real(e_
                                          value)<0
148    pe_vector=zeros(4*L,pcnt); % positive mode eigen_vector
149
150    pcnt=0;
151
152    for k=1:4*L
153
154        if real(e_value(k)) <= 0
155            pcnt=pcnt+1;
156            pe_value(pcnt)=e_value(k);
157            pe_vector(:,pcnt)=e_vector(:,k);
158        else
159         ;
160        end;
161
162    end;
163
```

```
164    % negative Bloch mode
165
166    [W, D]=eig(SBSb,SBSa);
167
168    for k=1:4*L
169    e_value(k)=-log(D(k,k))/d;
170    end;
171
172    pnl=0;
173    mnl=0;
174    for k=1:4*L
175        if abs(real(e_value(k))) >= abs(imag(e_value(k)))
176            pnl=pnl+1;pvalue(pnl)=e_
                value(k);pvec(:,pnl)=W(:,k);
177        else
178            mnl=mnl+1;mvalue(mnl)=e_
                value(k);mvec(:,mnl)=W(:,k);
179        end;
180    end;
181
182    if pnl ~= 0
183    [pevv pevn]=sort(abs(real(pvalue)));
184
185    for k=1:pnl
186    e_value(k)=pvalue(pevn(k));
187    e_vector(:,k)=pvec(:,pevn(k));
188    end;
189    end;
190    [mevv mevn]=sort(abs(imag(mvalue)));
191    for k=1:mnl
192        e_value(k+pnl)=mvalue(mevn(k));
193    e_vector(:,k+pnl)=mvec(:,mevn(k));
194
195    end;
196
197    mcnt=0;
198
199    for k=1:4*L
200        if real(e_value(k)) <=0
201          ;
202        else
203          mcnt=mcnt+1; % negative mode
204        end;
205    end;
206
```

```
207    me_value=zeros(1,mcnt);              % negative mode real(e_
                                               value)>0
208    me_vector=zeros(4*L,mcnt);           % negative mode eigen_
                                               vector
209
210    mcnt=0;
211
212    for k=1:4*L
213
214        if real(e_value(k)) <= 0
215          ;
216        else
217          mcnt=mcnt+1;
218          me_value(mcnt)=e_value(k);
219          me_vector(:,mcnt)=e_vector(:,k);
220        end;
221
222    end;
223
224    CCp=zeros(4*L,Nlay,pcnt); % positive Bloch mode
225    CCn=zeros(4*L,Nlay,mcnt); % negative Bloch mode
226
227          for k=1:pcnt
228              for lnt=1:Nlay
229
230    CCp(:,lnt,k)=Ca(:,:,lnt)*pe_vector(1:2*L,k)+Cb(:,:,lnt)
          *exp(pe_value(k)*d)*pe_vector(2*L+1:4*L,k);
231              end;
232            end;
233
234          for k=1:mcnt
235              for lnt=1:Nlay
236
237    CCn(:,lnt,k)=Ca(:,:,lnt)*me_vector(1:2*L,k)+Cb(:,:,lnt)*
          exp(me_value(k)*d)*me_vector(2*L+1:4*L,k);
238              end;
239            end;
```

Step 2.3: Pseudo-Fourier representation of Bloch eigenmodes. The pseudo-Fourier representation of the Bloch eigenmode, Equation (5.7), can be efficiently calculated with the use of the built-in DFT function of MATLAB. The field calculation of Bloch eigenmodes is implemented in the part of Step 2.3 in *AmodeBlochAnalysis.m*.

Step 3: S-matrix and coupling coefficient operator calculation of multi-super-block structures with extended Redheffer star product (lines 147~151). The S-matrix computation of Equations (5.17)~(5.18) and Equations (5.20)~(5.21) is performed in *SMM_2port_Amode.m*. The system matrix of Equation (5.21a) is coded in lines 88~93. The left-to-right and right-to-left characterizations in the extraction of S-matrix components are implemented in lines 95~107. The boundary S-matrices are calculated in lines 111~161.

MATLAB Code 5.3: *SMM_2port_Amode.m*

```
24      AWh=Iden;
25      AVh=KII*(1/(w0*mu0));
26
27      AWpm=zeros(2*L,pcnt); AVpm=zeros(2*L,pcnt);
28      AWnm=zeros(2*L,mcnt); AVnm=zeros(2*L,mcnt);
29
30      AWpp=zeros(2*L,pcnt); AVpp=zeros(2*L,pcnt);
31      AWnp=zeros(2*L,mcnt); AVnp=zeros(2*L,mcnt);
32
33      for pbm_ind=1:pcnt % AWpm, AVpm, AWpp, AVpp
34
35          for k=1:NBx
36              for l=1:NBy
37
38              [ey ex]=awp(zm-zm,k,l,pbm_ind,ApEFy,ApEFx,ap_
                    evalue);
39              AWpm(NBy*(k-1)+l,pbm_ind)=ey;
40              AWpm(L+NBy*(k-1)+l,pbm_ind)=ex;
41
42              [hy hx]=avp(zm-zm,k,l,pbm_ind,ApHFy,ApHFx,ap_
                    evalue);
43              AVpm(NBy*(k-1)+l,pbm_ind)=hy;
44              AVpm(L+NBy*(k-1)+l,pbm_ind)=hx;
45
46
47              [ey ex]=awp(zp-zm,k,l,pbm_ind,ApEFy,ApEFx,ap_
                    evalue);
48              AWpp(NBy*(k-1)+l,pbm_ind)=ey;
49              AWpp(L+NBy*(k-1)+l,pbm_ind)=ex;
50
51              [hy hx]=avp(zp-zm,k,l,pbm_ind,ApHFy,ApHFx,ap_
                    evalue);
52              AVpp(NBy*(k-1)+l,pbm_ind)=hy;
53              AVpp(L+NBy*(k-1)+l,pbm_ind)=hx;
54
```

```
55                  end;
56              end;
57
58      end;
59
60      for pbm_ind=1:mcnt % AWnm, AVnm, AWnp, AVnp
61
62              for k=1:NBx
63                  for l=1:NBy
64
65              [ey ex]=awn(zm-zp,k,l,pbm_ind,AnEFy,AnEFx,am_
                    evalue);
66              AWnm(NBy*(k-1)+l,pbm_ind)=ey;
67              AWnm(L+NBy*(k-1)+l,pbm_ind)=ex;
68
69              [hy hx]=avn(zm-zp,k,l,pbm_ind,AnHFy,AnHFx,am_
                    evalue);
70              AVnm(NBy*(k-1)+l,pbm_ind)=hy;
71              AVnm(L+NBy*(k-1)+l,pbm_ind)=hx;
72
73
74              [ey ex]=awn(zp-zp,k,l,pbm_ind,AnEFy,AnEFx,am_
                    evalue);
75              AWnp(NBy*(k-1)+l,pbm_ind)=ey;
76              AWnp(L+NBy*(k-1)+l,pbm_ind)=ex;
77
78              [hy hx]=avn(zp-zp,k,l,pbm_ind,AnHFy,AnHFx,am_
                    evalue);
79              AVnp(NBy*(k-1)+l,pbm_ind)=hy;
80              AVnp(L+NBy*(k-1)+l,pbm_ind)=hx;
81
82                  end;
83              end;
84
85      end;
86
87      % Block S-matrix
88      U=eye(2*NBx*NBy);
89      S11=inv(AWh)*AWpm+inv(AVh)*AVpm;
90      S12=inv(AWh)*AWnm+inv(AVh)*AVnm;
91      S21=inv(AWh)*AWpp-inv(AVh)*AVpp;
92      S22=inv(AWh)*AWnp-inv(AVh)*AVnp;
93      S=[S11 S12;S21 S22];
94
95      % left-to-right block S-matrix
96      ACa=inv(S)*[2*U;zeros(2*NBx*NBy)]; %% very good!!!
```

```
97    ACap=ACa(1:pcnt,:);
98    ACam=ACa(pcnt+1:pcnt+mcnt,:);
99    AR11=inv(AWh)*(AWpm*ACap+AWnm*ACam-AWh*U);
100   AT12=inv(AWh)*(AWpp*ACap+AWnp*ACam);
101
102   % right-to-left block S-matrix
103   ACb=inv(S)*[zeros(2*NBx*NBy);2*U];
104   ACbp=ACb(1:pcnt,:);
105   ACbm=ACb(pcnt+1:pcnt+mcnt,:);
106   AR22=inv(AWh)*(AWpp*ACbp+AWnp*ACbm-AWh*U);
107   AT21=inv(AWh)*(AWpm*ACbp+AWnm*ACbm);
108
109   % Boundary S-matrix
110
111   AWp=zeros(2*L,pcnt); AVp=zeros(2*L,pcnt);
112   AWn=zeros(2*L,mcnt); AVn=zeros(2*L,mcnt);
113
114
115   for pbm_ind=1:pcnt % AWpm, AVpm, AWpp, AVpp
116
117       for k=1:NBx
118           for l=1:NBy
119
120           [ey ex]=awp(zc,k,l,pbm_ind,ApEFy,ApEFx,ap_evalue);
121           AWp(NBy*(k-1)+1,pbm_ind)=ey;
122           AWp(L+NBy*(k-1)+1,pbm_ind)=ex;
123
124           [hy hx]=avp(zc,k,l,pbm_ind,ApHFy,ApHFx,ap_evalue);
125           AVp(NBy*(k-1)+1,pbm_ind)=hy;
126           AVp(L+NBy*(k-1)+1,pbm_ind)=hx;
127
128               end;
129           end;
130
131   end;
132
133   for pbm_ind=1:mcnt % AWnm, AVnm, AWnp, AVnp
134
135       for k=1:NBx
136           for l=1:NBy
137
138           [ey ex]=awn(zc,k,l,pbm_ind,AnEFy,AnEFx,am_evalue);
139           AWn(NBy*(k-1)+1,pbm_ind)=ey;
140           AWn(L+NBy*(k-1)+1,pbm_ind)=ex;
141
142           [hy hx]=avn(zc,k,l,pbm_ind,AnHFy,AnHFx,am_evalue);
```

```
143          AVn(NBy*(k-1)+l,pbm_ind)=hy;
144          AVn(L+NBy*(k-1)+l,pbm_ind)=hx;
145
146              end;
147          end;
148
149      end;
150
151      % left side semi-infinite waveguide
152      ALR11=-inv(inv(AWh)*AWn-inv(AVh)*AVn)*(inv(AWh)*AWp-
             inv(AVh)*AVp);
153      ALT12=inv(inv(AWn)*AWh-inv(AVn)*AVh)*(inv(AWn)*AWp-
             inv(AVn)*AVp);
154      ALT21=2*inv(inv(AWh)*AWn-inv(AVh)*AVn);
155      ALR22=-inv(inv(AWn)*AWh-inv(AVn)*AVh)*(inv(AWn)*AWh+inv(
             AVn)*AVh);
156
157      % right side semi-infinite waveguide
158      ARR11=-inv(inv(AWp)*AWh+inv(AVp)*AVh)*(inv(AWp)*AWh-
             inv(AVp)*AVh);
159      ART12=2*inv(inv(AWh)*AWp+inv(AVh)*AVp);
160      ART21=inv(inv(AWp)*AWh+inv(AVp)*AVh)*(inv(AWp)*AWn-
             inv(AVp)*AVn);
161      ARR22=-inv(inv(AWh)*AWp+inv(AVh)*AVp)*(inv(AWh)*AWn+inv(
             AVh)*AVn);
```

In the MATLAB code 5.3 of *SMM_2port_Amode.m*, the necessary variables to form the system matrix of Equation (5.21a) are prepared.

$\underline{\underline{W}}^+(0)$	AWpm	awp(zm-zm,k,l,pbm_ind,ApEFy,ApEFx,ap_evalue)
$\underline{\underline{W}}^-(z_- - z_+)$	AWnm	awn(zm-zp,k,l,pbm_ind,AnEFy,AnEFx,am_evalue)
$\underline{\underline{V}}^+(0)$	AVpm	avp(zm-zm,k,l,pbm_ind,ApHFy,ApHFx,ap_evalue)
$\underline{\underline{V}}^-(z_- - z_+)$	AVnm	avn(zm-zp,k,l,pbm_ind,AnHFy,AnHFx,am_evalue)
$\underline{\underline{W}}^+(z_+ - z_-)$	Awpp	awp(zp-zm,k,l,pbm_ind,ApEFy,ApEFx,ap_evalue)
$\underline{\underline{V}}^-(0)$	Awnp	awn(zp-zp,k,l,pbm_ind,AnEFy,AnEFx,am_evalue)
$\underline{\underline{V}}^+(z_+ - z_-)$	Avpp	avp(zp-zm,k,l,pbm_ind,ApHFy,ApHFx,ap_evalue)
$\underline{\underline{V}}^-(0)$	AVnp	avn(zp-zp,k,l,pbm_ind,AnHFy,AnHFx,am_evalue)

The outputs of *AmodeBlochAnalysis.m* and *SMM_2port_Amode.m* are given as

S-Matrix and Coupling Coefficient Operator		Positive Eigenmodes		Negative Eigenmodes	
$\vec{\underline{\underline{T}}}$	AT12	$E^{(g)+}_{x,m,n,p}$	ApEFx	$E^{(g)-}_{x,m,n,p}$	AmEFx
$\vec{\underline{\underline{R}}}$	AR11	$E^{(g)+}_{y,m,n,p}$	ApEFy	$E^{(g)-}_{y,m,n,p}$	AmEFy
$\overleftarrow{\underline{\underline{T}}}$	AT21	$E^{(g)+}_{z,m,n,p}$	ApEFz	$E^{(g)-}_{z,m,n,p}$	AmEFz
$\overleftarrow{\underline{\underline{R}}}$	AR22	$H^{(g)+}_{x,m,n,p}$	ApHFx	$H^{(g)-}_{x,m,n,p}$	AmHFx
$\underline{\underline{C}}_a$	ACa	$H^{(g)+}_{y,m,n,p}$	ApHFy	$H^{(g)-}_{y,m,n,p}$	AmHFy
$\underline{\underline{C}}_b$	ACb	$H^{(g)+}_{z,m,n,p}$	ApHFz	$H^{(g)-}_{z,m,n,p}$	AmHFz
		$k^{(g)+}_z$	pevalue	$k^{(g)-}_z$	mevalue

Step 4: Field visualization and data processing for specific purposes. The field profiles of the positive and negative Bloch eigenmodes are calculated in *Amode_visualization_fft.m*. The FFT built-in function is used here for the sake of efficiency. The six components of the positive Bloch eigenmodes in *Amode_visualization_fft.m* are saved in aPG_Ex_xz, aPG_Ey_xz, aPG_Ez_xz, aPG_Hx_xz, aPG_Hy_xz, and aPG_Hz_xz. The six components of the negative Bloch eigenmodes are saved in aNG_Ex_xz, aNG_Ey_xz, aNG_Ez_xz, aNG_Hx_xz, aNG_Hy_xz, and aNG_Hz_xz. As aforementioned, the field calculation analysis: (A) half-infinite homogeneous space–multiblock structure–half-infinite homogeneous space, (B) half-infinite homogeneous space–multiblock structure–half-infinite inhomogeneous space, and (C) half-infinite inhomogeneous space–multiblock structure–half-infinite homogeneous space are implemented in *LFMM_Smat_Amode_case1.m*, *LFMM_Smat_Amode_case2.m*, and *LFMM_Smat_Amode_case3.m*, respectively.

5.4 Applications

5.4.1 Field Localization in Photonic Crystals

The optical isolation of a block from other blocks in nanophotonic networks through tight field localization is a basic and important assumption. LFMA adopts the optical isolation assumption in analyzing nanophotonic networks (discussed further in Chapter 6), which means that optical energy can be

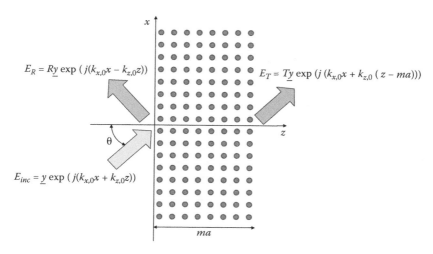

$E_R = R\underline{y} \exp\left(j(k_{x,0}x - k_{z,0}z)\right)$

$E_T = T\underline{y} \exp\left(j\left(k_{x,0}x + k_{z,0}\left(z - ma\right)\right)\right)$

$E_{inc} = \underline{y} \exp\left(j(k_{x,0}x + k_{z,0}z)\right)$

θ

ma

FIGURE 5.10
Transmission and reflection of a plane wave by finite photonic crystal slab. The transmission coefficient T is strongly dependent on the thickness of the photonic crystal slab and the incidence angle θ.

transferred only through predetermined waveguide channels in integrated nanophotonic networks.

Here, the optical field localization in a photonic crystal is examined with LFMA. First, through an analysis of the reflection and transmission characteristics of the finite-sized photonic crystal bandgap medium, the basic criteria for determining the thickness of the bandap medium validating the localization assumption of LFMA is addressed. Second, the characteristics of coupling and isolation between two parallel photonic crystal waveguides separated by a certain distance are analyzed.

Figure 5.10 shows that a y-polarization plane wave with an oblique incidence angle of θ, E_{inc}, is incident on a finite-sized photonic crystal slab with a thickness of $t = ma$, where m is the number of single-super-blocks. Since the period of the photonic crystal is on a subwavelength scale, the zeroth order diffraction fields, that is, reflection and transmission plane waves, E_R and E_T are generated in the left free space and right free space regions, respectively. With LFMA, the dependence of the transmission coefficient on the thickness of the photonic crystal and the incidence angle is inspected. Figure 5.11(a) shows the variation of the absolute value of the transmission coefficient with changes in the number of unit multiblocks, m, and the incidence angle θ. At the normal incidence ($\theta = 0$), the penetration depth of the optical field into the bandgap medium becomes the maximum. In Figure 5.11(b), the absolute value of the transmission coefficient obtained for the normal incidence is plotted for the number of single-super-block, m. Through a photonic crystal slab of a thickness larger than 10 single-super-blocks, almost all optical

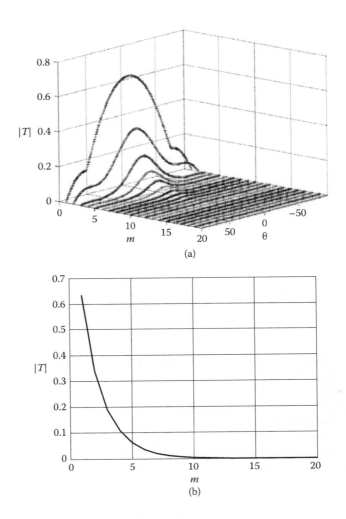

(a)

(b)

FIGURE 5.11
(a) Variation in transmission coefficients with respect to the incidence angle and the thickness of photonic crystal. (b) Variation in transmission coefficient of normal incidence case with respect to the thickness of the photonic crystal slab.

power is reflected and the transmission is negligible. It can be seen that the bandgap medium thickness should be larger than 10 single-super-blocks for effective isolation.

Next, let us consider two waveguides A and B, as shown in Figure 5.12. The optical mode A propagating in the waveguide A is cross-coupled to the waveguide B and transfers optical power to the optical mode B by evanescent field channel. The optical power variations in each waveguide show periodic patterns, and the sum of the optical powers of the optical mode A and optical

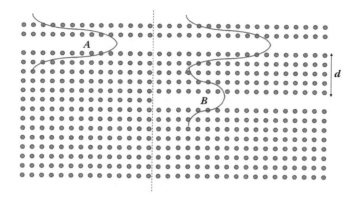

FIGURE 5.12

Optical mode propagation through waveguide A is coupled to waveguide B by evanescent field coupling. The coupling length, i.e., oscillation period of the optical power rate of the optical mode A and optical mode B, is strongly dependent on the separation distance d of two waveguides.

mode B is conserved. The coupling length, i.e., oscillation period of the optical rate of the optical mode A and optical mode B, is strongly dependent on the separation distance between the two waveguides. In Figure 5.13, the change in the coupling length with the variation of the separation distances $d = a \sim d = 6a$ is visualized and the condition for the effective isolation of two waveguides is addressed. As shown in Figure 5.13, the cross-coupling between two waveguides within the propagation length of $405a$ can be negligible when the separation distance d is larger than $6a$. A waveguide is assumed to be effectively isolated from the other waveguide within the propagation length of $405a$.

Here 2D photonic crystals are an example. However, the mathematical description in this book holds the form of 3D theory for real 3D photonic crystal-based nanophotonic networks. In the case of real 3D photonic crystal slabs, the criteria of the field localization assumption for the local pseudo-Fourier modal method is also important, but requires further study.

5.4.2 Tapered Photonic Crystal Resonator

A photonic crystal is an optical structure in which the dielectric constant exhibits periodic modulation [6]. Depending on the dimension of periodic function, it can be classified into a one-, two-, or three-dimensional photonic crystal. In mathematics, it is known that a partial derivative equation for a periodic boundary condition can be described by a linear combination of Floquet modes. This theory can also be applied to optics. A solution of the Helmholtz wave equation in such a structure yields a Bloch mode, and any field distribution in a photonic crystal is expressed as the superposition of Bloch modes. The most attractive feature in a photonic crystal

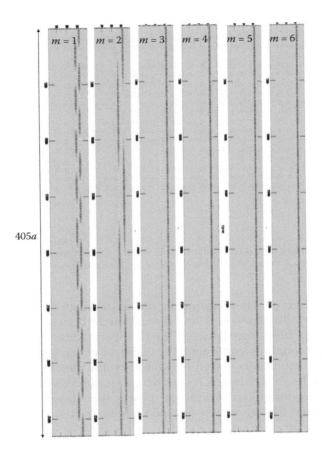

FIGURE 5.13
Change of the coupling length of the optical modes A and B with the variation of the separation
length: $d = a, d = 2a, d = 3a, d = 4a, d = 5a, d = 6a$.

is that there can be an energy level in which the wave cannot propagate
through the photonic crystal. These energy levels depend on the direction of propagation of light and can be obtained by solving the Helmholtz
equation. In particular, there can be a range of energy levels in which
propagation is inhibited regardless of the direction. This is referred to as
the photonic bandgap. This concept has its origin in solid-state physics.

Various applications have been proposed based on a photonic crystal.
Since photonic crystals do not allow the propagation of light, they can act
as mirrors that reflect light. If two photonic crystals are located at a fixed
distance, or line defect, then light is only allowed to propagate along the
defect line [7]. This suggests a photonic crystal waveguide. The modal size
of the guided mode in a photonic crystal waveguide is much smaller than
that in a conventional dielectric waveguide. Moreover, it turned out that

a bending loss of the photonic crystal waveguide can be decreased enormously [8]. Meanwhile, a point defect surrounded by photonic crystals can be used as an optical resonator. Due to low-loss property of dielectric materials, those resonators based on the photonic crystal exhibit extraordinary high Q-factors [9].

Photonic crystals have been a subject of a great deal of research. Many numerical algorithms have successfully been adopted for simulating various photonic crystal structures. However, it is inevitable that they are limited to the scale of the total structure of interest. This arises from the fact that most algorithms solve Maxwell's equations in the space domain rather than the spatial frequency domain. The total memory required to solve a photonic crystal is proportional to the size of the structure of interest. However, since LFMM approaches the same problem in the spatial frequency domain, the required memory in use can be dramatically reduced.

In this section, we implement an LFMM model of a triangular photonic crystal waveguide with an air hole and a photonic crystal resonator with chirped air holes [10]. Figure 5.14(a) and (b) shows a schematic diagram of a triangular lattice structure with an air hole and a photonic crystal waveguide. Throughout this section, it is assumed that both geometry and physical quantities do not vary along the out-of-plane direction, i.e., y-direction. The lattice constant a is 420 nm and the hole radius r is 108 nm (see lines 23 and 25 on *PRCWA_basic.m*). The refractive index of the surrounding dielectric media is 3.46 (see lines 26 to 27 on *PRCWA_basic.m*). The operating wavelength in free space is 1.7 μm (see line 24 on *PRCWA_basic.m*). In the photonic crystal waveguide structure, there is a line defect of missing holes of one row and we employed an offset ratio of 0.98 [10] (see line 30 on *PRCWA_basic.m*).

MATLAB Code 5.4: *PRCWA_basic.m*

```
23      a               = 0.42*um;
24      lambda          = 1.7*um;
25      rr              = 108 * nm;
26      n_air           = 1;
27      n_diel          = 3.46;
28      NN              = 30;
29
30      first_offset=0.98;
31      % first_offset=1.0; % to test the complete bandgap
32      PP=5;
33      QQ=PP;
34
35      Tx=(sqrt(3)*a*PP + sqrt(3)*a/2*first_offset) * 2;
36      Ty=Tx;
37      Tz=a;
38
```

```
39    % basic constant
40    % lambda=a/0.41;          % operating wavelength
41    c0=2.99792458*10^8;       % light speed
42    w0=2*pi*c0/lambda;        % angular frequency
43    eps0=1/(36*pi)*1e-9;% permittivity in free space
44    mu0=4*pi*10^-7;
45
46    % for 2D structure
47    nx=20;                    % x direction truncation order
48    ny=0;                     % y direction truncation order
49    nz=20;                    % z direction truncation order
```

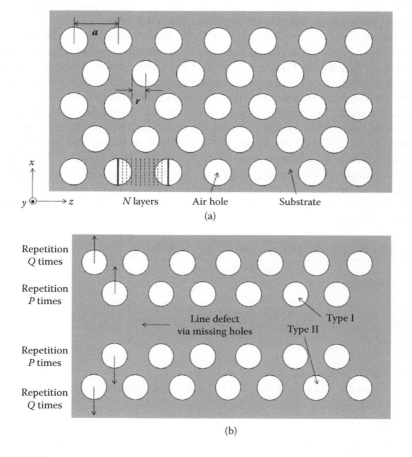

(a)

(b)

FIGURE 5.14
(a) Schematic diagram of a triangular photonic crystal structure with a lattice constant a = 420 nm and air hole radius r = 180 nm. (b) A line defect of missing holes results in the photonic crystal waveguide.

In LFMM code implementation, it is necessary to define computational parameters in addition to the aforementioned geometrical parameters. We adopted the staircase approximation with 30 subblocks for each longitudinal (*z*-direction) period. The repetition time of the very neighboring row of the missing holes toward the transverse direction (*x*-direction), i.e., the type I row, is set to be $P = 5$ (see line 32 on *PRCWA_basic.m*). The repetition time of the next row, i.e., the type II row, is $Q = 5$. Based on these parameters, the transverse period of the computation super-block *Tx* is calculated (see line 35 on *PRCWA_basic.m*). The longitudinal period is the same as the period of the photonic crystal, i.e., $Tz = a$ (see line 37 on *PRCWA_basic.m*). The numbers of the *x*- and *z*-directional Fourier harmonics are set as $nx = 20$ and $nz = 20$, respectively (see lines 47 to 49 on *PRCWA_basic.m*).

LFMM code implementation of the photonic crystal waveguide is presented in the file *Grating_gen_PhC_waveguide.m*. Note that the structure in a period has a mirror symmetry. Therefore, after obtaining the Fourier representation of the first half, this Fourier representation can be copied to the second half in reverse order (see lines 54 to 61 on *Grating_gen_PhC_waveguide.m*). The variable *NN* denotes the total number of subblocks N. The index *nn* running from 1 to N/2 is used to determine the location *xn* of each block in a period (see lines 14 and 16 on *Grating_gen_PhC_waveguide.m*). There are two kinds of regions: one corresponds to the very neighboring row of the missing holes and other holes having the same position in the *z*-direction (type I in Figure 5.14(b)), and the other corresponds to the next row and other holes having the same position in the *z*-direction (type II in Figure 5.14(b)). The former holes lie in the region $xn < rr$ (see line 20 on *Grating_gen_PhC_waveguide.m*), whereas the latter the region is $abs(xn - a/2) < rr$ (see line 34 on *Grating_gen_PhC_waveguide.m*). Each block is composed of rectangular structures representing the region of the air hole, and the size of those regions are calculated as p_size for type I and q_size for type II (see lines 21 and 35 on *Grating_gen_PhC_waveguide.m*). By using the function rect_2D_mesh, we define the region of air holes (see lines 25 to 43 on *Grating_gen_PhC_waveguide.m*). It should be pointed out that the air holes at the edge of the computation cell should cover only half of the region. This is because the computation cells are periodic and there exist other air hole structures across the interface between two computation super-blocks. In LFMM code implementation, this consideration is shown in lines 28 to 31 on *Grating_gen_PhC_waveguide.m*. If this condition is not satisfied, then the photonic crystal waveguide has another defect line located along the edge of a computation cell, which is undesirable.

At this point, we have calculated the Fourier representation of the air hole. This information is carried in the variable *hole_pa* declared in line 18 on *Grating_gen_PhC_waveguide.m*. The Fourier representation of the surrounding dielectric material is easily obtained by taking the inverse pattern of the air hole. A pattern for the rectangular structure covering the total computation cell (from $-Tx/2$ to $Tx/2$) was stored in the variable *full_pattern* (see line 11 on *Grating_gen_PhC_waveguide.m*). Consequently, the Fourier representation of the surrounding dielectric material is given by *inverse_pa = full_pattern - hole_pa*, which is shown in line 46 on *Grating_gen_PhC_waveguide.m*.

The material information expressed as dielectric constants is then multiplied to each geometric structure (see lines 48 to 51 on *Grating_gen_PhC_waveguide.m*). The variables *epra* and *eprd* denote the dielectric constants of the air and dielectric, receptively, and were defined in lines 5 and 6 on *PRCWA_Gen_PhC_waveguide.m*, respectively. It is noteworthy that, in the definition, we employed a very small loss (see lines 5 and 6 on *PRCWA_Gen_PhC_waveguide.m*). Although this value is negligible, it is known that these small values can dramatically enhance the numerical stability. The final process is to take a mirror symmetry for the second half (see lines 54 to 61 on *Grating_gen_PhC_waveguide.m*).

MATLAB Code 5.5: *Grating_gen_PhC_waveguide.m*

```
1     %% space determination
2     [m,n] = meshgrid([1:1:num_hx]-NBx,[1:1:num_hy]-NBy);
3     m = m';
4     n = n';
5
6     eps_L = zeros(num_hx,num_hy,NN);
7     aps_L = zeros(num_hx,num_hy,NN);
8     mu_L = zeros(num_hx,num_hy,NN);
9     bu_L = zeros(num_hx,num_hy,NN);
10
11    full_pattern = rect_2D_mesh(m,n,1,Tx,Ty,-Tx/2,Tx/2,-
      Ty/2,Ty/2);
12
13    %% pattern for the first half
14    for nn = 1 : NN/2
15
16     xn = a/NN * nn;
17
18     hole_pa = zeros(num_hx,num_hy);
19
20     if xn < rr
21     p_size = sqrt(rr^2 - xn^2);
22     for pp = 1 : PP
23     offset = sqrt(3)*a*(pp-1) + sqrt(3)*a*first_offset/2;
24     hole_pa = hole_pa + ...
25     rect_2D_mesh(m,n,1,Tx,Ty,+offset-p_size,+offset+p_
       size,-Ty/2,Ty/2) + ...
26     rect_2D_mesh(m,n,1,Tx,Ty,-offset-p_size,-offset+p_
       size,-Ty/2,Ty/2);
```

```
27     end
28     offset = sqrt(3)*a*((PP+1)-1) + sqrt(3)*a*first_
       offset/2; % should be equal to Tx/2
29     hole_pa = hole_pa + ...
30     rect_2D_mesh(m,n,1,Tx,Ty,+offset-p_size,+offset,-
       Ty/2,Ty/2) + ...
31     rect_2D_mesh(m,n,1,Tx,Ty,-offset,-offset+p_size,-
       Ty/2,Ty/2);
32     end
33
34     if abs(xn - a/2) < rr
35     q_size = sqrt(rr^2 - (xn - a/2)^2);
36     for qq = 1 : QQ
37     offset = sqrt(3)*a*(qq-1) + sqrt(3)*a*first_offset/2 +
       sqrt(3)*a/2;
38     hole_pa = hole_pa + ...
39     rect_2D_mesh(m,n,1,Tx,Ty,+offset-q_size,+offset+q_
       size,-Ty/2,Ty/2) + ...
40     rect_2D_mesh(m,n,1,Tx,Ty,-offset-q_size,-offset+q_
       size,-Ty/2,Ty/2);
41     end
42     % hole_pa = hole_pa + ... % to test the complete bandgap
43     % rect_2D_mesh(m,n,1,Tx,Ty,-q_size,+q_size,-Ty/2,Ty/2);
44     end
45
46     inverse_pa = full_pattern - hole_pa;
47
48     eps_L(:,:,nn) = epra * hole_pa + eprd * inverse_pa;
49     aps_L(:,:,nn) = 1/epra * hole_pa + 1/eprd * inverse_pa;
50     mu_L (:,:,nn) = mur0 * full_pattern;
51     bu_L (:,:,nn) = 1/mur0 * full_pattern;
52     end
53
54     %% copy to the other half side
55     for nn = NN : -1 : NN/2+1
56     eps_L(:,:,nn) = eps_L(:,:,NN+1-nn);
57     aps_L(:,:,nn) = aps_L(:,:,NN+1-nn);
58     mu_L(:,:,nn) = mu_L (:,:,NN+1-nn);
59     bu_L(:,:,nn) = bu_L (:,:,NN+1-nn);
60
61     end
```

In order to check the result of the Fourier representation of the photonic crystal waveguide, we calculate the distribution of the dielectric constant from the Fourier representation. This process is shown in lines 79 to 105 on *PRCWA_Gen_PhC_waveguide.m*. The key process is the inverse Fourier

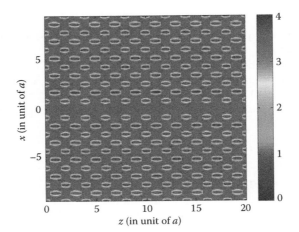

FIGURE 5.15
Dielectric constant distribution of the photonic crystal waveguide inversely calculated from the Fourier representation.

transform shown in line 94 on *PRCWA_Gen_PhC_waveguide.m*. The result is shown in Figure 5.15. It can clearly be seen that there is the line defect of missing holes along the middle of the photonic crystal. Note that this figure is generated by 20 repetitions of the basic Fourier representation (see line 82 on *PRCWA_Gen_PhC_waveguide.m*). This value can be arbitrarily changed.

MATLAB Code 5.6: *PRCWA_Gen_PhC_waveguide.m*

```
79    xx=1*(-Tx/2:Tx*0.002:Tx/2);
80    yy=0;%yy=3*(-Ty/2:Ty*0.002:Ty/2);
81
82    z_repetition=20;
83    zz=linspace(0,Tz*z_repetition,NN*z_repetition);;
84
85
86
87    Gr_str=zeros(length(xx),Nlay);
88    [ya xa]=meshgrid(yy,xx);
89
90    for nn=1:Nlay
91     Gr_str_=zeros(length(xx),length(yy));
92     for k=-2*nx:2*nx
93     for l=-2*ny:2*ny
94     Gr_str_=Gr_str_+Aeps_xx(k+NBx,l+NBy,nn)*exp(j*(k*xa*2
          *pi/Tx+l*ya*2*pi/Ty));
95     end
```

```
96     end
97     Gr_str(:,nn)=Gr_str_(:,1);
98     end
99
100    figure 60);set(gca,'fontsize',16);set(gca,'fontname','ti
       mes new roman');
101    imagesc(zz/a,xx/a,sqrt(abs(repmat(Gr_str,[1 10]))));set
       (gca,'ydir','normal');set(gca,'fontname','times new
       roman');
102    xlabel('z (in unit of a)');set(gca,'fontname','times new
       roman');
103    ylabel('x (in unit of a)');set(gca,'fontname','times new
       roman');
104    axis equal;axis([zz(1)/a zz(end)/a xx(1)/a xx(end)/a]);
105    colorbar;caxis([0 4]);set(gca,'fontsize',16);set(gca,
       'fontname','times new roman');
```

The next process is the Bloch mode analysis. This is a common and shared process, so that it is not necessary for users to modify it. The Bloch mode analysis can be classified into two categories: the *A* Bloch mode analysis and the *B* Bloch mode analysis. The former corresponds to the case where light is incident from the left (negative *z*) domain, whereas the latter is from the right (positive *z*) domain. Depending on the simulation condition, users need to choose an appropriate analysis. Here, we choose the *A* Bloch mode analysis (see lines 121 and 147 on *LFMM_pcwg.m*). The *A* Bloch mode analysis is composed of four steps; the block S-matrix calculation of each subblock (see lines 15 to 36 on *AmodeBlochAnalysis.m*), the total S-matrix calculation based on the extended Redheffer star product (see lines 38 to 93 on *AmodeBlochAnalysis.m*), the positive and negative Bloch mode analysis (see lines 94 to 274 on *AmodeBlochAnalysis.m*), and the pseudo-Fourier representation of Bloch modes (see lines 276 to 624 on *AmodeBlochAnalysis.m*). A detailed description of each step is introduced in Section 5.3.

Figure 5.16 shows the electromagnetic field distribution of the photonic crystal waveguide with the fundamental mode. This field distribution is calculated in the file *LFMM_Smat_Amode_case1.m*, which is called in line 244 on *LFMM_pcwg.m*. The *x*-polarized plane wave is incident from the left side. The photonic crystal waveguide with 10 periods along the *z*-direction is located in the middle of the figure. The *x*-component of the electric field (E_x) and the *y*-component of the magnetic field (H_y) exhibit the symmetric field distributions, while the *z*-component of the electric field (E_z) shows the antisymmetric field distribution.

The Fourier representation of the complete bandgap can be easily obtained by modifying a few code lines. First, the first offset value of 0.98 needs to be changed to 1.0 (see lines 30 and 31 on *PRCWA_basic.m*). This slight change

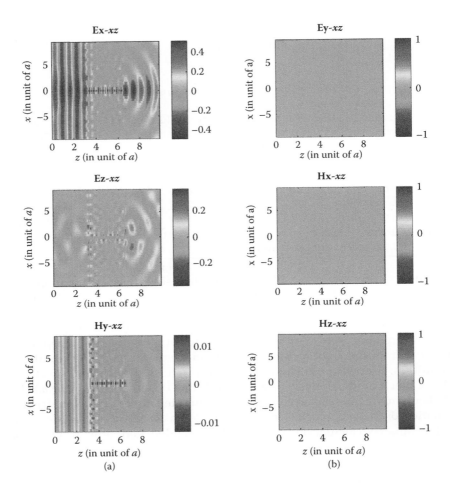

FIGURE 5.16
Electromagnetic field distribution of the photonic crystal waveguide with the fundamental mode.

was adopted for the photonic crystal waveguide. Any irregular lattice structure may play a role of a defect, resulting in a guided mode through the defect line. The missing hole row is then recovered by uncomment lines 42 and 43 on *Grating_gen_PhC_waveguide.m*. The result of the electromagnetic field distribution for the complete bandgap test is presented in Figure 5.17. It is seen that the incident electromagnetic field is totally reflected and cannot propagate through the photonic crystal structure.

Now, we are led to a discussion of photonic crystal resonators. This function can be accomplished by various methods such as local index modulation on a photonic crystal waveguide, introducing an additional hole in the line defect via missing hole array in a photonic crystal waveguide, or lateral shifting of the hole array near the line defect via missing hole. In this section,

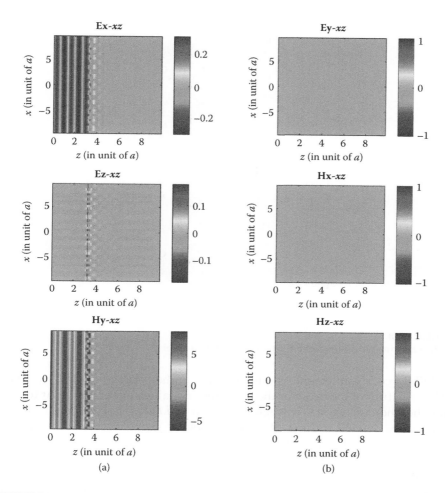

FIGURE 5.17
Electromagnetic field distribution of the photonic crystal bandgap.

by tapering the position of the air hole around the line defect of the photonic crystal waveguide discussed above, we implement the photonic crystal resonator. In addition, the cascaded structure, i.e., the photonic crystal waveguide–the photonic crystal resonator–the photonic crystal waveguide, will be implemented in LFMM MATLAB code.

Figure 5.18 shows a schematic diagram of the tapered photonic crystal waveguide. Through seven periods along the z-direction, air holes near the core are adiabatically shifted. The offset distance varies gradually from 8 nm (type C) through 16 nm (type B) to 24 nm (type A). Considering that the lattice constant a is 420 nm, the offset distance is very short. However, it is sufficient to act as a resonator, as will be shown below.

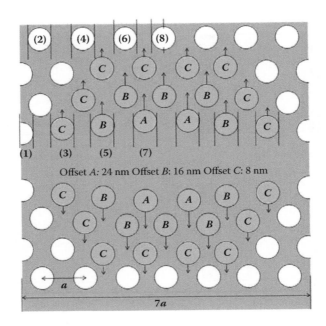

FIGURE 5.18
Schematic diagram of the tapered photonic crystal waveguide.

LFMM code implementation is provided in the package *LFMM_resonator*. The structure of the code for the photonic crystal resonator is similar to that for the photonic crystal waveguide shown in *LFMM_pcwg*. The offset distances are defined in lines 14 to 16 on *Grating_gen_PhC_resonator.m*. As in the photonic crystal waveguide, the structure has a mirror symmetry. The Fourier representation for the second half can thus be obtained from the first half, and we only iterate the first half (see line 19 on on *Grating_gen_ PhC_resonator.m*). The region can be classified into eight subregions; each of them is a number with parentheses from (1) to (8), as shown in Figure 5.18. In accordance with the subregions, the MATLAB code branches off as shown in lines 25, 39, 49, 67, 81, 103, 117, and 139 on *Grating_gen_PhC_resonator.m*. Let us first consider the region (7), where the first row experiences a shift with offset A and the second row a shift with offset C. In line 119, the offset of the first row is calculated regarding the offset A. Then, in line 123, the offset of the second row is obtained by using the offset C. Next, through lines 127 to 132, the rest of the rows are expressed using for-loop with the variable pp iterating from 3 to PP. Finally, lines 133 to 136 describe the final row that lies across the interface of the computation super-blocks. The overall information for air holes is stored in the variable *hole_pa*, and the Fourier representation of the surrounding dielectric material can be easily obtained by taking the inverse (see line 153 on *Grating_gen_PhC_resonator.m*). Figure 5.19 shows the dielectric constant distribution calculated from the Fourier representation of

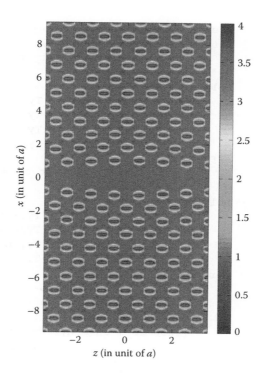

FIGURE 5.19
Dielectric constant distribution of the photonic crystal resonator inversely calculated from the Fourier representation.

the photonic crystal resonator. In order to put an emphasis on the shift of the holes, an exaggeration is adopted; the offset distances of types A, B, and C are set to be 90, 60, and 30 nm, respectively.

MATLAB Code 5.7: *Grating_gen_PhC_resonator.m*

```
14      offset_C      = 8 * nm;
15      offset_B      = 16 * nm;
16      offset_A      = 24 * nm;
19      for nn        = 1 : Nlay/2
21          xn        = cell_num*a/Nlay * nn;
23          hole_pa   = zeros(num_hx,num_hy);
25          if abs(xn - 0*a/2) < rr          % region (1)
39          if abs(xn - 1*a/2) < rr          % region (2)
49          if abs(xn - 2*a/2) < rr          % region (3)
67          if abs(xn - 3*a/2) < rr          % region (4)
81          if abs(xn - 4*a/2) < rr          % region (5)
103         if abs(xn - 5*a/2) < rr          % region (6)
117         if abs(xn - 6*a/2) < rr          % region (7)
```

```
118              p_size   = sqrt(rr^2 - (xn - 6*a/2)^2);
119              offset   = sqrt(3)*a*( 1-1) +
                             sqrt(3)*a*first_offset/2 +
                             offset_A;
120              hole_pa  = hole_pa + ...
121                  rect_2D_mesh(m,n,1,Tx,Ty,+offset-p_
                         size,+offset+p_size,-Ty/2,Ty/2) + ...
122                  rect_2D_mesh(m,n,1,Tx,Ty,-offset-p_
                         size,-offset+p_size,-Ty/2,Ty/2);
123              offset   = sqrt(3)*a*( 2-1) +
                             sqrt(3)*a*first_offset/2 +
                             offset_C;
124              hole_pa  = hole_pa + ...
125                  rect_2D_mesh(m,n,1,Tx,Ty,+offset-p_
                         size,+offset+p_size,-Ty/2,Ty/2) + ...
126                  rect_2D_mesh(m,n,1,Tx,Ty,-offset-p_
                         size,-offset+p_size,-Ty/2,Ty/2);
127          for pp = 3 : PP
128                  offset = sqrt(3)*a*(pp-1) +
                         sqrt(3)*a*first_offset/2;
129                  hole_pa = hole_pa + ...
130                      rect_2D_mesh(m,n,1,Tx,Ty,+offset-p_
                             size,+offset+p_size,-Ty/2,Ty/2) + ...
131                      rect_2D_mesh(m,n,1,Tx,Ty,-offset-p_
                             size,-offset+p_size,-Ty/2,Ty/2);
132          end
133                  offset = sqrt(3)*a*((PP+1)-1) +
                         sqrt(3)*a*first_offset/2; % should be
                         equal to Tx/2
                     hole_pa = hole_pa + ...
134                      rect_2D_mesh(m,n,1,Tx,Ty,+offset-p_
                             size,+offset,-Ty/2,Ty/2) + ...
135                      rect_2D_mesh(m,n,1,Tx,Ty,-offset,-
                             offset+p_size,-Ty/2,Ty/2);
136      end
137
138      if abs(xn - 7*a/2) < rr % region (8)
139      inverse_pa = full_pattern - hole_pa;
153
154      eps_L(:,:,nn) = epra * hole_pa + eprd * inverse_pa;
155      aps_L(:,:,nn) = 1/epra * hole_pa + 1/eprd *
          inverse_pa;
156      mu_L (:,:,nn) = mur0 * full_pattern;
157      bu_L (:,:,nn) = 1/mur0 * full_pattern;
158  end
159
```

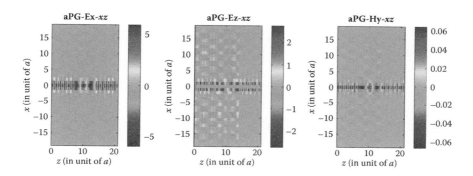

FIGURE 5.20
Electromagnetic field distribution in the cascaded structure of the photonic crystal waveguide-resonator-waveguide.

The rest of the analysis for implementing the photonic crystal resonator is similar to that for the photonic crystal waveguide discussed above. However, users should exercise care in setting the z-directional period of the computation super-block, Tz (see line 36 on *PRCWA_basic.m*).

After establishing the photonic crystal waveguide and resonator, we can build up a cascaded structure by using the extended Redheffer star product. This implementation is presented in *main.m*. Each simulation result for the photonic crystal waveguide and resonator was saved in the files *2portDATA_pcwg.mat* and *2portDATA_resonator.mat*. The first part of the code is loading that information (see lines 12 and 52 on *main.m*). Through lines 85 to 117 on *main.m*, the S-matrix for the multi-super-block of the input region (the photonic crystal waveguide) and the body (the photonic crystal resonator) is calculated via the extended Redheffer star product. After that, the S-matrix for the total super-block is obtained by taking the output region (the photonic crystal waveguide) into account (see lines 119 to 143 on *main.m*). The overall field distribution is calculated in the file *LFMM_Amode_field_visual_xz_case4_Lwg_Rwg_leftright.m*, in which the fundamental propagating mode in the photonic crystal waveguide is excited (see lines 4 to 6 on *LFMM_Amode_field_visual_xz_case4_Lwg_Rwg_leftright.m*). Figure 5.20 shows the electromagnetic field distribution of the cascaded structure of the photonic crystal waveguide and resonator. Significant field enhancement inside the photonic crystal resonator is observed. Since the group velocity through the photonic crystal resonator is enormously reduced, this slow light property can be utilized for an optical buffer.

6

Perspectives on the Fourier Modal Method

In the previous chapters, FMM and LFMM were established for use in the modal analysis of collinear multiblock and multi-super-block structures. Using the perfect matched layer (PML) approach, nonperiodic structures were successfully modeled and analyzed by FMM. At this stage, it might be concluded that FMM is now nearly mature and that the focus of future issues of FMM should be applications. However, as this chapter will show, by the idea of local Fourier analysis, the theoretical territory of FMM can be expanded to the development of novel theories. The present FMM was applied to collinearly cascaded multiblock or multi-super-block structures that were actually considered to be limitations for FMM. However, the concept of localization enables LFMM to be exploited for the general topology of nanophotonic devices, i.e., nanophotonic networks. In this chapter, for this topic, the generalized S-matrix method (GSMM) is proposed and the unification of LFMM and GSMM is developed for modeling nanophotonic networks.

6.1 Nanophotonic Network Modeling

The collective system of such nanophotonic devices, more generally photonic blocks, is presented in Figure 6.1. The functional photonic blocks are represented in the form of geometric figures: circles, rectangles, and other geometric figures. The interaction between blocks is denoted by bidirectional arrows between photonic blocks such as two-port blocks, four-port cross-blocks, two-port light sources, and general blocks having more ports. Let us call this type of system a nanophotonic network. Nanophotonic networks are composed of a wide variety of photonic devices. To study the collective dynamics of such nanophotonic networks, an understanding of the theory of nanophotonic networks is necessary.

It is not possible to correctly analyze such networks or global phenomena without access to huge computing resources. Although the optical fields act in accordance with Maxwell's equations, the information needed to represent the optical field is too huge to permit a direct analysis. The development of rigorous and efficient mathematical modeling of nanophotonic networks is an urgent and important theme at present.

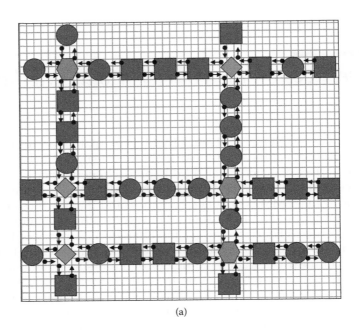

(a)

Basic Elements

2 port block

4 port block   2 light source

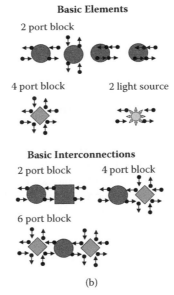

Basic Interconnections

2 port block   4 port block

6 port block

(b)

FIGURE 6.1
Schematic of nanophotonic network and basic elements and interconnections. (From H. Kim and B. Lee, *J. Opt. Soc. Am. B*, 25(4), 2008. With permission.)

For this, a technical approach can be used to deal with such global structures with an intuitive assumption. First, in practical nanophotonic networks, optical fields are usually localized on engineered structures or devices such as waveguides and resonators. Thus we can obtain a reasonable solution for Maxwell's equations around these local areas, where the optical field energy is not negligible. It should then be possible to develop a systematic mathematical model that describes the relationship and interactions between each locally analyzed region.

This concept is the motivation for the development of the linear system theory of nanophotonic networks. The mathematical model of a nanophotonic network is made up of two key subtheories, LFMM and GSMM. The LFMA is an extended FMM of solving electromagnetic fields within four-port cross-block structures with four boundaries. LFMM addressed in Chapter 5 is a specific case of the general scheme of LFMM that will be elucidated in this chapter.

Conventional FMM is only applied to 1D, cascaded multiblocks. This type of structure can be considered to be a two-port block structure. The conventional SMM is also restricted to applications to only two-port block structures in combination with FMM. This structural restriction has been considered to be an inherent limitation of FMMs. However, the local Fourier modal analysis (LFMA or LFMM) overcomes this limitation and can successfully be used to analyze four-port cross-blocks with four boundaries. Furthermore, the interconnections between a two-port block and a two-port block, a four-port block and a two-port block, and a four-port block and a four-port block can be systematically described by GSMM. The LFMA is composed of two main subtheories: (1) LFMA for analyzing internal eigenmodes of four-port cross-blocks and (2) GSMM for modeling a nanophotonic network by interconnecting several two-port blocks and four-port blocks.

This model can be recursively extended to characterize a general $N \times M$ port nanophotonic network composed of several four-port blocks and two-port blocks. As a result, the linear system model of the $N \times M$ port nanophotonic network based on the GSMM can be built. The linear system theory of nanophotonic networks with LFMM and GSMM is expected to provide an elegant numerical framework of design and analysis of complex large-scale integrated nanophotonic networks. This topic is beyond the scope of this book and requires further research.

From the standpoint of methodology, the LFMM scheme is efficient, since the local field analysis is performed with the reasonable and practical assumption of field localization in nanophotonic networks. From the theoretical point of view, the proposed methodology can be considered to be a linear system theory of nanophotonic networks based on a local spectral analysis of electromagnetic field distributions. The basic assumption of this local analysis is the field localization around nanophotonic devices such as waveguides, resonators, mirrors, and so on. This assumption is a reasonable

premise in integrated nanophotonic networks of interest. Instead of computing the whole structure of a network, local regions that are occupied by
functional photonic blocks are characterized. The proposed theory provides
a mathematical framework of design and analysis of complex large-scale
integrated nanophotonic networks. The contents of this chapter are based on
the original paper [1].

In this section, as a prerequisite step for constructing a general linear system theory of nanophotonic networks, the electromagnetic analysis on the
basic element, a four-port crossed nanophotonic structure, is described. The
most important feature of LFMA is the pseudo-Fourier representation of
electromagnetic fields.

Figure 6.2 shows a schematic diagram of a four-port cross-block. As an
analysis example, the 2D photonic crystal cross-waveguide structure shown
in Figure 6.2(a) is chosen. This cross-waveguide structure is composed of
five subparts: port 1, port 2, port 3, port 4, and the intersection cross-block.
The complete characterizations of the four-port cross-block and the two-port
block are represented by the 4×4 S-matrix and the 2×2 S-matrix, respectively. For convenience, let the S-matrices of two-port blocks placed along
the transverse direction and those placed along the longitudinal direction be
distinguished, respectively, by

$$\mathbf{S} = \begin{pmatrix} \vec{T} & \vec{R} \\ \bar{R} & \bar{T} \end{pmatrix} \tag{6.1a}$$

$$\mathbf{S} = \begin{pmatrix} T_\uparrow & R_\downarrow \\ R_\uparrow & T_\downarrow \end{pmatrix} \tag{6.1b}$$

The S-matrix of a four-port cross-block is defined by

$$\mathbf{S} = \begin{pmatrix} \mathbf{S}_{11} & \mathbf{S}_{21} & \mathbf{S}_{31} & \mathbf{S}_{41} \\ \mathbf{S}_{12} & \mathbf{S}_{22} & \mathbf{S}_{32} & \mathbf{S}_{42} \\ \mathbf{S}_{13} & \mathbf{S}_{23} & \mathbf{S}_{33} & \mathbf{S}_{43} \\ \mathbf{S}_{14} & \mathbf{S}_{24} & \mathbf{S}_{34} & \mathbf{S}_{44} \end{pmatrix} \tag{6.2}$$

In 2D geometry, four vertex points, (x_+, z_-), (x_+, z_+), (x_-, z_-), and (x_-, z_+), define
the four boundaries where the cross-block contacts port 1, port 2, port 3,
and port 4. Figure 6.2(b) shows the S-matrix diagram of the photonic crystal
cross-waveguide structure.

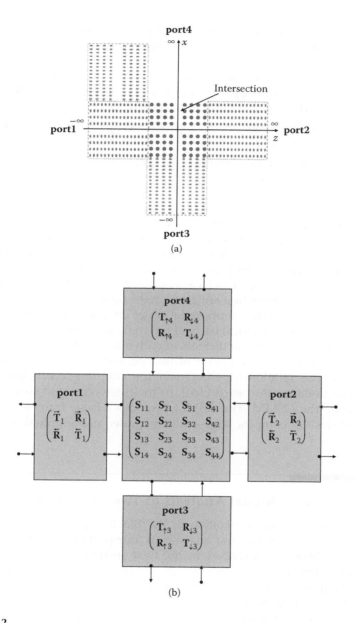

FIGURE 6.2

(a) Photonic crystal cross-waveguide structure. (b) S-matrix diagram of the photonic crystal cross-waveguide structure. (From H. Kim and B. Lee, *J. Opt. Soc. Am. B*, 25(4), 2008. With permission.)

For building the mathematical modeling of the described nanophotonic structures, the following basic elements should be prepared:

1. S-matrices, Bloch eigenmodes represented by pseudo-Fourier series, coupling coefficients operators of two-port blocks placed along the transverse direction and the longitudinal direction
2. S-matrices, Bloch eigenmodes represented by pseudo-Fourier series, coupling coefficients operators of four-port blocks
3. Generalized recursion formulas for interconnecting the S-matrices of composing blocks and updating internal coupling coefficient operators of each block

The analysis will progress in a stepwise manner. After completing the analysis of the 4×4 S-matrix of the four-port intersection block, ports 1, 2, 3, and 4 of the four-port block will be consecutively connected to the half-infinite horizontal two-port blocks and vertical two-port blocks. The S-matrices of the combined blocks are updated at each stage of the process. Details of this interconnection are given in the following section. In the case of LFMA, the local Fourier representation of internal Bloch eigenmodes is identified as the mathematical basis of internal electromagnetic field distributions, and the coupling coefficients of each eigenmode are determined so as to satisfy the transverse field continuation conditions at the four given boundaries. The relationships between the coupling coefficients of the Bloch eigenmodes within each block in the interconnected structures are described by the GSMM. The GSMM provides a basic framework for the general analysis of complex large-scale integrated nanophotonic networks.

6.2 Local Fourier Modal Analysis of Two-Port Block Structures

The pseudo-Fourier representation of the Bloch eigenmodes is an essential factor in the modeling and analysis of four-port crossed nanophotonic structures, which are described in the next section. Hereafter, the Bloch eigenmodes represented by the pseudo-Fourier series are adopted as the mathematical basis for the electromagnetic field distributions. Let the two-port blocks be placed along the transverse direction and those placed along the longitudinal direction be distinguished by the prefixes α- and β-, respectively. Figure 6.3 illustrates three kinds of α-blocks: two-port α-block with finite size, two-port half-infinite α-block with right boundary, and two-port half-infinite α-block with left boundary. On the other hand, Figure 6.4 illustrates three kinds of β-blocks: two-port β-block with finite size, two-port half-infinite β-block with upper boundary, and two-port half-infinite β-block with lower boundary.

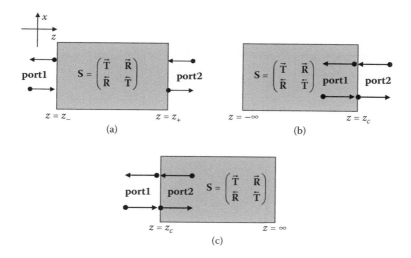

FIGURE 6.3
(a) Two-port α-block with finite size, (b) two-port half-infinite α-block with right boundary, and (c) two-port half-infinite α-block with left boundary. (From H. Kim and B. Lee, *J. Opt. Soc. Am. B*, 25(4), 2008. With permission.)

In the representation of the field distributions in α-blocks, the subscript α is also used. The reciprocal vectors of α-modes, $(k_{\alpha,x,m}, k_{\alpha,y,n}, k_{\alpha,z,p}^{(g)})$, are defined by

$$\left(k_{\alpha,x,m}, k_{\alpha,y,n}, k_{\alpha,z,p}^{(g)}\right) = \left(k_{x,0} + \frac{2\pi}{T_x} m, k_{y,0} + \frac{2\pi}{T_y} n, k_{\alpha,z}^{(g)} + \frac{2\pi}{T_z} p\right) \quad (6.3)$$

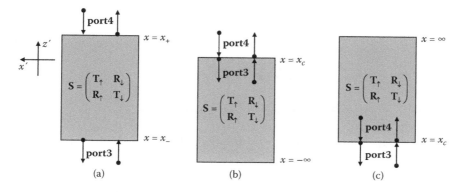

FIGURE 6.4
(a) Two-port β-block with finite size, (b) two-port half-infinite β-block with right boundary, and (c) two-port half-infinite β-block with left boundary. (From H. Kim and B. Lee, *J. Opt. Soc. Am. B*, 25(4), 2008. With permission.)

Then let us represent the Bloch eigenmodes of two-port α-blocks as

$$
\begin{pmatrix} \mathbf{E}_{\alpha,(g)}^{\pm} \\ \mathbf{H}_{\alpha,(g)}^{\pm} \end{pmatrix} = \sum_{m=-M}^{M} \sum_{n=-N}^{N} \sum_{p=-H}^{H} \begin{pmatrix} E_{\alpha,x,m,n,p}^{(g)\pm}, E_{\alpha,y,m,n,p}^{(g)\pm}, E_{\alpha,z,m,n,p}^{(g)\pm} \\ H_{\alpha,x,m,n,p}^{(g)\pm}, H_{\alpha,y,m,n,p}^{(g)\pm}, H_{\alpha,z,m,n,p}^{(g)\pm} \end{pmatrix} e^{\left[j\left(k_{\alpha,x,m}x + k_{\alpha,y,n}y + k_{\alpha,z,p}^{(g)\pm}(z-z_{\mp}) \right) \right]}
$$

$$(6.4)$$

In the representation of the field distributions in β-blocks, the subscript β is used. The reciprocal vectors of β-modes are denoted by $(k_{\beta,x,m}^{(g)}, k_{\beta,y,n}, k_{\beta,z,p})$ and defined by

$$
\left(k_{\beta,x,m}^{(g)}, k_{\beta,y,n}, k_{\beta,z,p} \right) = \left(k_{\beta,x,0}^{(g)} + \frac{2\pi}{T_x}m, k_{y,0} + \frac{2\pi}{T_y}n, k_{z,0} + \frac{2\pi}{T_z}p \right)
$$

$$(6.5)$$

The Bloch eigenmodes of two-port β-blocks are represented by the pseudo-Fourier series with the subscript β as

$$
\begin{pmatrix} \mathbf{E}_{\beta,(g)}^{\pm} \\ \mathbf{H}_{\beta,(g)}^{\pm} \end{pmatrix} = \sum_{m=-M}^{M} \sum_{n=-N}^{N} \sum_{p=-H}^{H} \begin{pmatrix} E_{\beta,x,m,n,p}^{(g)\pm}, E_{\beta,y,m,n,p}^{(g)\pm}, E_{\beta,z,m,n,p}^{(g)\pm} \\ H_{\beta,x,m,n,p}^{(g)\pm}, H_{\beta,y,m,n,p}^{(g)\pm}, H_{\beta,z,m,n,p}^{(g)\pm} \end{pmatrix} e^{j\left(k_{\beta,x,m}^{(g)\pm}(x-x_{\mp}) + k_{\beta,y,n}y + k_{\beta,z,p}z \right)}
$$

$$(6.6)$$

Let us note the meanings of the local coordinate of the β-block, (x',y',z'), shown in Figure 6.4(a). In practical computation, β-blocks are analyzed on the local coordinate (x',y',z') for convenience. When converting the information obtained in the local coordinate, (x',y',z'), to those in the default coordinate, (x',y',z'), a careful consideration on the following relations is required:

$$
\varepsilon'(x',y',z') = \varepsilon(z',y',-x') \tag{6.7a}
$$

$$
\mu'(x',y',z') = \mu(z',y',-x') \tag{6.7b}
$$

$$
(E_{x,m,n,p}, E_{y,m,n,p}, E_{z,m,n,p}) = (E_{z',p,n,-m}, E_{y',p,n,-m}, -E_{x',p,n,-m}) \tag{6.7c}
$$

$$
(H_{x,m,n,p}, H_{y,m,n,p}, H_{z,m,n,p}) = (H_{z',p,n,-m}, H_{y',p,n,-m}, -H_{x',p,n,-m}) \tag{6.7d}
$$

Next, the S-matrices of two-port blocks are derived. The S-matrix components are obtained by the following bidirectional characterization procedure. Let us consider the left-to-right directional characterization of the multiblock for obtaining the block S-matrix of this structure, **S**. The left-to-right and right-to-left characterizations of the two-port α-blocks read, respectively, as

$$
\begin{pmatrix} \underline{\underline{W}}_{\alpha,h} & \underline{\underline{W}}_{\alpha,h} \\ \underline{\underline{V}}_{\alpha,h} & -\underline{\underline{V}}_{\alpha,h} \end{pmatrix} \begin{pmatrix} \vec{\underline{U}} \\ \vec{\underline{R}} \end{pmatrix} = \begin{pmatrix} \underline{\underline{W}}_\alpha^+(0) & \underline{\underline{W}}_\alpha^-(z_- - z_+) \\ \underline{\underline{V}}_\alpha^+(0) & \underline{\underline{V}}_\alpha^-(z_- - z_+) \end{pmatrix} \begin{pmatrix} \underline{\underline{C}}_{\alpha,a}^+ \\ \underline{\underline{C}}_{\alpha,a}^- \end{pmatrix} \quad \text{at } z = z_- \tag{6.8a}
$$

$$
\begin{pmatrix} \underline{\underline{W}}_{\alpha,h} & \underline{\underline{W}}_{\alpha,h} \\ \underline{\underline{V}}_{\alpha,h} & -\underline{\underline{V}}_{\alpha,h} \end{pmatrix} \begin{pmatrix} \vec{\underline{T}} \\ 0 \end{pmatrix} = \begin{pmatrix} \underline{\underline{W}}_\alpha^+(z_+ - z_-) & \underline{\underline{W}}_\alpha^-(0) \\ \underline{\underline{V}}_\alpha^+(z_+ - z_-) & \underline{\underline{V}}_\alpha^-(0) \end{pmatrix} \begin{pmatrix} \underline{\underline{C}}_{\alpha,a}^+ \\ \underline{\underline{C}}_{\alpha,a}^- \end{pmatrix} \quad \text{at } z = z_+ \tag{6.8b}
$$

and

$$
\begin{pmatrix} \underline{\underline{W}}_{\alpha,h} & \underline{\underline{W}}_{\alpha,h} \\ \underline{\underline{V}}_{\alpha,h} & -\underline{\underline{V}}_{\alpha,h} \end{pmatrix} \begin{pmatrix} 0 \\ \vec{\underline{T}} \end{pmatrix} = \begin{pmatrix} \underline{\underline{W}}_\alpha^+(0) & \underline{\underline{W}}_\alpha^-(z_- - z_+) \\ \underline{\underline{V}}_\alpha^+(0) & \underline{\underline{V}}_\alpha^-(z_- - z_+) \end{pmatrix} \begin{pmatrix} \underline{\underline{C}}_{\alpha,b}^+ \\ \underline{\underline{C}}_{\alpha,b}^- \end{pmatrix} \quad \text{at } z = z_- \tag{6.8c}
$$

$$
\begin{pmatrix} \underline{\underline{W}}_{\alpha,h} & \underline{\underline{W}}_{\alpha,h} \\ \underline{\underline{V}}_{\alpha,h} & -\underline{\underline{V}}_{\alpha,h} \end{pmatrix} \begin{pmatrix} \vec{\underline{R}} \\ \vec{\underline{U}} \end{pmatrix} = \begin{pmatrix} \underline{\underline{W}}_\alpha^+(z_+ - z_-) & \underline{\underline{W}}_\alpha^-(0) \\ \underline{\underline{V}}_\alpha^+(z_+ - z_-) & \underline{\underline{V}}_\alpha^-(0) \end{pmatrix} \begin{pmatrix} \underline{\underline{C}}_{\alpha,b}^+ \\ \underline{\underline{C}}_{\alpha,b}^- \end{pmatrix} \quad \text{at } z = z_+ \tag{6.8d}
$$

where $\underline{\underline{W}}_{\alpha,h}$ and $\underline{\underline{V}}_{\alpha,h}$ are $[2(2M+1)(2N+1)]\times[2(2M+1)(2N+1)]$ matrices given, respectively, by

$$
\underline{\underline{W}}_{\alpha,h} = \begin{pmatrix} \underline{\underline{I}} & 0 \\ 0 & \underline{\underline{I}} \end{pmatrix} \tag{6.9a}
$$

$$
\underline{\underline{V}}_{\alpha,h} = \begin{pmatrix} \left[\dfrac{1}{\omega\mu_0} \dfrac{k_{\alpha,x,m}k_{\alpha,y,n}}{k_{\alpha,z,m,n}} \right] & \left[\dfrac{1}{\omega\mu_0} \dfrac{\left(k_{\alpha,z,m,n}^2 + k_{\alpha,x,m}^2\right)}{k_{\alpha,z,m,n}} \right] \\ \left[-\dfrac{1}{\omega\mu_0} \dfrac{\left(k_{\alpha,y,n}^2 + k_{\alpha,z,m,n}^2\right)}{k_{\alpha,z,m,n}} \right] & \left[-\dfrac{1}{\omega\mu_0} \dfrac{k_{\alpha,y,n}k_{\alpha,x,m}}{k_{\alpha,z,m,n}} \right] \end{pmatrix} \tag{6.9b}
$$

$\underline{\underline{W}}_\alpha^+(z)$ and $\underline{\underline{V}}_\alpha^+(z)$ are $[2(2M+1)(2N+1)]\times M^+$ matrices indicating the part of the positive modes given, respectively, by

$$
\underline{\underline{W}}_\alpha^+(z) = \begin{pmatrix} \left[\displaystyle\sum_{p=-H}^{H} \tilde{E}_{\alpha,y,m,n,p}^{(1)+} e^{jk_{\alpha,z,p}^{(1)+}z} \right] & \cdots & \left[\displaystyle\sum_{p=-H}^{H} \tilde{E}_{\alpha,y,m,n,p}^{(M^+)+} e^{jk_{\alpha,z,p}^{(M^+)+}z} \right] \\ \left[\displaystyle\sum_{p=-H}^{H} \tilde{E}_{\alpha,x,m,n,p}^{(1)+} e^{jk_{\alpha,z,p}^{(1)+}z} \right] & \cdots & \left[\displaystyle\sum_{p=-H}^{H} \tilde{E}_{\alpha,x,m,n,p}^{(M^+)+} e^{jk_{\alpha,z,p}^{(M^+)+}z} \right] \end{pmatrix} \tag{6.9c}
$$

$$
\underline{\underline{V}}_\alpha^+(z) = \left(
\begin{bmatrix}
\displaystyle\sum_{p=-H}^{H} \tilde{H}_{\alpha,y,m,n,p}^{(1)+} e^{jk_{\alpha,z,p}^{(1)+}z}
\end{bmatrix}
\quad \cdots \quad
\begin{bmatrix}
\displaystyle\sum_{p=-H}^{H} \tilde{H}_{\alpha,y,m,n,p}^{(M^+)+} e^{jk_{\alpha,z,p}^{(M^+)+}z}
\end{bmatrix} \\[2em]
\begin{bmatrix}
\displaystyle\sum_{p=-H}^{H} \tilde{H}_{\alpha,x,m,n,p}^{(1)+} e^{jk_{\alpha,z,p}^{(1)+}z}
\end{bmatrix}
\quad \cdots \quad
\begin{bmatrix}
\displaystyle\sum_{p=-H}^{H} \tilde{H}_{\alpha,x,m,n,p}^{(M^+)+} e^{jk_{\alpha,z,p}^{(M^+)+}z}
\end{bmatrix}
\end{cases}
\right) \tag{6.9d}
$$

$\underline{\underline{W}}_\alpha^-(z)$, and $\underline{\underline{V}}_\alpha^-(z)$ are $[2(2M+1)(2N+1)] \times M^-$ matrices indicating the part of the negative modes, given, respectively, by

$$
\underline{\underline{W}}_\alpha^-(z) = \left(
\begin{bmatrix}
\displaystyle\sum_{p=-H}^{H} \tilde{E}_{\alpha,y,m,n,p}^{(1)-} e^{jk_{\alpha,z,p}^{(1)-}z}
\end{bmatrix}
\quad \cdots \quad
\begin{bmatrix}
\displaystyle\sum_{p=-H}^{H} \tilde{E}_{\alpha,y,m,n,p}^{(M-)-} e^{jk_{\alpha,z,p}^{(M-)-}z}
\end{bmatrix} \\[2em]
\begin{bmatrix}
\displaystyle\sum_{p=-H}^{H} \tilde{E}_{\alpha,x,m,n,p}^{(1)-} e^{jk_{\alpha,z,p}^{(1)-}z}
\end{bmatrix}
\quad \cdots \quad
\begin{bmatrix}
\displaystyle\sum_{p=-H}^{H} \tilde{E}_{\alpha,x,m,n,p}^{(M-)-} e^{jk_{\alpha,z,p}^{(M-)-}z}
\end{bmatrix}
\right) \tag{6.9e}
$$

$$
\underline{\underline{V}}_\alpha^-(z) = \left(
\begin{bmatrix}
\displaystyle\sum_{p=-H}^{H} \tilde{H}_{\alpha,y,m,n,p}^{(1)-} e^{jk_{\alpha,z,p}^{(1)-}z}
\end{bmatrix}
\quad \cdots \quad
\begin{bmatrix}
\displaystyle\sum_{p=-H}^{H} \tilde{H}_{\alpha,y,m,n,p}^{(M^-)-} e^{jk_{\alpha,z,p}^{(M^-)-}z}
\end{bmatrix} \\[2em]
\begin{bmatrix}
\displaystyle\sum_{p=-H}^{H} \tilde{H}_{\alpha,x,m,n,p}^{(1)-} e^{jk_{\alpha,z,p}^{(1)-}z}
\end{bmatrix}
\quad \cdots \quad
\begin{bmatrix}
\displaystyle\sum_{p=-H}^{H} \tilde{H}_{\alpha,x,m,n,p}^{(M^-)-} e^{jk_{\alpha,z,p}^{(M^-)-}z}
\end{bmatrix}
\right) \tag{6.9f}
$$

$\underline{\underline{U}}$ and $\underline{\underline{\tilde{U}}}$ are the input operators, $[2(2M+1)(2N+1)] \times [2(2M+1)(2N+1)]$ identity matrices. $\underline{\underline{\tilde{R}}}$, $\underline{\underline{R}}$ and $\underline{\underline{\tilde{T}}}$, $\underline{\underline{T}}$ are the reflection coefficient matrix operators and transmission coefficient matrix operators, respectively.

The coupling coefficient operator S-matrix components, $\underline{\underline{C}}_{\alpha,a}^+$, $\underline{\underline{C}}_{\alpha,a}^-$, $\underline{\underline{C}}_{\alpha,b}^+$, and $\underline{\underline{C}}_{\alpha,b}$, are obtained from Equations (6.8a) to (6.8d),

$$
\begin{pmatrix} \underline{\underline{C}}_{\alpha,a}^+ \\ \underline{\underline{C}}_{\alpha,a}^- \end{pmatrix} = \left(
\begin{array}{cc}
\underline{\underline{W}}_{\alpha,h}^{-1}\underline{\underline{W}}_{\alpha,+}(0) + \underline{\underline{V}}_{\alpha,h}^{-1}\underline{\underline{V}}_{\alpha,+}(0) & \underline{\underline{W}}_{\alpha,h}^{-1}\underline{\underline{W}}_{\alpha,-}(z_- - z_+) + \underline{\underline{V}}_{\alpha,h}^{-1}\underline{\underline{V}}_{\alpha,-}(z_- - z_+) \\
\underline{\underline{W}}_{\alpha,h}^{-1}\underline{\underline{W}}_{\alpha,+}(z_+ - z_-) - \underline{\underline{V}}_{\alpha,h}^{-1}\underline{\underline{V}}_{\alpha,+}(z_+ - z_-) & \underline{\underline{W}}_{\alpha,h}^{-1}\underline{\underline{W}}_{\alpha,-}(0) - \underline{\underline{V}}_{\alpha,h}^{-1}\underline{\underline{V}}_{\alpha,-}(0)
\end{array}
\right)^{-1}
$$
$$
\times \begin{pmatrix} 2\underline{\underline{\tilde{U}}} \\ \underline{\underline{0}} \end{pmatrix}, \tag{6.10a}
$$

$$\begin{pmatrix} \underline{C}^+_{\alpha,b} \\ \underline{C}^-_{\alpha,b} \end{pmatrix} = \begin{pmatrix} \underline{W}^{-1}_{\alpha,h}\underline{W}_{\alpha,+}(0) + \underline{V}^{-1}_{\alpha,h}\underline{V}_{\alpha,+}(0) & \underline{W}^{-1}_{\alpha,h}\underline{W}_{\alpha,-}(z_- - z_+) + \underline{V}^{-1}_{\alpha,h}\underline{V}_{\alpha,-}(z_- - z_+) \\ \underline{W}^{-1}_{\alpha,h}\underline{W}_{\alpha,+}(z_+ - z_-) - \underline{V}^{-1}_{\alpha,h}\underline{V}_{\alpha,+}(z_+ - z_-) & \underline{W}^{-1}_{\alpha,h}\underline{W}_{\alpha,-}(0) - \underline{V}^{-1}_{\alpha,h}\underline{V}_{\alpha,-}(0) \end{pmatrix}^{-1}$$

$$\times \begin{pmatrix} \underline{0} \\ 2\underline{\tilde{U}} \end{pmatrix}. \tag{6.10b}$$

Then, the block S-matrix components, $\vec{\underline{R}}$, $\vec{\underline{T}}$, $\overleftarrow{\underline{R}}$, and $\overleftarrow{\underline{T}}$, are given by

$$\vec{\underline{R}} = \underline{W}^{-1}_{\alpha,h}[\underline{W}^+_\alpha(0)\underline{C}^+_{\alpha,a} + \underline{W}^-_\alpha(z_- - z_+)\underline{C}^-_{\alpha,a} - \underline{W}_{\alpha,h}] \tag{6.11a}$$

$$\vec{\underline{T}} = \underline{W}^{-1}_{\alpha,h}[\underline{W}^+_\alpha(z_+ - z_-)\underline{C}^+_{\alpha,a} + \underline{W}^-_\alpha(0)\underline{C}^-_{\alpha,a}] \tag{6.11b}$$

$$\overleftarrow{\underline{R}} = \underline{W}^{-1}_{\alpha,h}[\underline{W}^+_\alpha(z_+ - z_-)\underline{C}^+_{\alpha,b} + \underline{W}^-_\alpha(0)\underline{C}^-_{\alpha,b} - \underline{W}_{\alpha,h}] \tag{6.11c}$$

$$\overleftarrow{\underline{T}} = \underline{W}^{-1}_{\alpha,h}[\underline{W}^+_\alpha(0)\underline{C}^+_{\alpha,b} + \underline{W}^-_\alpha(z_- - z_+)\underline{C}^-_{\alpha,b}] \tag{6.11d}$$

The boundary S-matrix components, $\vec{\underline{R}}$, $\vec{\underline{T}}$, $\overleftarrow{\underline{R}}$, and $\overleftarrow{\underline{T}}$, of the half-infinite α-block with a right boundary are given by

$$\vec{\underline{R}} = -\left[(\underline{W}_{\alpha,h})^{-1}\underline{W}^-_\alpha(z_c) - (\underline{V}_{\alpha,h})^{-1}\underline{V}^-_\alpha(z_c)\right]^{-1}\left[(\underline{W}_{\alpha,h})^{-1}\underline{W}^+_\alpha(z_c) - (\underline{V}_{\alpha,h})^{-1}\underline{V}^+_\alpha(z_c)\right]$$

$$\tag{6.12a}$$

$$\vec{\underline{T}} = \left[(\underline{W}^-_\alpha(z_c))^{-1}\underline{W}_{\alpha,h} - (\underline{V}^-_\alpha(z_c))^{-1}\underline{V}_{\alpha,h}\right]^{-1}\left[(\underline{W}^-_\alpha(z_c))^{-1}\underline{W}^+_\alpha(z_c) - (\underline{V}^-_\alpha(z_c))^{-1}\underline{V}^+_\alpha(z_c)\right]$$

$$\tag{6.12b}$$

$$\overleftarrow{\underline{R}} = -\left[(\underline{W}^-_\alpha(z_c))^{-1}\underline{W}_{\alpha,h} - (\underline{V}^-_\alpha(z_c))^{-1}\underline{V}_{\alpha,h}\right]^{-1}\left[(\underline{W}^-_\alpha(z_c))^{-1}\underline{W}_{\alpha,h} + (\underline{V}^-_\alpha(z_c))^{-1}\underline{V}_{\alpha,h}\right]$$

$$\tag{6.12c}$$

$$\overleftarrow{\underline{T}} = 2\left[(\underline{W}_{\alpha,h})^{-1}\underline{W}^-_\alpha(z_c) - (\underline{V}_{\alpha,h})^{-1}\underline{V}^-_\alpha(z_c)\right]^{-1} \tag{6.12d}$$

The boundary S-matrix components, $\overleftarrow{\underline{\underline{R}}}$, $\vec{\underline{\underline{T}}}$, $\vec{\underline{\underline{R}}}$, and $\overleftarrow{\underline{\underline{T}}}$, of the half-infinite α-block with left boundary are given by

$$\overleftarrow{\underline{\underline{R}}} = -\left[(\underline{\underline{W}}_\alpha^+(z_c))^{-1} \underline{\underline{W}}_{\alpha,h} + (\underline{\underline{V}}_\alpha^+(z_c))^{-1} \underline{\underline{V}}_{\alpha,h} \right]^{-1} \left[(\underline{\underline{W}}_\alpha^+(z_c))^{-1} \underline{\underline{W}}_{\alpha,h} - (\underline{\underline{V}}_\alpha^+(z_c))^{-1} \underline{\underline{V}}_{\alpha,h} \right]$$

(6.13a)

$$\vec{\underline{\underline{T}}} = 2\left[(\underline{\underline{W}}_{\alpha,h})^{-1} \underline{\underline{W}}_\alpha^+(z_c) + (\underline{\underline{V}}_{\alpha,h})^{-1} \underline{\underline{V}}_\alpha^+(z_c) \right]^{-1}$$

(6.13b)

$$\vec{\underline{\underline{R}}} = -\left[(\underline{\underline{W}}_{\alpha,h})^{-1} \underline{\underline{W}}_\alpha^+(z_c) + (\underline{\underline{V}}_{\alpha,h})^{-1} \underline{\underline{V}}_\alpha^+(z_c) \right]^{-1} \left[(\underline{\underline{W}}_{\alpha,h})^{-1} \underline{\underline{W}}_\alpha^-(z_c) + (\underline{\underline{V}}_{\alpha,h})^{-1} \underline{\underline{V}}_\alpha^-(z_c) \right]$$

(6.13c)

$$\overleftarrow{\underline{\underline{T}}} = \left[(\underline{\underline{W}}_\alpha^+(z_c))^{-1} \underline{\underline{W}}_{\alpha,h} + (\underline{\underline{V}}_\alpha^+(z_c))^{-1} \underline{\underline{V}}_{\alpha,h} \right]^{-1} \left[(\underline{\underline{W}}_\alpha^+(z_c))^{-1} \underline{\underline{W}}_\alpha^-(z_c) - (\underline{\underline{V}}_\alpha^+(z_c))^{-1} \underline{\underline{V}}_\alpha^-(z_c) \right]$$

(6.13d)

By a similar manner, we can derive the block S-matrix and the boundary S-matrices from the transverse field continuation boundary conditions of the two-port β-blocks. The boundary conditions of the down-to-up characterization and the up-to-down characterization of the two-port β-blocks read, respectively, as

$$\begin{pmatrix} \underline{\underline{Y}}_{\beta,h} & \underline{\underline{Y}}_{\beta,h} \\ \underline{\underline{Z}}_{\beta,h} & -\underline{\underline{Z}}_{\beta,h} \end{pmatrix} \begin{pmatrix} \underline{U}_\uparrow \\ \underline{R}_\downarrow \end{pmatrix} = \begin{pmatrix} \underline{\underline{Y}}_\beta^+(0) & \underline{\underline{Y}}_\beta^-(x_- - x_+) \\ \underline{\underline{Z}}_\beta^+(0) & \underline{\underline{Z}}_\beta^-(x_- - x_+) \end{pmatrix} \begin{pmatrix} \underline{C}_{\beta,a}^+ \\ \underline{C}_{\beta,a}^- \end{pmatrix} \quad \text{at } x = x_- \quad (6.14\text{a})$$

$$\begin{pmatrix} \underline{\underline{Y}}_{\beta,h} & \underline{\underline{Y}}_{\beta,h} \\ \underline{\underline{Z}}_{\beta,h} & -\underline{\underline{Z}}_{\beta,h} \end{pmatrix} \begin{pmatrix} \underline{T}_\uparrow \\ \underline{0} \end{pmatrix} = \begin{pmatrix} \underline{\underline{Y}}_\beta^+(x_+ - x_-) & \underline{\underline{Y}}_\beta^-(0) \\ \underline{\underline{Z}}_\beta^+(x_+ - x_-) & \underline{\underline{Z}}_\beta^-(0) \end{pmatrix} \begin{pmatrix} \underline{C}_{\beta,a}^+ \\ \underline{C}_{\beta,a}^- \end{pmatrix} \quad \text{at } x = x_+ \quad (6.14\text{b})$$

and

$$\begin{pmatrix} \underline{\underline{Y}}_{\beta,h} & \underline{\underline{Y}}_{\beta,h} \\ \underline{\underline{Z}}_{\beta,h} & -\underline{\underline{Z}}_{\beta,h} \end{pmatrix} \begin{pmatrix} \underline{0} \\ \underline{T}_\downarrow \end{pmatrix} = \begin{pmatrix} \underline{\underline{Y}}_\beta^+(0) & \underline{\underline{Y}}_\beta^-(x_- - x_+) \\ \underline{\underline{Z}}_\beta^+(0) & \underline{\underline{Z}}_\beta^-(x_- - x_+) \end{pmatrix} \begin{pmatrix} \underline{C}_{\beta,b}^+ \\ \underline{C}_{\beta,b}^- \end{pmatrix} \quad \text{at } x = x_- \quad (6.14\text{c})$$

$$\begin{pmatrix} \underline{\underline{Y}}_{\beta,h} & \underline{\underline{Y}}_{\beta,h} \\ \underline{\underline{Z}}_{\beta,h} & -\underline{\underline{Z}}_{\beta,h} \end{pmatrix} \begin{pmatrix} \underline{R}_\uparrow \\ \underline{U}_\downarrow \end{pmatrix} = \begin{pmatrix} \underline{\underline{Y}}_\beta^+(x_+ - x_-) & \underline{\underline{Y}}_\beta^-(0) \\ \underline{\underline{Z}}_\beta^+(x_+ - x_-) & \underline{\underline{Z}}_\beta^-(0) \end{pmatrix} \begin{pmatrix} \underline{C}_{\beta,b}^+ \\ \underline{C}_{\beta,b}^- \end{pmatrix} \quad \text{at } x = x_+ \quad (6.14\text{d})$$

where $\underline{\underline{Y}}_{\beta,h}$ and $\underline{\underline{Z}}_{\beta,h}$ are $[2(2M+1)(2N+1)]\times[2(2M+1)(2N+1)]$ matrices given, respectively, by

$$\underline{\underline{Y}}_{\beta,h} = \begin{pmatrix} \underline{\underline{I}} & 0 \\ 0 & \underline{\underline{I}} \end{pmatrix} \qquad (6.15a)$$

$$\underline{\underline{Z}}_{\beta,h} = \begin{pmatrix} \left[-\dfrac{1}{\omega\mu_0}\dfrac{k_{\beta,z,m}k_{\beta,y,n}}{k_{\beta,x,m,n}} \right] & \left[-\dfrac{1}{\omega\mu_0}\dfrac{\left(k_{\beta,x,m,n}^2 + k_{\beta,z,m}^2\right)}{k_{\beta,x,m,n}} \right] \\[3ex] \left[\dfrac{1}{\omega\mu_0}\dfrac{\left(k_{\beta,y,n}^2 + k_{\beta,x,m,n}^2\right)}{k_{\beta,x,m,n}} \right] & \left[\dfrac{1}{\omega\mu_0}\dfrac{k_{\beta,y,n}k_{\beta,z,m}}{k_{\beta,x,m,n}} \right] \end{pmatrix} \qquad (6.15b)$$

$\underline{Y}_\beta^+(z)$ and $\underline{Z}_\beta^+(z)$ are $[2(2M+1)(2N+1)]\times M^+$ matrices indicating the part of the positive modes given, respectively, by

$$\underline{Y}_\beta^+(x) = \begin{pmatrix} \left[\displaystyle\sum_{m=-M}^{M} \tilde{E}_{\beta,y,m,n,s}^{(1)+} e^{jk_{x,m}^{(1)+}x} \right] & \cdots & \left[\displaystyle\sum_{m=-M}^{M} \tilde{E}_{\beta,y,m,n,s}^{(M^+)+} e^{jk_{x,m}^{(M^+)+}x} \right] \\[3ex] \left[\displaystyle\sum_{m=-M}^{M} \tilde{E}_{\beta,z,m,n,s}^{(1)+} e^{jk_{x,m}^{(1)+}x} \right] & \cdots & \left[\displaystyle\sum_{m=-M}^{M} \tilde{E}_{\beta,z,m,n,s}^{(M^+)+} e^{jk_{x,m}^{(M^+)+}x} \right] \end{pmatrix} \qquad (6.15c)$$

$$\underline{Z}_\beta^+(x) = \begin{pmatrix} \left[\displaystyle\sum_{m=-M}^{M} \tilde{H}_{\beta,y,m,n,s}^{(1)+} e^{jk_{x,m}^{(1)+}x} \right] & \cdots & \left[\displaystyle\sum_{m=-M}^{M} \tilde{H}_{\beta,y,m,n,s}^{(M^+)+} e^{jk_{x,m}^{(M^+)+}x} \right] \\[3ex] \left[\displaystyle\sum_{m=-M}^{M} \tilde{H}_{\beta,z,m,n,s}^{(1)+} e^{jk_{x,m}^{(1)+}x} \right] & \cdots & \left[\displaystyle\sum_{m=-M}^{M} \tilde{H}_{\beta,z,m,n,s}^{(M^+)+} e^{jk_{x,m}^{(M^+)+}x} \right] \end{pmatrix} \qquad (6.15d)$$

$\underline{Y}_\beta^-(z)$, and $\underline{Z}_\beta^-(z)$ are $[2(2M+1)(2N+1)]\times M^-$ matrices indicating the part of the negative modes, given, respectively, by

$$\underline{Y}_\beta^-(x) = \begin{pmatrix} \left[\displaystyle\sum_{m=-M}^{M} \tilde{E}_{\beta,y,m,n,s}^{(1)-} e^{jk_{x,m}^{(1)-}x} \right] & \cdots & \left[\displaystyle\sum_{m=-M}^{M} \tilde{E}_{\beta,y,m,n,s}^{(M^-)-} e^{jk_{x,m}^{(M^-)-}x} \right] \\[3ex] \left[\displaystyle\sum_{m=-M}^{M} \tilde{E}_{\beta,z,m,n,s}^{(1)-} e^{jk_{x,m}^{(1)-}x} \right] & \cdots & \left[\displaystyle\sum_{m=-M}^{M} \tilde{E}_{\beta,z,m,n,s}^{(M^-)-} e^{jk_{x,m}^{(M^-)-}x} \right] \end{pmatrix} \qquad (6.15e)$$

$$
\underline{\underline{Z}}_\beta^-(x) = \left(
\begin{array}{ccc}
\left[\displaystyle\sum_{m=-M}^{M} \tilde{H}_{\beta,y,m,n,s}^{(1)-} e^{jk_{x,m}^{(1)-}x}\right] & \cdots & \left[\displaystyle\sum_{m=-M}^{M} \tilde{H}_{\beta,y,m,n,s}^{(M^-)-} e^{jk_{x,m}^{(M^-)-}x}\right] \\[3ex]
\left[\displaystyle\sum_{m=-M}^{M} \tilde{H}_{\beta,z,m,n,s}^{(1)-} e^{jk_{x,m}^{(1)-}x}\right] & \cdots & \left[\displaystyle\sum_{m=-M}^{M} \tilde{H}_{\beta,z,m,n,s}^{(M^-)-} e^{jk_{x,m}^{(M^-)-}x}\right]
\end{array}
\right) \tag{6.15f}
$$

The coupling coefficient operators, $\underline{\underline{C}}_{\beta,a}^+$, $\underline{\underline{C}}_{\beta,a}^-$, $\underline{\underline{C}}_{\beta,b}^+$, and $\underline{\underline{C}}_{\beta,b}^-$, are obtained, from Equations (6.12a) to (6.12d), by

$$
\begin{bmatrix} \underline{\underline{C}}_{\beta,a}^+ \\[1ex] \underline{\underline{C}}_{\beta,a}^- \end{bmatrix} = \left(
\begin{array}{cc}
\left(\underline{\underline{Y}}_{\beta,h}^{-1}\underline{\underline{Y}}_\beta^+(0) + \underline{\underline{Z}}_{\beta,h}^{-1}\underline{\underline{Z}}_\beta^+(0)\right) & \left(\underline{\underline{Y}}_{\beta,h}^{-1}\underline{\underline{Y}}_\beta^-(x_- - x_+) + \underline{\underline{Z}}_{\beta,h}^{-1}\underline{\underline{Z}}_\beta^-(x_- - x_+)\right) \\[2ex]
\left(\underline{\underline{Y}}_{\beta,h}^{-1}\underline{\underline{Y}}_\beta^+(x_+ - x_-) - \underline{\underline{Z}}_{\beta,h}^{-1}\underline{\underline{Z}}_\beta^+(x_+ - x_-)\right) & \left(\underline{\underline{Y}}_{\beta,h}^{-1}\underline{\underline{Y}}_\beta^-(0) - \underline{\underline{Z}}_{\beta,h}^{-1}\underline{\underline{Z}}_\beta^-(0)\right)
\end{array}
\right)^{-1}
$$
$$
\times \begin{bmatrix} 2\underline{\underline{U}}_\uparrow \\[1ex] \underline{\underline{0}} \end{bmatrix}, \tag{6.16a}
$$

$$
\begin{bmatrix} \underline{\underline{C}}_{\beta,b}^+ \\[1ex] \underline{\underline{C}}_{\beta,b}^- \end{bmatrix} = \left(
\begin{array}{cc}
\left(\underline{\underline{Y}}_{\beta,h}^{-1}\underline{\underline{Y}}_\beta^+(0) + \underline{\underline{Z}}_{\beta,h}^{-1}\underline{\underline{Z}}_\beta^+(0)\right) & \left(\underline{\underline{Y}}_{\beta,h}^{-1}\underline{\underline{Y}}_\beta^-(x_- - x_+) + \underline{\underline{Z}}_{\beta,h}^{-1}\underline{\underline{Z}}_\beta^-(x_- - x_+)\right) \\[2ex]
\left(\underline{\underline{Y}}_{\beta,h}^{-1}\underline{\underline{Y}}_\beta^+(x_+ - x_-) - \underline{\underline{Z}}_{\beta,h}^{-1}\underline{\underline{Z}}_\beta^+(x_+ - x_-)\right) & \left(\underline{\underline{Y}}_{\beta,h}^{-1}\underline{\underline{Y}}_\beta^-(0) - \underline{\underline{Z}}_{\beta,h}^{-1}\underline{\underline{Z}}_\beta^-(0)\right)
\end{array}
\right)^{-1}
$$
$$
\times \begin{bmatrix} \underline{\underline{0}} \\[1ex] 2\underline{\underline{U}}_\downarrow \end{bmatrix}. \tag{6.16b}
$$

The block S-matrix components, $\underline{\underline{R}}_\downarrow$, $\underline{\underline{T}}_\uparrow$, $\underline{\underline{R}}_\uparrow$, and $\underline{\underline{T}}_\downarrow$, of the two-port β-block are given by

$$
\underline{\underline{R}}_\downarrow = \underline{\underline{Y}}_{\beta,h}^{-1}[\underline{\underline{Y}}_\beta^+(0)\underline{\underline{C}}_{\beta,a}^+ + \underline{\underline{Y}}_\beta^-(x_- - x_+)\underline{\underline{C}}_{\beta,a}^- - \underline{\underline{Y}}_{\beta,h}] \tag{6.17a}
$$

$$
\underline{\underline{T}}_\uparrow = \underline{\underline{Y}}_{\beta,h}^{-1}[\underline{\underline{Y}}_\beta^+(x_+ - x_-)\underline{\underline{C}}_{\beta,a}^+ + \underline{\underline{Y}}_\beta^-(0)\underline{\underline{C}}_{\beta,a}^-] \tag{6.17b}
$$

$$
\underline{\underline{R}}_\uparrow = \underline{\underline{Y}}_{\beta,h}^{-1}[\underline{\underline{Y}}_\beta^+(x_+ - x_-)\underline{\underline{C}}_{\beta,b}^+ + \underline{\underline{Y}}_\beta^-(0)\underline{\underline{C}}_{\beta,b}^- - \underline{\underline{Y}}_{\beta,h}] \tag{6.17c}
$$

$$
\underline{\underline{T}}_\downarrow = \underline{\underline{Y}}_{\beta,h}^{-1}[\underline{\underline{Y}}_\beta^+(0)\underline{\underline{C}}_{\beta,b}^+ + \underline{\underline{Y}}_\beta^-(x_- - x_+)\underline{\underline{C}}_{\beta,b}^-] \tag{6.17d}
$$

The boundary S-matrix components, $\underline{R}_\downarrow$, $\underline{T}_\uparrow$, $\underline{R}_\uparrow$, and $\underline{T}_\downarrow$, of the half-infinite β-block with an upper boundary are given by

$$\underline{R}_\downarrow = -\left[(\underline{\underline{Y}}_{\beta,h})^{-1}\underline{\underline{Y}}_\beta^-(x_c) - (\underline{\underline{Z}}_{\beta,h})^{-1}\underline{\underline{Z}}_\beta^-(x_c) \right]^{-1} \left[(\underline{\underline{Y}}_{\beta,h})^{-1}\underline{\underline{Y}}_\beta^+(x_c) - (\underline{\underline{Z}}_{\beta,h})^{-1}\underline{\underline{Z}}_\beta^+(x_c) \right] \quad (6.18a)$$

$$\underline{T}_\uparrow = \left[(\underline{\underline{Y}}_\beta^-(x_c))^{-1}\underline{\underline{Y}}_{\beta,h} - (\underline{\underline{Z}}_\beta^-(x_c))^{-1}\underline{\underline{Z}}_{\beta,h} \right]^{-1} \left[(\underline{\underline{Y}}_\beta^-(x_c))^{-1}\underline{\underline{Y}}_\beta^+(x_c) - (\underline{\underline{Z}}_\beta^-(x_c))^{-1}\underline{\underline{Z}}_\beta^+(x_c) \right]$$

$$(6.18b)$$

$$\underline{R}_\uparrow = -\left[(\underline{\underline{Y}}_\beta^-(x_c))^{-1}\underline{\underline{Y}}_{\beta,h} - (\underline{\underline{Z}}_\beta^-(x_c))^{-1}\underline{\underline{Z}}_{\beta,h} \right]^{-1} \left[(\underline{\underline{Y}}_\beta^-(x_c))^{-1}\underline{\underline{Y}}_{\beta,h} + (\underline{\underline{Z}}_\beta^-(x_c))^{-1}\underline{\underline{Z}}_{\beta,h} \right]$$

$$(6.18c)$$

$$\underline{T}_\downarrow = 2\left[(\underline{\underline{Y}}_{\beta,h})^{-1}\underline{\underline{Y}}_\beta^-(x_c) - (\underline{\underline{Z}}_{\beta,h})^{-1}\underline{\underline{Z}}_\beta^-(x_c) \right]^{-1} \qquad (6.18d)$$

The boundary S-matrix components, $\underline{R}_\downarrow$, $\underline{T}_\uparrow$, $\underline{R}_\uparrow$, and $\underline{T}_\downarrow$, of the half-infinite β-block with a lower boundary are given by

$$\underline{R}_\downarrow = -\left[(\underline{\underline{Y}}_\beta^+(x_c))^{-1}\underline{\underline{Y}}_{\beta,h} + (\underline{\underline{Z}}_\beta^+(x_c))^{-1}\underline{\underline{Z}}_{\beta,h} \right]^{-1} \left[(\underline{\underline{Y}}_\beta^+(x_c))^{-1}\underline{\underline{Y}}_{\beta,h} - (\underline{\underline{Z}}_\beta^+(x_c))^{-1}\underline{\underline{Z}}_{\beta,h} \right] \quad (6.19a)$$

$$\underline{T}_\uparrow = 2\left[(\underline{\underline{Y}}_{\beta,h})^{-1}\underline{\underline{Y}}_\beta^+(x_c) + (\underline{\underline{Z}}_{\beta,h})^{-1}\underline{\underline{Z}}_\beta^+(x_c) \right]^{-1} \qquad (6.19b)$$

$$\underline{R}_\uparrow = -\left[(\underline{\underline{Y}}_{\beta,h})^{-1}\underline{\underline{Y}}_\beta^+(x_c) + (\underline{\underline{Z}}_{\beta,h})^{-1}\underline{\underline{Z}}_\beta^+(x_c) \right]^{-1} \left[(\underline{\underline{Y}}_{\beta,h})^{-1}\underline{\underline{Y}}_\beta^-(x_c) + (\underline{\underline{Z}}_{\beta,h})^{-1}\underline{\underline{Z}}_\beta^-(x_c) \right] \quad (6.19c)$$

$$\underline{T}_\downarrow = \left[(\underline{\underline{Y}}_\beta^+(x_c))^{-1}\underline{\underline{Y}}_{\beta,h} + (\underline{\underline{Z}}_\beta^+(x_c))^{-1}\underline{\underline{Z}}_{\beta,h} \right]^{-1} \left[(\underline{\underline{Y}}_\beta^+(x_c))^{-1}\underline{\underline{Y}}_\beta^- - (\underline{\underline{Z}}_\beta^+(x_c))^{-1}\underline{\underline{Z}}_\beta^- \right] \quad (6.19d)$$

6.3 Local Fourier Modal Analysis of Four-Port Cross-Block Structures

As stated in the previous section, the ultimate objective of this paper is the complete mathematical modeling of four-port crossed nanophotonic structures. The crossed photonic crystal waveguide structure shown in Figure 6.2(a) is composed of five subparts: ports 1 to 4 and the intersection cross-block. In this section, the LFMA for obtaining the Bloch eigenmodes and S-matrix of the four-port intersection block is described. The four-port cross-block interconnecting two-port α-blocks and two-port β-blocks is characterized by 4 × 4 S-matrix. Figure 6.5(a) and (b) shows the separated intersection cross-block

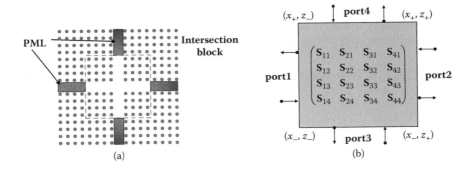

FIGURE 6.5
(a) Intersection block model with PML placed within four waveguide branches, and (b) diagram of 4×4 S-matrix. (From H. Kim and B. Lee, *J. Opt. Soc. Am. B*, 25(4), 2008. With permission.)

of the photonic crystal cross-waveguide structure and its S-matrix diagram, respectively. In the LFMA, the intersection block must be embedded into a larger block with an absorbing medium (or perfect matched layer (PML)) block placed within the waveguide branches connected to the ports as shown in Figure 6.5(a). The internal part, indicated by the dash-lined rectangle, is the intersection part of the analyzed cross-waveguide structure. The dash-lined rectangle is the actual boundary of the intersection block. The PML within each waveguide branch is necessary to model nonperiodic structures and prevent the eigenmode profile from undergoing deterioration by interference that is induced by periodicity. With the PML, the power flow through each waveguide branch is outward without nonphysical reflection at the interface of the cross-block. The basic intuitive assumption of the LFMA is the electromagnetic isolation among individual blocks by field localization on nanophotonic structures.

The field representation of the Bloch eigenmode within the four-port intersection block should be prepared. The pseudo-Fourier representations of the Bloch modes in the α-blocks and β-blocks of the four-port intersection block are taken, respectively, as the same forms of Equations (6.4) and (6.6). The total electromagnetic fields in the four-port cross-block can be represented by the superposition of the obtained α-Bloch eigenmodes and β-Bloch eigenmodes:

$$
\begin{pmatrix} \mathbf{E} \\ \mathbf{H} \end{pmatrix} = \sum_{g=1}^{M^+} C_{\alpha,g}^+ \begin{pmatrix} \mathbf{E}_\alpha^{(g)+} \\ \mathbf{H}_\alpha^{(g)+} \end{pmatrix} + \sum_{g=1}^{M^-} C_{\alpha,g}^- \begin{pmatrix} \mathbf{E}_\alpha^{(g)-} \\ \mathbf{H}_\alpha^{(g)-} \end{pmatrix} + \sum_{g=1}^{M^+} C_{\beta,g}^+ \begin{pmatrix} \mathbf{E}_\beta^{(g)+} \\ \mathbf{H}_\beta^{(g)+} \end{pmatrix} + \sum_{g=1}^{M^-} C_{\beta,g}^- \begin{pmatrix} \mathbf{E}_\beta^{(g)-} \\ \mathbf{H}_\beta^{(g)-} \end{pmatrix}
$$

(6.20)

where $C_{\alpha,g}^+$ and $C_{\alpha,g}^-$ are the coupling coefficients of the positive α-Bloch eigenmodes and the negative α-Bloch eigenmodes, respectively. $C_{\beta,g}^+$ and $C_{\beta,g}^-$ are the coupling coefficients of the positive β-Bloch eigenmodes and the negative β-Bloch eigenmodes, respectively.

In addition, we take the Fourier series approximations of the exponential functions of the eigenvalues in Equations (6.4) and (6.6) as follows:

$$\exp\left(jk_{\alpha,z,0}^{(g)\pm}z\right) \approx \sum_{q=-H}^{H} \zeta_{\alpha,q}^{(g)\pm} \exp\left[j\frac{2\pi q}{T_z}z\right] \tag{6.21a}$$

$$\exp\left(jk_{\beta,x,0}^{(g)\pm}x\right) \approx \sum_{q=-H}^{H} \zeta_{\beta,q}^{(g)\pm} \exp\left[j\frac{2\pi q}{T_x}x\right] \tag{6.21b}$$

where $\zeta_{\alpha,q}^{(g)\pm}$ and $\zeta_{\beta,q}^{(g)\pm}$ are given, respectively, by

$$\zeta_{\alpha,q}^{(g)\pm} = \mathrm{sinc}\left(\frac{k_{z,0}^{(g)\pm}T_z}{2\pi} - q\right) \tag{6.22a}$$

$$\zeta_{\beta,q}^{(g)\pm} = \mathrm{sinc}\left(\frac{k_{x,0}^{(g)\pm}T_x}{2\pi} - q\right) \tag{6.22b}$$

By substituting Equations (6.21a) and (6.21b) into the pseudo-Fourier representations of the Bloch eigenmodes, Equations (6.4) and (6.6), we can obtain the Fourier approximation representation of the Bloch eigenmodes. The resultant α-Bloch eigenmode representations read as

$$\begin{pmatrix} \mathbf{E}_\alpha^{(g)+} \\ \mathbf{H}_\alpha^{(g)+} \end{pmatrix} = \begin{pmatrix} \mathbf{E}_{\alpha,x}^{(g)+}, \mathbf{E}_{\alpha,y}^{(g)+}, \mathbf{E}_{\alpha,z}^{(g)+} \\ \mathbf{H}_{\alpha,x}^{(g)+}, \mathbf{H}_{\alpha,y}^{(g)+}, \mathbf{H}_{\alpha,z}^{(g)+} \end{pmatrix}$$

$$= e^{j(k_{x,0}x+k_{y,0}y)} \sum_{m=-M}^{M} \sum_{n=-N}^{N} \sum_{s=-H}^{H} \begin{pmatrix} \tilde{E}_{\alpha,x,m,n,s}^{(g)\pm}, \tilde{E}_{\alpha,y,m,n,s}^{(g)\pm}, \tilde{E}_{\alpha,z,m,n,s}^{(g)\pm} \\ \tilde{H}_{\alpha,x,m,n,s}^{(g)\pm}, \tilde{H}_{\alpha,y,m,n,s}^{(g)\pm}, \tilde{H}_{\alpha,z,m,n,s}^{(g)\pm} \end{pmatrix} \tag{6.23a}$$

$$\times e^{j(G_{x,m}x+G_{y,n}y+G_{z,s}z)}$$

where the Fourier coefficients of the representations are obtained by

$$\begin{pmatrix} \tilde{E}_{\alpha,x,m,n,s}^{(g)\pm}, \tilde{E}_{\alpha,y,m,n,s}^{(g)\pm}, \tilde{E}_{\alpha,z,m,n,s}^{(g)\pm} \\ \tilde{H}_{\alpha,x,m,n,s}^{(g)\pm}, \tilde{H}_{\alpha,y,m,n,s}^{(g)\pm}, \tilde{H}_{\alpha,z,m,n,s}^{(g)\pm} \end{pmatrix}$$

$$= \sum_{p=-H}^{H} \begin{pmatrix} \zeta_{\alpha,s-p}^{(g)\pm} E_{\alpha,x,m,n,p}^{(g)\pm}, \zeta_{\alpha,s-p}^{(g)\pm} E_{\alpha,y,m,n,p}^{(g)\pm}, \zeta_{\alpha,s-p}^{(g)\pm} E_{\alpha,z,m,n,p}^{(g)\pm} \\ \zeta_{\alpha,s-p}^{(g)\pm} H_{\alpha,x,m,n,p}^{(g)\pm}, \zeta_{\alpha,s-p}^{(g)\pm} H_{\alpha,y,m,n,p}^{(g)\pm}, \zeta_{\alpha,s-p}^{(g)\pm} H_{\alpha,z,m,n,p}^{(g)\pm} \end{pmatrix} e^{\left(-jk_{z,p}^{(g)\pm}z_\mp\right)} \tag{6.23b}$$

The resultant β-Bloch eigenmode representations read as

$$
\begin{pmatrix} \mathbf{E}_\beta^{(g)\pm} \\ \mathbf{H}_\beta^{(g)\pm} \end{pmatrix} = e^{j(k_{z,0}z+k_{y,0}y)} \sum_{s=-M}^{M} \sum_{n=-N}^{N} \sum_{p=-H}^{H} \begin{pmatrix} \tilde{E}_{\beta,s,m,n,p}^{(g)\pm}, \tilde{E}_{\beta,y,s,n,p}^{(g)\pm}, \tilde{E}_{\beta,z,s,n,p}^{(g)\pm} \\ \tilde{H}_{\beta,x,s,n,p}^{(g)\pm}, \tilde{H}_{\beta,y,s,n,p}^{(g)\pm}, \tilde{H}_{\beta,z,s,n,p}^{(g)\pm} \end{pmatrix} e^{j(G_{x,s}x+G_{y,n}y+G_{z,p}z)}
$$

$$(6.24a)$$

where the Fourier coefficients of the representations are obtained by

$$
\begin{pmatrix} \tilde{E}_{\beta,x,s,n,p}^{(g)\pm}, \tilde{E}_{\beta,y,s,n,p}^{(g)\pm}, \tilde{E}_{\beta,z,s,n,p}^{(g)\pm} \\ \tilde{H}_{\beta,x,s,n,p}^{(g)+}, \tilde{H}_{\beta,y,s,n,p}^{(g)+}, \tilde{H}_{\beta,z,s,n,p}^{(g)+} \end{pmatrix}
$$

$$(6.24b)$$

$$
= \sum_{m=-M}^{M} \begin{pmatrix} \zeta_{\beta,s-m}^{(g)\pm} E_{\beta,x,m,n,p}^{(g)\pm}, \zeta_{\beta,s-m}^{(g)\pm} E_{\beta,y,m,n,p}^{(g)\pm}, \zeta_{\beta,s-m}^{(g)\pm} E_{\beta,z,m,n,p}^{(g)\pm} \\ \zeta_{\beta,s-m}^{(g)+} H_{\beta,x,m,n,p}^{(g)+}, \zeta_{\beta,s-m}^{(g)+} H_{\beta,y,m,n,p}^{(g)+}, \zeta_{\beta,s-m}^{(g)+} H_{\beta,z,m,n,p}^{(g)+} \end{pmatrix} e^{-jk_{x,m}^{(g)\pm}x_\mp}.
$$

With the Fourier approximation of the pseudo-Fourier Bloch eigenmode representations, we can find the appropriate boundary condition equations for obtaining the S-matrix of the four-port intersection block.

Let us examine the α-Bloch eigenmodes and β-Bloch eigenmodes of three examples: 2D photonic crystal cross-waveguide structure (reference [44] in Chapter 1), 2D photonic crystal T-branch structure (reference [47] in Chapter 1), and 2D photonic crystal 90° bending structure (reference [48] in Chapter 1), the permittivity profiles of which are shown in Figure 6.6(a) to (c), respectively. The PML blocks are placed in the waveguide channels within the dummy region. The mode profiles of the dominant α-Bloch eigenmode and β-Bloch eigenmode transferring electromagnetic power are analyzed by the LFMA with the Fourier truncation order of $M = 14$, $N = 0$, and $P = 14$.

Figures 6.7 to 6.9 show the dominant eigenmode profiles of the intersection blocks of the 2D photonic crystal cross-waveguide structure, 2D photonic crystal T-branch structure, and 2D photonic crystal 90° bending structure, respectively.

The 4×4 S-matrix is derived by solving four boundary conditions at four boundaries of the intersection block. Hereafter mathematical terms related to four-port characterization are denoted by bold font for convenience. Let us denote four excitation fields at ports 1 to 4 as $\mathbf{U}_1$, $\mathbf{U}_2$, $\mathbf{U}_3$, and $\mathbf{U}_4$, which are represented, respectively, by

$$
\mathbf{U}_1 = \sum_{m=-M}^{M} \sum_{n=-N}^{N} (u_{1,x,m,n}, u_{1,y,m,n}, u_{1,z,m,n}) e^{j(k_{\alpha,x,m}x+k_{\alpha,y,n}y+k_{\alpha,z,m,n}(z-z_-))}, \quad (6.25a)
$$

$$
\mathbf{U}_2 = \sum_{m=-M}^{M} \sum_{n=-N}^{N} (u_{2,x,m,n}, u_{2,y,m,n}, u_{2,z,m,n}) e^{j(k_{\alpha,x,m}x+k_{\alpha,y,n}y-k_{\alpha,z,m,n}(z+z_+))}, \quad (6.25b)
$$

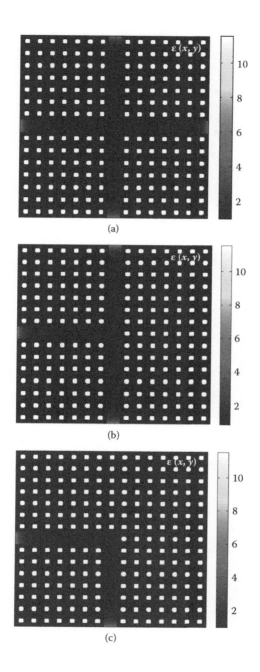

FIGURE 6.6
Permittivity profiles of (a) the intersection block of the 2D photonic crystal cross-waveguide structure, (b) the intersection block of the 2D photonic crystal T-branch structure, (c) the intersection block of the 2D photonic crystal 90° bending structure. (From H. Kim and B. Lee, *J. Opt. Soc. Am. B*, 25(4), 2008. With permission.)

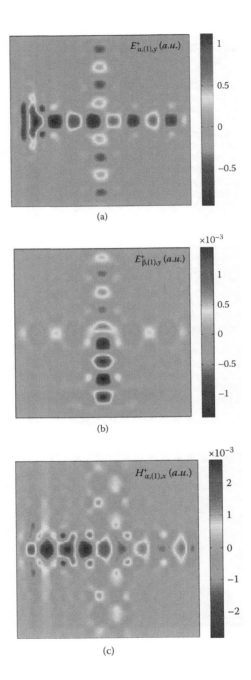

(a)

(b)

(c)

FIGURE 6.7
Dominant eigenmode profiles of the intersection block of the cross-waveguide structure:
(a) $E^+_{\alpha,(1),y}$, (b) $E^+_{\beta,(1),y}$, (c) $H^+_{\alpha,(1),x}$, (d) $H^+_{\beta,(1),x}$, (e) $H^+_{\alpha,(1),z}$, and (f) $H^+_{\beta,(1),z}$. (From H. Kim and B. Lee, *J. Opt. Soc. Am. B*, 25(4), 2008. With permission.)

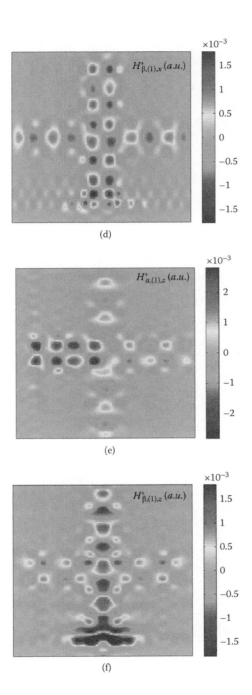

FIGURE 6.7
(Continued).

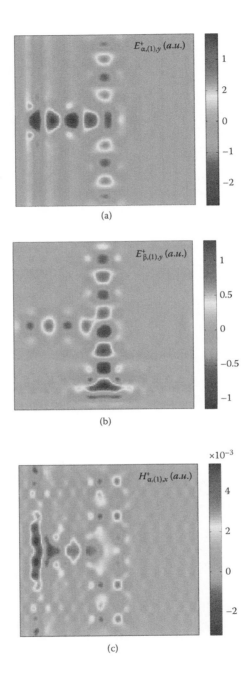

FIGURE 6.8
Dominant eigenmode profiles of the intersection block of the photonic crystal T-branch structure: (a) $E_{\alpha,(1),y}^+$, (b) $E_{\beta,(1),y}^+$, (c) $H_{\alpha,(1),x}^+$, (d) $H_{\beta,(1),x}^+$, (e) $H_{\alpha,(1),z}^+$, and (f) $H_{\beta,(1),z}^+$. (From H. Kim and B. Lee, *J. Opt. Soc. Am. B*, 25(4), 2008. With permission.)

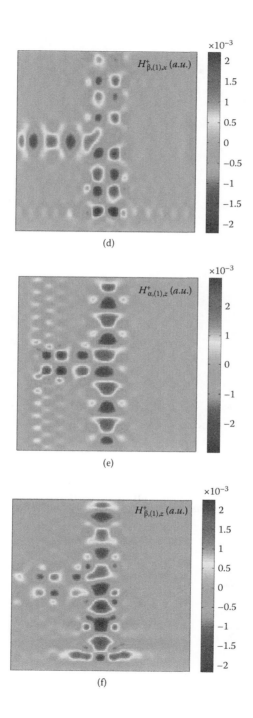

(d)

(e)

(f)

FIGURE 6.8
(Continued).

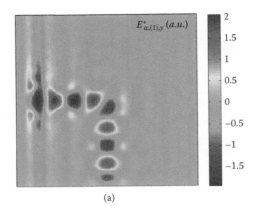

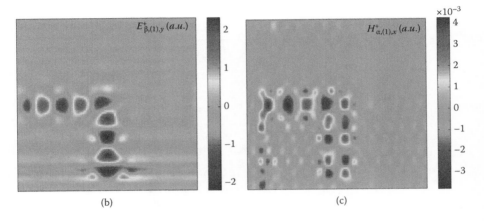

FIGURE 6.9
Dominant eigenmode profiles of the intersection block of the photonic crystal 90° bending structure: (a) $E^+_{\alpha,(1),y}$, (b) $E^+_{\beta,(1),y}$, (c) $H^+_{\alpha,(1),x}$, (d) $H^+_{\beta,(1),x}$, (e) $H^+_{\alpha,(1),z}$, and (f) $H^+_{\beta,(1),z}$. (From H. Kim and B. Lee, *J. Opt. Soc. Am. B*, 25(4), 2008. With permission.)

$$\mathbf{U}_3 = \sum_{m=-M}^{M}\sum_{n=-N}^{N}(u_{3,x,m,n}, u_{3,y,m,n}, u_{3,z,m,n})e^{j(k_{\beta,x,m,n}(x-x_-)+k_{\beta,y,n}y+k_{\beta,z,m}z)}, \quad (6.25c)$$

$$\mathbf{U}_4 = \sum_{m=-M}^{M}\sum_{n=-N}^{N}(u_{4,x,m,n}, u_{4,y,m,n}, u_{4,z,m,n})e^{j(-k_{\beta,x,m,n}(x-x_+)+k_{\beta,y,n}y+k_{\beta,z,m}z)}. \quad (6.25d)$$

Let us denote the radiation fields (transmission and reflection fields) at ports 1, 2, 3, and 4 induced by the excitation of port i (for $= 1, 2, 3, 4$) as $\mathbf{T}_{i1}$, $\mathbf{T}_{i2}$,

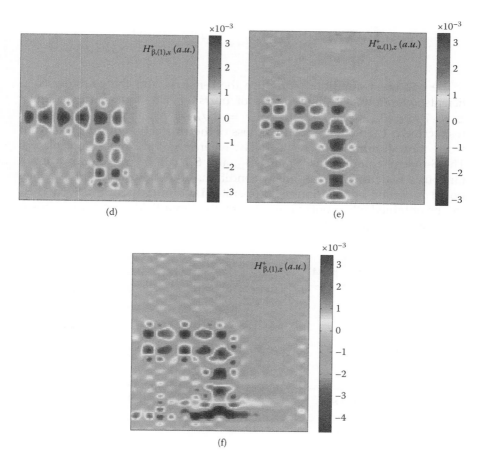

FIGURE 6.9
(Continued).

$\mathbf{T}_{i3}$, and $\mathbf{T}_{i4}$ that are given, respectively, by

$$\mathbf{T}_{i1} = \sum_{m=-M}^{M} \sum_{n=-N}^{N} (t_{i1,x,m,n}, t_{i1,y,m,n}, t_{i1,z,m,n}) e^{[j(k_{\alpha,x,m}x + k_{\alpha,y,n}y - k_{\alpha,z,m,n}(z-z_-))]}, \quad (6.26a)$$

$$\mathbf{T}_{i2} = \sum_{m=-M}^{M} \sum_{n=-N}^{N} (t_{i2,x,m,n}, t_{i2,y,m,n}, t_{i2,z,m,n}) e^{[j(k_{\alpha,x,m}x + k_{\alpha,y,n}y + k_{\alpha,z,m,n}(z-z_+))]}, \quad (6.26b)$$

$$\mathbf{T}_{i3} = \sum_{m=-M}^{M} \sum_{n=-N}^{N} (t_{i3,x,m,n}, t_{i3,y,m,n}, t_{i3,z,m,n}) e^{[j(-k_{\beta,x,m,n}(x-x_-) + k_{\beta,y,n}y + k_{\beta,z,m}z)]}, \quad (6.26c)$$

$$\mathbf{T}_{i4} = \sum_{m=-M}^{M} \sum_{n=-N}^{N} (t_{i4,x,m,n}, t_{i4,y,m,n}, t_{i4,z,m,n}) e^{[j(k_{\beta,x,m,n}(x-x_+)+k_{\beta,y,n}y+k_{\beta,z,m}z)]}. \quad (6.26d)$$

The boundary conditions produce the S-matrix and the coupling coefficients operators. The boundaries of ports 1, 2, 3, and 4 are set up at $z_- = -T_z/2 + \Delta T_z$, $z_+ = T_z/2 - \Delta T_z$, $x_- = -T_x/2 + \Delta T_x$, and $x_+ = T_x/2 - \Delta T_x$, respectively. T_x and T_z are the thickness of the x-direction dummy area and z-direction dummy area, respectively, where the PML blocks are placed. The S-matrix components, $\mathbf{S}_{11}$, $\mathbf{S}_{12}$, $\mathbf{S}_{13}$, and $\mathbf{S}_{14}$, can be obtained by simultaneously matching the boundary conditions at the four boundaries, when port 1 is excited by the input field operator $\mathbf{U}_1$. Similarly $\mathbf{S}_{i1}$, $\mathbf{S}_{i2}$, $\mathbf{S}_{i3}$, and $\mathbf{S}_{i4}$ can be obtained when port i is excited by the input field operator $\mathbf{U}_i$. This is illustrated in Figure 6.10.

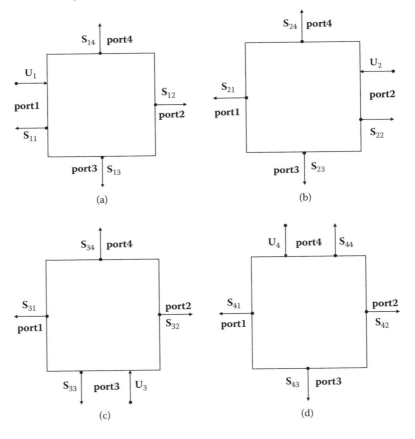

FIGURE 6.10
(a) Excitation of port 1, (b) excitation of port 2, (c) excitation of port 3, and (d) excitation of port 4. (From H. Kim and B. Lee, *J. Opt. Soc. Am. B*, 25(4), 2008. With permission.)

The continuity conditions of the transverse fields at four boundaries can be expressed by the following matrix equations:

1. At $z = z_-$

$$
\begin{pmatrix} \mathbf{W}_{\alpha,h} & \mathbf{W}_{\alpha,h} \\ \mathbf{V}_{\alpha,h} & -\mathbf{V}_{\alpha,h} \end{pmatrix} \begin{pmatrix} \mathbf{U}_1 & 0 & 0 & 0 \\ \mathbf{S}_{11} & \mathbf{S}_{21} & \mathbf{S}_{31} & \mathbf{S}_{41} \end{pmatrix}
$$
$$
= \begin{pmatrix} \tilde{\mathbf{W}}_{\alpha,+}(z_-) & \tilde{\mathbf{W}}_{\alpha,-}(z_-) \\ \tilde{\mathbf{V}}_{\alpha,+}(z_-) & \tilde{\mathbf{V}}_{\alpha,-}(z_-) \end{pmatrix} \begin{pmatrix} \mathbf{C}_{\alpha,1}^+ & \mathbf{C}_{\alpha,2}^+ & \mathbf{C}_{\alpha,3}^+ & \mathbf{C}_{\alpha,4}^+ \\ \mathbf{C}_{\alpha,1}^- & \mathbf{C}_{\alpha,2}^- & \mathbf{C}_{\alpha,3}^- & \mathbf{C}_{\alpha,4}^- \end{pmatrix}
$$
$$
+ \begin{pmatrix} \tilde{\mathbf{W}}_{\beta,+}(z_-) & \tilde{\mathbf{W}}_{\beta,-}(z_-) \\ \tilde{\mathbf{V}}_{\beta,+}(z_-) & \tilde{\mathbf{V}}_{\beta,-}(z_-) \end{pmatrix} \begin{pmatrix} \mathbf{C}_{\beta,1}^+ & \mathbf{C}_{\beta,2}^+ & \mathbf{C}_{\beta,3}^+ & \mathbf{C}_{\beta,4}^+ \\ \mathbf{C}_{\beta,1}^- & \mathbf{C}_{\beta,2}^- & \mathbf{C}_{\beta,3}^- & \mathbf{C}_{\beta,4}^- \end{pmatrix},
$$

(6.27a)

2. At $z = z_t$

$$
\begin{pmatrix} \mathbf{W}_{\alpha,h} & \mathbf{W}_{\alpha,h} \\ \mathbf{V}_{\alpha,h} & -\mathbf{V}_{\alpha,h} \end{pmatrix} \begin{pmatrix} \mathbf{S}_{12} & \mathbf{S}_{22} & \mathbf{S}_{32} & \mathbf{S}_{42} \\ 0 & \mathbf{U}_2 & 0 & 0 \end{pmatrix}
$$
$$
= \begin{pmatrix} \tilde{\mathbf{W}}_{\alpha,+}(z_+) & \tilde{\mathbf{W}}_{\alpha,-}(z_+) \\ \tilde{\mathbf{V}}_{\alpha,+}(z_+) & \tilde{\mathbf{V}}_{\alpha,-}(z_+) \end{pmatrix} \begin{pmatrix} \mathbf{C}_{\alpha,1}^+ & \mathbf{C}_{\alpha,2}^+ & \mathbf{C}_{\alpha,3}^+ & \mathbf{C}_{\alpha,4}^+ \\ \mathbf{C}_{\alpha,1}^- & \mathbf{C}_{\alpha,2}^- & \mathbf{C}_{\alpha,3}^- & \mathbf{C}_{\alpha,4}^- \end{pmatrix}
$$
$$
+ \begin{pmatrix} \tilde{\mathbf{W}}_{\beta,+}(z_+) & \tilde{\mathbf{W}}_{\beta,-}(z_+) \\ \tilde{\mathbf{V}}_{\beta,+}(z_+) & \tilde{\mathbf{V}}_{\beta,-}(z_+) \end{pmatrix} \begin{pmatrix} \mathbf{C}_{\beta,1}^+ & \mathbf{C}_{\beta,2}^+ & \mathbf{C}_{\beta,3}^+ & \mathbf{C}_{\beta,4}^+ \\ \mathbf{C}_{\beta,1}^- & \mathbf{C}_{\beta,2}^- & \mathbf{C}_{\beta,3}^- & \mathbf{C}_{\beta,4}^- \end{pmatrix},
$$

(6.27b)

3. At $x = x_-$

$$
\begin{pmatrix} \mathbf{Y}_{\beta,h} & \mathbf{Y}_{\beta,h} \\ \mathbf{Z}_{\beta,h} & -\mathbf{Z}_{\beta,h} \end{pmatrix} \begin{pmatrix} 0 & 0 & 0 & \mathbf{U}_3 & 0 \\ \mathbf{S}_{13} & \mathbf{S}_{13} & \mathbf{S}_{23} & \mathbf{S}_{33} & \mathbf{S}_{43} \end{pmatrix}
$$
$$
= \begin{pmatrix} \tilde{\mathbf{Y}}_{\alpha,+}(x_-) & \tilde{\mathbf{Y}}_{\alpha,-}(x_-) \\ \tilde{\mathbf{Z}}_{\alpha,+}(x_-) & \tilde{\mathbf{Z}}_{\alpha,-}(x_-) \end{pmatrix} \begin{pmatrix} \mathbf{C}_{\alpha,1}^+ & \mathbf{C}_{\alpha,2}^+ & \mathbf{C}_{\alpha,3}^+ & \mathbf{C}_{\alpha,4}^+ \\ \mathbf{C}_{\alpha,1}^- & \mathbf{C}_{\alpha,2}^- & \mathbf{C}_{\alpha,3}^- & \mathbf{C}_{\alpha,4}^- \end{pmatrix}
$$
$$
+ \begin{pmatrix} \tilde{\mathbf{Y}}_{\beta,+}(x_-) & \tilde{\mathbf{Y}}_{\beta,-}(x_-) \\ \tilde{\mathbf{Z}}_{\beta,+}(x_-) & \tilde{\mathbf{Z}}_{\beta,-}(x_-) \end{pmatrix} \begin{pmatrix} \mathbf{C}_{\beta,1}^+ & \mathbf{C}_{\beta,2}^+ & \mathbf{C}_{\beta,3}^+ & \mathbf{C}_{\beta,4}^+ \\ \mathbf{C}_{\beta,1}^- & \mathbf{C}_{\beta,2}^- & \mathbf{C}_{\beta,3}^- & \mathbf{C}_{\beta,4}^- \end{pmatrix},
$$

(6.27c)

4. At $x = x_T$

$$
\begin{pmatrix} \mathbf{Y}_{\beta,h} & \mathbf{Y}_{\beta,h} \\ \mathbf{Z}_{\beta,h} & -\mathbf{Z}_{\beta,h} \end{pmatrix} \begin{pmatrix} \mathbf{S}_{14} & \mathbf{S}_{24} & \mathbf{S}_{34} & \mathbf{S}_{44} \\ \mathbf{0} & \mathbf{0} & \mathbf{0} & \mathbf{U}_4 \end{pmatrix}
$$

$$
= \begin{pmatrix} \tilde{\mathbf{Y}}_{\alpha,+}(x_+) & \tilde{\mathbf{Y}}_{\alpha,-}(x_+) \\ \tilde{\mathbf{Z}}_{\alpha,+}(x_+) & \tilde{\mathbf{Z}}_{\alpha,-}(x_+) \end{pmatrix} \begin{pmatrix} \mathbf{C}_{\alpha,1}^+ & \mathbf{C}_{\alpha,2}^+ & \mathbf{C}_{\alpha,3}^+ & \mathbf{C}_{\alpha,4}^+ \\ \mathbf{C}_{\alpha,1}^- & \mathbf{C}_{\alpha,2}^- & \mathbf{C}_{\alpha,3}^- & \mathbf{C}_{\alpha,4}^- \end{pmatrix} \tag{6.27d}
$$

$$
+ \begin{pmatrix} \tilde{\mathbf{Y}}_{\beta,+}(x_+) & \tilde{\mathbf{Y}}_{\beta,-}(x_+) \\ \tilde{\mathbf{Z}}_{\beta,+}(x_+) & \tilde{\mathbf{Z}}_{\beta,-}(x_+) \end{pmatrix} \begin{pmatrix} \mathbf{C}_{\beta,1}^+ & \mathbf{C}_{\beta,2}^+ & \mathbf{C}_{\beta,3}^+ & \mathbf{C}_{\beta,4}^+ \\ \mathbf{C}_{\beta,1}^- & \mathbf{C}_{\beta,2}^- & \mathbf{C}_{\beta,3}^- & \mathbf{C}_{\beta,4}^- \end{pmatrix},
$$

where $\tilde{\mathbf{W}}_{\alpha,+}(z)$, $\tilde{\mathbf{W}}_{\alpha,-}(z)$, $\tilde{\mathbf{V}}_{\alpha,+}(z)$, $\tilde{\mathbf{V}}_{\alpha,-}(z)$, $\tilde{\mathbf{W}}_{\beta,+}(z)$, $\tilde{\mathbf{W}}_{\beta,-}(z)$, $\tilde{\mathbf{V}}_{\beta,+}(z)$, and $\tilde{\mathbf{V}}_{\beta,-}(z)$ are defined, respectively, by

$$
\tilde{\mathbf{W}}_{\alpha,\pm}(z) = \begin{pmatrix} \left[\displaystyle\sum_{s=-H}^{H} \tilde{E}_{\alpha,y,m,n,s}^{(1)\pm} e^{jG_{z,s}z} \right] & \cdots & \left[\displaystyle\sum_{s=-H}^{H} \tilde{E}_{\alpha,y,m,n,s}^{(M^\pm)\pm} e^{jG_{z,s}z} \right] \\ \left[\displaystyle\sum_{s=-H}^{H} \tilde{E}_{\alpha,x,m,n,s}^{(1)\pm} e^{jG_{z,s}z} \right] & \cdots & \left[\displaystyle\sum_{s=-H}^{H} \tilde{E}_{\alpha,x,m,n,s}^{(M^\pm)\pm} e^{jG_{z,s}z} \right] \end{pmatrix} \tag{6.28a}
$$

$$
\tilde{\mathbf{V}}_{\alpha,\pm}(z) = \begin{pmatrix} \left[\displaystyle\sum_{s=-H}^{H} \tilde{H}_{\alpha,y,m,n,s}^{(1)\pm} e^{jG_{z,s}z} \right] & \cdots & \left[\displaystyle\sum_{s=-H}^{H} \tilde{H}_{\alpha,y,m,n,s}^{(M^\pm)\pm} e^{jG_{z,s}z} \right] \\ \left[\displaystyle\sum_{s=-H}^{H} \tilde{H}_{\alpha,x,m,n,s}^{(1)\pm} e^{jG_{z,s}z} \right] & \cdots & \left[\displaystyle\sum_{s=-H}^{H} \tilde{H}_{\alpha,x,m,n,s}^{(M^\pm)\pm} e^{jG_{z,s}z} \right] \end{pmatrix} \tag{6.28b}
$$

$$
\tilde{\mathbf{W}}_{\beta,\pm}(z) = \begin{pmatrix} \left[\displaystyle\sum_{s=-H}^{H} \tilde{E}_{\beta,y,m,n,s}^{(1)\pm} e^{jG_{z,s}z} \right] & \cdots & \left[\displaystyle\sum_{s=-H}^{H} \tilde{E}_{\beta,y,m,n,s}^{(M^\pm)\pm} e^{jG_{z,s}z} \right] \\ \left[\displaystyle\sum_{s=-H}^{H} \tilde{E}_{\beta,x,m,n,s}^{(1)\pm} e^{jG_{z,s}z} \right] & \cdots & \left[\displaystyle\sum_{s=-H}^{H} \tilde{E}_{\beta,x,m,n,s}^{(M^\pm)\pm} e^{jG_{z,s}z} \right] \end{pmatrix} \tag{6.28c}
$$

$$\tilde{\mathbf{V}}_{\beta,\pm}(z) = \left(\begin{bmatrix} \sum\limits_{s=-H}^{H} \tilde{H}^{(1)\pm}_{\beta,y,m,n,s}e^{jG_{z,s}z} \\ \sum\limits_{s=-H}^{H} \tilde{H}^{(1)\pm}_{\beta,x,m,n,s}e^{jG_{z,s}z} \end{bmatrix} \cdots \begin{bmatrix} \sum\limits_{s=-H}^{H} \tilde{H}^{(M^{\pm})\pm}_{\beta,y,m,n,s}e^{jG_{z,s}z} \\ \sum\limits_{s=-H}^{H} \tilde{H}^{(M^{\pm})\pm}_{\beta,x,m,n,s}e^{jG_{z,s}z} \end{bmatrix} \right) \tag{6.28d}$$

and $\tilde{\mathbf{Y}}_{\beta,+}(x)$, $\tilde{\mathbf{Y}}_{\beta,-}(x)$, $\tilde{\mathbf{Z}}_{\beta,+}(x)$, $\tilde{\mathbf{Z}}_{\beta,-}(x)$, $\tilde{\mathbf{Y}}_{\alpha,+}(x)$, $\tilde{\mathbf{Y}}_{\alpha,-}(x)$, $\tilde{\mathbf{Z}}_{\alpha,+}(x)$, and $\tilde{\mathbf{Z}}_{\alpha,-}(x)$ are defined, respectively, by

$$\tilde{\mathbf{Y}}_{\beta,\pm}(x) = \left(\begin{bmatrix} \sum\limits_{m=-M}^{M} \tilde{E}^{(1)\pm}_{\beta,y,m,n,s}e^{jG_{x,m}x} \\ \sum\limits_{m=-M}^{M} \tilde{E}^{(1)\pm}_{\beta,z,m,n,s}e^{jG_{x,m}x} \end{bmatrix} \cdots \begin{bmatrix} \sum\limits_{m=-M}^{M} \tilde{E}^{(M^{\pm})\pm}_{\beta,y,m,n,s}e^{jG_{x,m}x} \\ \sum\limits_{m=-M}^{M} \tilde{E}^{(M^{\pm})\pm}_{\beta,z,m,n,s}e^{jG_{x,m}x} \end{bmatrix} \right) \tag{6.29a}$$

$$\tilde{\mathbf{Z}}_{\beta,+}(x) = \left(\begin{bmatrix} \sum\limits_{m=-M}^{M} \tilde{H}^{(1)+}_{\beta,y,m,n,s}e^{jG_{x,m}x} \\ \sum\limits_{m=-M}^{M} \tilde{H}^{(1)+}_{\beta,z,m,n,s}e^{jG_{x,m}x} \end{bmatrix} \cdots \begin{bmatrix} \sum\limits_{m=-M}^{M} \tilde{H}^{(M^{+})+}_{\beta,y,m,n,s}e^{jG_{x,m}x} \\ \sum\limits_{m=-M}^{M} \tilde{H}^{(M^{+})+}_{\beta,z,m,n,s}e^{jG_{x,m}x} \end{bmatrix} \right) \tag{6.29b}$$

$$\tilde{\mathbf{Y}}_{\alpha,\pm}(x) = \left(\begin{bmatrix} \sum\limits_{m=-M}^{M} \tilde{E}^{(1)\pm}_{\alpha,y,m,n,s}e^{jG_{x,m}x} \\ \sum\limits_{m=-M}^{M} \tilde{E}^{(1)\pm}_{\alpha,z,m,n,s}e^{jG_{x,m}x} \end{bmatrix} \cdots \begin{bmatrix} \sum\limits_{m=-M}^{M} \tilde{E}^{(M^{\pm})\pm}_{\alpha,y,m,n,s}e^{jG_{x,m}x} \\ \sum\limits_{m=-M}^{M} \tilde{E}^{(M^{+})\pm}_{\alpha,z,m,n,s}e^{jG_{x,m}x} \end{bmatrix} \right) \tag{6.29c}$$

$$\tilde{\mathbf{Z}}_{\alpha,\pm}(x) = \left(\begin{bmatrix} \sum\limits_{m=-M}^{M} \tilde{H}^{(1)\pm}_{\alpha,y,m,n,s}e^{jG_{x,m}x} \\ \sum\limits_{m=-M}^{M} \tilde{H}^{(1)\pm}_{\alpha,z,m,n,s}e^{jG_{x,m}x} \end{bmatrix} \cdots \begin{bmatrix} \sum\limits_{m=-M}^{M} \tilde{H}^{(M^{\pm})\pm}_{\alpha,y,m,n,s}e^{jG_{x,m}x} \\ \sum\limits_{m=-M}^{M} \tilde{H}^{(M^{\pm})\pm}_{\alpha,z,m,n,s}e^{jG_{x,m}x} \end{bmatrix} \right) \tag{6.29d}$$

From Equations (6.27a) to (6.27d), the coupling coefficient matrix operators $C_{\alpha,i}^+$, $C_{\alpha,i}^-$, $C_{\beta,i}^+$, and $C_{\beta,i}^-$ for $i = 1, 2, 3, 4$ are obtained by

$$
\begin{bmatrix} C_{\alpha,i}^+ \\ C_{\alpha,i}^- \\ C_{\beta,i}^+ \\ C_{\beta,i}^- \end{bmatrix} =
$$

$$
\begin{pmatrix}
\left(\tilde{W}_{\alpha,h}^{-1}\tilde{W}_{\alpha,+}(z_-) + \tilde{V}_{\alpha,h}^{-1}\tilde{V}_{\alpha,+}(z_-) \right) & \left(\tilde{W}_{\alpha,h}^{-1}\tilde{W}_{\alpha,-}(z_-) + \tilde{V}_{\alpha,h}^{-1}\tilde{V}_{\alpha,-}(z_-) \right) & \left(\tilde{W}_{\alpha,h}^{-1}\tilde{W}_{\beta,+}(z_-) + \tilde{V}_{\alpha,h}^{-1}\tilde{V}_{\beta,+}(z_-) \right) & \left(\tilde{W}_{\alpha,h}^{-1}\tilde{W}_{\beta,-}(z_-) + \tilde{V}_{\alpha,h}^{-1}\tilde{V}_{\beta,-}(z_-) \right) \\
\left(\tilde{W}_{\alpha,h}^{-1}\tilde{W}_{\alpha,+}(z_+) - \tilde{V}_{\alpha,h}^{-1}\tilde{V}_{\alpha,+}(z_+) \right) & \left(\tilde{W}_{\alpha,h}^{-1}\tilde{W}_{\alpha,-}(z_+) - \tilde{V}_{\alpha,h}^{-1}\tilde{V}_{\alpha,-}(z_+) \right) & \left(\tilde{W}_{\alpha,h}^{-1}\tilde{W}_{\beta,+}(z_+) - \tilde{V}_{\alpha,h}^{-1}\tilde{V}_{\beta,+}(z_+) \right) & \left(\tilde{W}_{\alpha,h}^{-1}\tilde{W}_{\beta,-}(z_+) - \tilde{V}_{\alpha,h}^{-1}\tilde{V}_{\beta,-}(z_+) \right) \\
\left(\tilde{Y}_{\beta,h}^{-1}\tilde{Y}_{\alpha,+}(x_-) + \tilde{Z}_{\beta,h}^{-1}\tilde{Z}_{\alpha,+}(x_-) \right) & \left(\tilde{Y}_{\beta,h}^{-1}\tilde{Y}_{\alpha,-}(x_-) + \tilde{Z}_{\beta,h}^{-1}\tilde{Z}_{\alpha,-}(x_-) \right) & \left(\tilde{Y}_{\beta,h}^{-1}\tilde{Y}_{\beta,+}(x_-) + \tilde{Z}_{\beta,h}^{-1}\tilde{Z}_{\beta,+}(x_-) \right) & \left(\tilde{Y}_{\beta,h}^{-1}\tilde{Y}_{\beta,-}(x_-) + \tilde{Z}_{\beta,h}^{-1}\tilde{Z}_{\beta,-}(x_-) \right) \\
\left(\tilde{Y}_{\beta,h}^{-1}\tilde{Y}_{\alpha,+}(x_+) - \tilde{Z}_{\beta,h}^{-1}\tilde{Z}_{\alpha,+}(x_+) \right) & \left(\tilde{Y}_{\beta,h}^{-1}\tilde{Y}_{\alpha,-}(x_+) - \tilde{Z}_{\beta,h}^{-1}\tilde{Z}_{\alpha,-}(x_+) \right) & \left(\tilde{Y}_{\beta,h}^{-1}\tilde{Y}_{\beta,+}(x_+) - \tilde{Z}_{\beta,h}^{-1}\tilde{Z}_{\beta,+}(x_+) \right) & \left(\tilde{Y}_{\beta,h}^{-1}\tilde{Y}_{\beta,-}(x_+) - \tilde{Z}_{\beta,h}^{-1}\tilde{Z}_{\beta,-}(x_+) \right)
\end{pmatrix}^{-1}
$$

$$
\times \begin{bmatrix} 2U_1\delta_{1i} \\ 2U_2\delta_{2i} \\ 2U_3\delta_{3i} \\ 2U_4\delta_{4i} \end{bmatrix}.
$$

$$(6.30)$$

The S-matrix components, S_{i1}, S_{i2}, S_{i3}, and S_{i4}, are also obtained, respectively, from Equations (6.27a) to (6.27d),

$$
S_{i1} = W_{\alpha,h}^{-1}[\tilde{W}_{\alpha,+}(z_-)C_{\alpha,i}^+ + \tilde{W}_{\alpha,-}(z_-)C_{\alpha,i}^- + \tilde{W}_{\beta,+}(z_-)C_{\beta,i}^+ + \tilde{W}_{\beta,-}(z_-)C_{\beta,i}^- - W_{\alpha,h}\delta_{1i}],
$$

$$(6.31a)$$

$$
S_{i2} = W_{\alpha,h}^{-1}[\tilde{W}_{\alpha,+}(z_+)C_{\alpha,i}^+ + \tilde{W}_{\alpha,-}(z_+)C_{\alpha,i}^- + \tilde{W}_{\beta,+}(z_+)C_{\beta,i}^+ + \tilde{W}_{\beta,-}(z_+)C_{\beta,i}^- - W_{\alpha,h}\delta_{2i}]
$$

$$(6.31b)$$

$$
S_{i3} = Y_{\beta,h}^{-1}[\tilde{Y}_{\alpha,+}(x_-)C_{\alpha,i}^+ + \tilde{Y}_{\alpha,-}(x_-)C_{\alpha,i}^- + \tilde{Y}_{\beta,+}(x_-)C_{\beta,i}^+ + \tilde{Y}_{\beta,-}(x_-)C_{\beta,i}^- - Y_{\beta,h}\delta_{3i}]
$$

$$(6.31c)$$

$$
S_{i4} = Y_{\beta,h}^{-1}[\tilde{Y}_{\alpha,+}(x_+)C_{\alpha,i}^+ + \tilde{Y}_{\alpha,-}(x_+)C_{\alpha,i}^- + \tilde{Y}_{\beta,+}(x_+)C_{\beta,i}^+ + \tilde{Y}_{\beta,-}(x_+)C_{\beta,i}^- - Y_{\beta,h}\delta_{4i}]
$$

$$(6.31d)$$

We can easily understand that the equations for the S-matrix and coupling coefficients for the two-port blocks of Equations (6.10a) and (6.10b), (6.11a) to (6.11d), (6.16a) to (6.16b), and (6.17a) and (6.17b) are special cases of the above-stated equations for the four-port cross-blocks.

The validity of the derived formulas is examined by visualizing the field distributions when a normally incident plane wave impinges on the port 1 interface of the intersection block. Figure 6.11(a) to (c) illustrates the LFMA

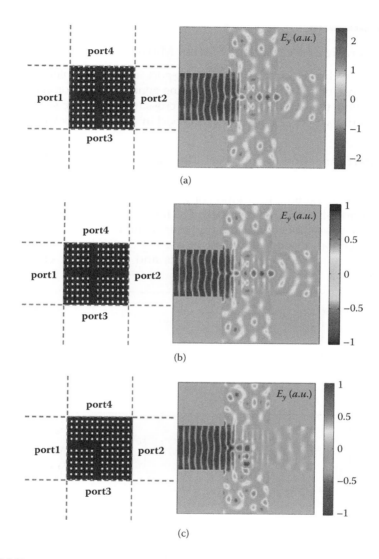

FIGURE 6.11
LFMA results of the S-matrix characterization of (a) the intersection block of the photonic crystal cross-waveguide structure, (b) the photonic crystal T-branch structure, and (c) the photonic crystal 90° bending structure.

results of the S-matrix characterization of the intersection block of the photonic crystal cross-waveguide structure, the T-branch structure, and the 90° bending structure, respectively. Here, the regions of ports 1 to 4 are free space. We can see that the field boundary conditions are well matched and that field continuity is conserved.

6.4 Generalized Scattering Matrix Method

The S-matrix characterization of the four-port intersection block provides a basis for constructing a GSMM for modeling general nanophotonic networks. In this section, the GSMM for a four-port cross nanophotonic structure composed of four two-port blocks and a four-port intersection block is developed.

6.4.1 Interconnection of Four-Port Block and Two-Port Blocks

The interconnection of two-port blocks can be described by the 2×2 S-matrix. In Figures 6.12 and 6.13, four kinds of interconnection of two-port α-blocks and two-port β-blocks are illustrated. The interconnection algorithm of two-port blocks was established in the previous section.

In this section, the development of a 4×4 S-matrix model for a four-port block composed of two-port blocks and a four-port cross-block is

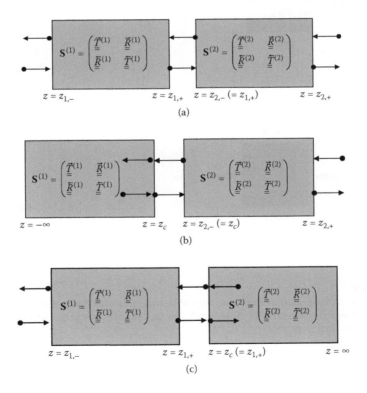

FIGURE 6.12
Two-port α-block interconnection: (a) left finite-sized block and right finite-sized block, (b) left half-infinite block and right finite-sized block, (c) left finite-sized block and right half-infinite block, and (d) left half-infinite block and right half-infinite block.

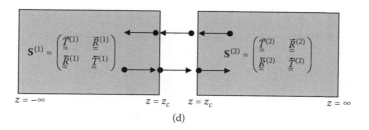

$$z = -\infty \qquad\qquad z = z_c \quad z = z_c \qquad\qquad z = \infty$$

(d)

FIGURE 6.12
(Continued)

described. Figure 6.14 shows the S-matrix model of the four-port crossed nanophotonic structure. A stepwise approach can be taken to obtain the total S-matrix of this four-port cross nanophotonic structure. Before obtaining the complete S-matrix, it is neccessary to manifest the interconnection of a two-port block and a four-port cross-block. In Figure 6.15, the interconnection of two-port block and four-port cross-block through port i is illustrated. The S-matrix components of the two-port block are denoted, with the subscript indicating the connected port number, by $\underline{\tilde{R}}_i$, $\underline{\tilde{T}}_i$, $\underline{\vec{R}}_i$, $\underline{\vec{T}}_i$ $(i = 1, 2)$ and $\underline{R}_{\uparrow i}$, $\underline{T}_{\downarrow i}$, $\underline{R}_{\downarrow i}$, $\underline{T}_{\uparrow i}$ $(i = 3, 4)$, respectively. The internal coupling coefficient matrix operators of the two-port block are denoted, with the superscript indicating the connected port number, by $\underline{C}_a^{(i)}$ and $\underline{C}_b^{(i)}$, respectively.

The internal infinite multiple reflections through the four-port block of the intersection block and a two-port block are intuitively formulated by a recursion equation of the S-matrix components and the coupling coefficient matrix operators. In the derived formulas, the tilde notation is used for denoting the components of the total S-matrix. Hence, $\tilde{C}_{p,1}^{(1)}$, $\tilde{C}_{p,2}^{(1)}$, $\tilde{C}_{p,3}^{(1)}$, and $\tilde{C}_{p,4}^{(1)}$ indicate the coupling coefficient matrix operators of the two-port block induced by the excitations of ports 1 to 4, respectively. The superscript means the port index through which the two-port block is connected to the four-port cross-block. The derived S-matrix recursion formulas of the interconnection of the two-port block and the four-port cross-block through port 1 are listed as follows:

$$
\begin{pmatrix}
\tilde{S}_{11} & \tilde{S}_{21} & \tilde{S}_{31} & \tilde{S}_{41} \\
\tilde{S}_{12} & \tilde{S}_{22} & \tilde{S}_{32} & \tilde{S}_{42} \\
\tilde{S}_{13} & \tilde{S}_{23} & \tilde{S}_{33} & \tilde{S}_{43} \\
\tilde{S}_{14} & \tilde{S}_{24} & \tilde{S}_{34} & \tilde{S}_{44}
\end{pmatrix}
$$

$$
=
\begin{pmatrix}
\underline{\vec{R}}_1 + \underline{\vec{T}}_1 S_{11}(I - \underline{\tilde{R}}_1 S_{11})^{-1}\underline{\tilde{T}}_1 & \underline{\vec{T}}_1(I - S_{11}\underline{\tilde{R}}_1)^{-1}S_{21} & \underline{\vec{T}}_1(I - S_{11}\underline{\tilde{R}}_1)^{-1}S_{31} & \underline{\vec{T}}_1(I - S_{11}\underline{\tilde{R}}_1)^{-1}S_{41} \\
S_{12}(I - \underline{\tilde{R}}_1 S_{11})^{-1}\underline{\tilde{T}}_1 & S_{22} + S_{12}\underline{\tilde{R}}_1(I - S_{11}\underline{\tilde{R}}_1)^{-1}S_{21} & S_{32} + S_{12}\underline{\tilde{R}}_1(I - S_{11}\underline{\tilde{R}}_1)^{-1}S_{31} & S_{42} + S_{12}\underline{\tilde{R}}_1(I - S_{11}\underline{\tilde{R}}_1)^{-1}S_{41} \\
S_{13}(I - \underline{\tilde{R}}_1 S_{11})^{-1}\underline{\tilde{T}}_1 & S_{23} + S_{13}\underline{\tilde{R}}_1(I - S_{11}\underline{\tilde{R}}_1)^{-1}S_{21} & S_{33} + S_{13}\underline{\tilde{R}}_1(I - S_{11}\underline{\tilde{R}}_1)^{-1}S_{31} & S_{43} + S_{13}\underline{\tilde{R}}_1(I - S_{11}\underline{\tilde{R}}_1)^{-1}S_{41} \\
S_{14}(I - \underline{\tilde{R}}_1 S_{11})^{-1}\underline{\tilde{T}}_1 & S_{24} + S_{14}\underline{\tilde{R}}_1(I - S_{11}\underline{\tilde{R}}_1)^{-1}S_{21} & S_{34} + S_{14}\underline{\tilde{R}}_1(I - S_{11}\underline{\tilde{R}}_1)^{-1}S_{31} & S_{44} + S_{14}\underline{\tilde{R}}_1(I - S_{11}\underline{\tilde{R}}_1)^{-1}S_{41}
\end{pmatrix},
$$

(6.32a)

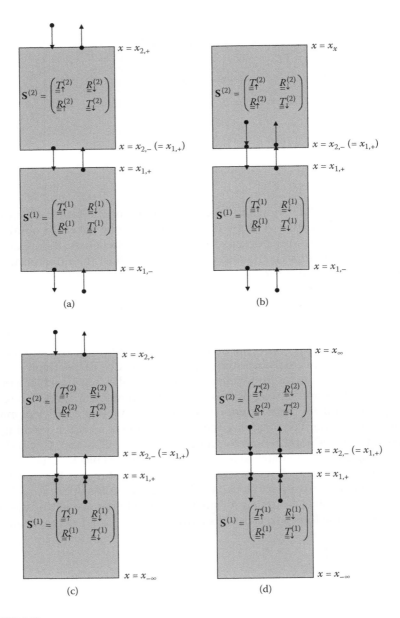

FIGURE 6.13
Two-port β-block interconnection: (a) upper finite-sized block and lower finite-sized block, (b) upper half-infinite block and lower finite-sized block, (c) upper finite-sized block and lower half-infinite block, and (d) upper half-infinite block and lower half-infinite block.

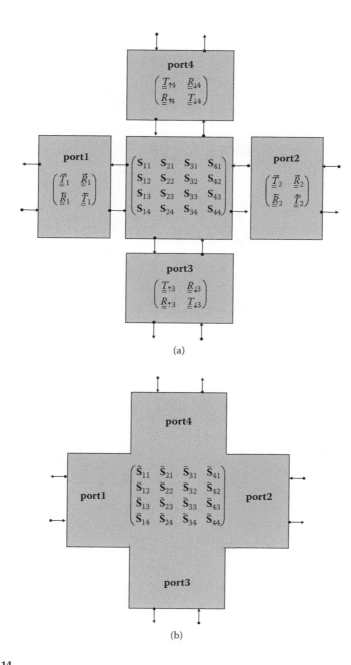

FIGURE 6.14

(a) Interconnection of four two-port blocks and a four-port cross-block, and (b) extended four-port cross-block composed of four two-port blocks and a four-port intersection block.

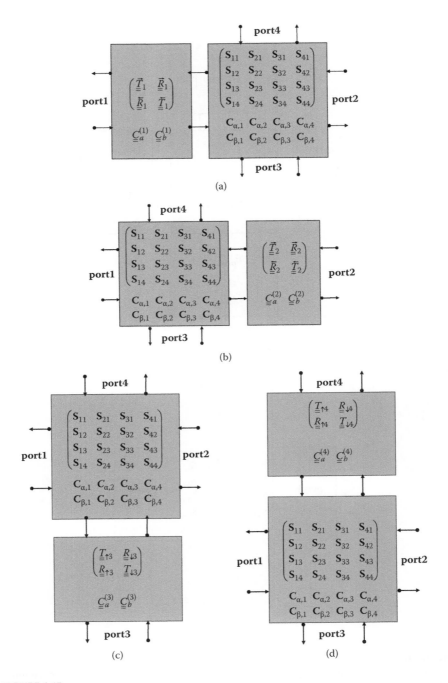

FIGURE 6.15
(a) Interconnection through port 1, (b) interconnection through port 2, (c) interconnection through port 3, and (d) interconnection through port 4.

$$
\begin{pmatrix} \tilde{\mathbf{C}}_{p,1}^{(1)} \\ \tilde{\mathbf{C}}_{p,2}^{(1)} \\ \tilde{\mathbf{C}}_{p,3}^{(1)} \\ \tilde{\mathbf{C}}_{p,4}^{(1)} \end{pmatrix} = \begin{pmatrix} \underline{\underline{C}}_a^{(1)} + \underline{\underline{C}}_b^{(1)} \mathbf{S}_{11} (\mathbf{I} - \underline{\underline{\vec{R}}}_1 \mathbf{S}_{11})^{-1} \underline{\vec{T}}_1 \\ \underline{\underline{C}}_b^{(1)} (\mathbf{I} - \mathbf{S}_{11} \underline{\underline{\vec{R}}}_1)^{-1} \mathbf{S}_{21} \\ \underline{\underline{C}}_b^{(1)} (\mathbf{I} - \mathbf{S}_{11} \underline{\underline{\vec{R}}}_1)^{-1} \mathbf{S}_{31} \\ \underline{\underline{C}}_b^{(1)} (\mathbf{I} - \mathbf{S}_{11} \underline{\underline{\vec{R}}}_1)^{-1} \mathbf{S}_{41} \end{pmatrix},
$$
(6.32b)

$$
\begin{pmatrix} \tilde{\mathbf{C}}_{\alpha,1} & \tilde{\mathbf{C}}_{\alpha,2} & \tilde{\mathbf{C}}_{\alpha,3} & \tilde{\mathbf{C}}_{\alpha,4} \\ \tilde{\mathbf{C}}_{\beta,1} & \tilde{\mathbf{C}}_{\beta,2} & \tilde{\mathbf{C}}_{\beta,3} & \tilde{\mathbf{C}}_{\beta,4} \end{pmatrix}
$$

$$
= \begin{pmatrix} \mathbf{C}_{\alpha,1}(\mathbf{I}-\underline{\underline{\vec{R}}}_1\mathbf{S}_{11})^{-1}\underline{\vec{T}} & \mathbf{C}_{\alpha,2}+\mathbf{C}_{\alpha,1}\underline{\underline{\vec{R}}}_1(\mathbf{I}-\mathbf{S}_{11}\underline{\underline{\vec{R}}}_1)^{-1}\mathbf{S}_{21} & \mathbf{C}_{\alpha,3}+\mathbf{C}_{\alpha,1}\underline{\underline{\vec{R}}}_1(\mathbf{I}-\mathbf{S}_{11}\underline{\underline{\vec{R}}}_1)^{-1}\mathbf{S}_{31} & \mathbf{C}_{\alpha,4}+\mathbf{C}_{\alpha,1}\underline{\underline{\vec{R}}}_1(\mathbf{I}-\mathbf{S}_{11}\underline{\underline{\vec{R}}}_1)^{-1}\mathbf{S}_{41} \\ \mathbf{C}_{\beta,1}(\mathbf{I}-\underline{\underline{\vec{R}}}_1\mathbf{S}_{11})^{-1}\underline{\vec{T}} & \mathbf{C}_{\beta,2}+\mathbf{C}_{\beta,1}\underline{\underline{\vec{R}}}_1(\mathbf{I}-\mathbf{S}_{11}\underline{\underline{\vec{R}}}_1)^{-1}\mathbf{S}_{21} & \mathbf{C}_{\beta,3}+\mathbf{C}_{\beta,1}\underline{\underline{\vec{R}}}_1(\mathbf{I}-\mathbf{S}_{11}\underline{\underline{\vec{R}}}_1)^{-1}\mathbf{S}_{31} & \mathbf{C}_{\beta,4}+\mathbf{C}_{\beta,1}(\mathbf{I}-\mathbf{S}_{11}\underline{\underline{\vec{R}}}_1)^{-1}\mathbf{S}_{41} \end{pmatrix}.
$$
(6.32c)

The derived S-matrix recursion formulas of the interconnection of the two-port block and the four-port cross-block through port 2 are listed as follows:

$$
\begin{pmatrix} \tilde{\mathbf{S}}_{11} & \tilde{\mathbf{S}}_{21} & \tilde{\mathbf{S}}_{31} & \tilde{\mathbf{S}}_{41} \\ \tilde{\mathbf{S}}_{12} & \tilde{\mathbf{S}}_{22} & \tilde{\mathbf{S}}_{32} & \tilde{\mathbf{S}}_{42} \\ \tilde{\mathbf{S}}_{13} & \tilde{\mathbf{S}}_{23} & \tilde{\mathbf{S}}_{33} & \tilde{\mathbf{S}}_{43} \\ \tilde{\mathbf{S}}_{14} & \tilde{\mathbf{S}}_{24} & \tilde{\mathbf{S}}_{34} & \tilde{\mathbf{S}}_{44} \end{pmatrix}
$$

$$
= \begin{pmatrix} \mathbf{S}_{11}+\mathbf{S}_{21}\underline{\underline{\vec{R}}}_2(\mathbf{I}-\mathbf{S}_{22}\underline{\underline{\vec{R}}}_2)^{-1}\mathbf{S}_{12} & \mathbf{S}_{21}(\mathbf{I}-\underline{\underline{\vec{R}}}_2\mathbf{S}_{22})^{-1}\underline{\vec{T}}_2 & \mathbf{S}_{31}+\mathbf{S}_{21}\underline{\underline{\vec{R}}}_2(\mathbf{I}-\mathbf{S}_{22}\underline{\underline{\vec{R}}}_2)^{-1}\mathbf{S}_{32} & \mathbf{S}_{41}+\mathbf{S}_{21}\underline{\underline{\vec{R}}}_2(\mathbf{I}-\mathbf{S}_{22}\underline{\underline{\vec{R}}}_2)^{-1}\mathbf{S}_{42} \\ \underline{\vec{T}}_2(\mathbf{I}-\mathbf{S}_{22}\underline{\underline{\vec{R}}}_2)^{-1}\mathbf{S}_{12} & \underline{\underline{\vec{R}}}_2+\underline{\vec{T}}_2\mathbf{S}_{22}(\mathbf{I}-\underline{\underline{\vec{R}}}_2\mathbf{S}_{22})^{-1}\underline{\vec{T}}_2 & \underline{\vec{T}}_2(\mathbf{I}-\mathbf{S}_{22}\underline{\underline{\vec{R}}}_2)^{-1}\mathbf{S}_{32} & \underline{\vec{T}}_2(\mathbf{I}-\mathbf{S}_{22}\underline{\underline{\vec{R}}}_2)^{-1}\mathbf{S}_{42} \\ \mathbf{S}_{13}+\mathbf{S}_{23}\underline{\underline{\vec{R}}}_2(\mathbf{I}-\mathbf{S}_{22}\underline{\underline{\vec{R}}}_2)^{-1}\mathbf{S}_{12} & \mathbf{S}_{23}(\mathbf{I}-\underline{\underline{\vec{R}}}_2\mathbf{S}_{22})^{-1}\underline{\vec{T}}_2 & \mathbf{S}_{33}+\mathbf{S}_{23}\underline{\underline{\vec{R}}}_2(\mathbf{I}-\mathbf{S}_{22}\underline{\underline{\vec{R}}}_2)^{-1}\mathbf{S}_{32} & \mathbf{S}_{43}+\mathbf{S}_{23}\underline{\underline{\vec{R}}}_2(\mathbf{I}-\mathbf{S}_{22}\underline{\underline{\vec{R}}}_2)^{-1}\mathbf{S}_{42} \\ \mathbf{S}_{14}+\mathbf{S}_{24}\underline{\underline{\vec{R}}}_2(\mathbf{I}-\mathbf{S}_{22}\underline{\underline{\vec{R}}}_2)^{-1}\mathbf{S}_{12} & \mathbf{S}_{24}(\mathbf{I}-\underline{\underline{\vec{R}}}_2\mathbf{S}_{22})^{-1}\underline{\vec{T}}_2 & \mathbf{S}_{34}+\mathbf{S}_{24}\underline{\underline{\vec{R}}}_2(\mathbf{I}-\mathbf{S}_{22}\underline{\underline{\vec{R}}}_2)^{-1}\mathbf{S}_{32} & \mathbf{S}_{44}+\mathbf{S}_{24}\underline{\underline{\vec{R}}}_2(\mathbf{I}-\mathbf{S}_{22}\underline{\underline{\vec{R}}}_2)^{-1}\mathbf{S}_{42} \end{pmatrix},
$$
(6.33a)

$$
\begin{pmatrix} \tilde{\mathbf{C}}_{p,1}^{(2)} \\ \tilde{\mathbf{C}}_{p,2}^{(2)} \\ \tilde{\mathbf{C}}_{p,3}^{(2)} \\ \tilde{\mathbf{C}}_{p,4}^{(2)} \end{pmatrix} = \begin{pmatrix} \mathbf{C}_a^{(2)}(\mathbf{I}-\mathbf{S}_{22}\underline{\underline{\vec{R}}}_2)^{-1}\mathbf{S}_{12} \\ \mathbf{C}_b^{(2)}+\mathbf{C}_a^{(2)}\mathbf{S}_{22}(\mathbf{I}-\underline{\underline{\vec{R}}}_2\mathbf{S}_{22})^{-1}\underline{\vec{T}}_2 \\ \mathbf{C}_a^{(2)}(\mathbf{I}-\mathbf{S}_{22}\underline{\underline{\vec{R}}}_2)^{-1}\mathbf{S}_{32} \\ \mathbf{C}_a^{(2)}(\mathbf{I}-\mathbf{S}_{22}\underline{\underline{\vec{R}}}_2)^{-1}\mathbf{S}_{42} \end{pmatrix},
$$
(6.33b)

$$
\begin{pmatrix}
\tilde{C}_{\alpha,1} & \tilde{C}_{\alpha,2} & \tilde{C}_{\alpha,3} & \tilde{C}_{\alpha,4} \\
\tilde{C}_{\beta,1} & \tilde{C}_{\beta,2} & \tilde{C}_{\beta,3} & \tilde{C}_{\beta,4}
\end{pmatrix}
$$

$$
=\begin{pmatrix}
C_{\alpha,1}+C_{\alpha,2}\underline{\underline{R}}_2(I-S_{22}\underline{\underline{R}}_2)^{-1}S_{12} & C_{\alpha,2}(I-\underline{\underline{R}}_2S_{22})^{-1}\underline{T}_2 & C_{\alpha,3}+C_{\alpha,2}\underline{\underline{R}}_2(I-S_{22}\underline{\underline{R}}_2)^{-1}S_{32} & C_{\alpha,4}+C_{\alpha,2}\underline{\underline{R}}_2(I-S_{22}\underline{\underline{R}}_2)^{-1}S_{42} \\
C_{\beta,1}+C_{\beta,2}\underline{\underline{R}}_2(I-S_{22}\underline{\underline{R}}_2)^{-1}S_{12} & C_{\beta,2}(I-\underline{\underline{R}}_2S_{22})^{-1}\underline{T}_2 & C_{\beta,3}+C_{\beta,2}\underline{\underline{R}}_2(I-S_{22}\underline{\underline{R}}_2)^{-1}S_{32} & C_{\beta,4}+C_{\beta,2}\underline{\underline{R}}_2(I-S_{22}\underline{\underline{R}}_2)^{-1}S_{42}
\end{pmatrix}.
$$

$$(6.33c)$$

The derived S-matrix recursion formulas of the interconnection of the two-port block and the four-port cross-block through port 3 are listed as follows:

$$
\begin{pmatrix}
\tilde{S}_{11} & \tilde{S}_{21} & \tilde{S}_{31} & \tilde{S}_{41} \\
\tilde{S}_{12} & \tilde{S}_{22} & \tilde{S}_{32} & \tilde{S}_{42} \\
\tilde{S}_{13} & \tilde{S}_{23} & \tilde{S}_{33} & \tilde{S}_{43} \\
\tilde{S}_{14} & \tilde{S}_{24} & \tilde{S}_{34} & \tilde{S}_{44}
\end{pmatrix}
$$

$$
=\begin{pmatrix}
S_{11}+S_{31}\underline{R}_{\uparrow3}(I-S_{33}\underline{R}_{\uparrow3})^{-1}S_{13} & S_{21}+S_{31}\underline{R}_{\uparrow3}(I-S_{33}\underline{R}_{\uparrow3})^{-1}S_{23} & S_{31}(I-\underline{R}_{\uparrow3}S_{33})^{-1}T_{\uparrow3} & S_{41}+S_{31}\underline{R}_{\uparrow3}(I-S_{33}\underline{R}_{\uparrow3})^{-1}S_{43} \\
S_{12}+S_{32}\underline{R}_{\uparrow3}(I-S_{33}\underline{R}_{\uparrow3})^{-1}S_{13} & S_{22}+S_{32}\underline{R}_{\uparrow3}(I-S_{33}\underline{R}_{\uparrow3})^{-1}S_{23} & S_{32}(I-\underline{R}_{\uparrow3}S_{33})^{-1}T_{\uparrow3} & S_{42}+S_{32}\underline{R}_{\uparrow3}(I-S_{33}\underline{R}_{\uparrow3})^{-1}S_{43} \\
\underline{T}_{\downarrow3}(I-S_{33}\underline{R}_{\uparrow3})^{-1}S_{13} & \underline{T}_{\downarrow3}(I-S_{33}\underline{R}_{\uparrow3})^{-1}S_{23} & \underline{R}_{\downarrow3}+\underline{T}_{\downarrow3}S_{33}(I-\underline{R}_{\uparrow3}S_{33})^{-1}T_{\uparrow3} & \underline{T}_{\downarrow3}(I-S_{33}\underline{R}_{\uparrow3})^{-1}S_{43} \\
S_{14}+S_{34}\underline{R}_{\uparrow3}(I-S_{33}\underline{R}_{\uparrow3})^{-1}S_{13} & S_{24}+S_{34}\underline{R}_{\uparrow3}(I-S_{33}\underline{R}_{\uparrow3})^{-1}S_{23} & S_{34}(I-\underline{R}_{\uparrow3}S_{33})^{-1}T_{\uparrow3} & S_{44}+S_{34}\underline{R}_{\uparrow3}(I-S_{33}\underline{R}_{\uparrow3})^{-1}S_{43}
\end{pmatrix},
$$

$$(6.34a)$$

$$
\begin{pmatrix}
\tilde{C}_{p,1} \\
\tilde{C}_{p,2} \\
\tilde{C}_{p,3} \\
\tilde{C}_{p,4}
\end{pmatrix}
=\begin{pmatrix}
C_b^{(3)}(I-S_{33}\underline{\underline{R}}_{\uparrow3})^{-1}S_{13} \\
C_b^{(3)}(I-S_{33}\underline{\underline{R}}_{\uparrow3})^{-1}S_{23} \\
C_a^{(3)}+C_b^{(3)}S_{33}(I-\underline{R}_{\uparrow3}S_{33})^{-1}\underline{T}_{\uparrow3} \\
C_b^{(3)}(I-S_{33}\underline{\underline{R}}_{\uparrow3})^{-1}S_{43}
\end{pmatrix},
$$

$$(6.34b)$$

$$
\begin{pmatrix}
\tilde{C}_{\alpha,1} & \tilde{C}_{\alpha,2} & \tilde{C}_{\alpha,3} & \tilde{C}_{\alpha,4} \\
\tilde{C}_{\beta,1} & \tilde{C}_{\beta,2} & \tilde{C}_{\beta,3} & \tilde{C}_{\beta,4}
\end{pmatrix}
$$

$$
=\begin{pmatrix}
C_{\alpha,1}+C_{\alpha,3}\underline{R}_{\uparrow3}(I-S_{33}\underline{R}_{\uparrow3})^{-1}S_{13} & C_{\alpha,2}+C_{\alpha,3}\underline{R}_{\uparrow3}(I-S_{33}\underline{R}_{\uparrow3})^{-1}S_{23} & C_{\alpha,3}(I-\underline{R}_{\uparrow3}S_{33})^{-1}\underline{T}_{\uparrow3} & C_{\alpha,4}+C_{\alpha,3}\underline{R}_{\uparrow3}(I-S_{33}\underline{R}_{\uparrow3})^{-1}S_{43} \\
C_{\beta,1}+C_{\beta,3}\underline{R}_{\uparrow3}(I-S_{33}\underline{R}_{\uparrow3})^{-1}S_{13} & C_{\beta,2}+C_{\beta,3}\underline{R}_{\uparrow3}(I-S_{33}\underline{R}_{\uparrow3})^{-1}S_{23} & C_{\beta,3}(I-\underline{R}_{\uparrow3}S_{33})^{-1}\underline{T}_{\uparrow3} & C_{\beta,4}+C_{\beta,3}\underline{R}_{\uparrow3}(I-S_{33}\underline{R}_{\uparrow3})^{-1}S_{43}
\end{pmatrix}.
$$

$$(6.34c)$$

The derived S-matrix recursion formulas of the interconnection of the two-port block and the four-port cross-block through port 4 are listed as follows:

$$
\begin{pmatrix}
\tilde{S}_{11} & \tilde{S}_{21} & \tilde{S}_{31} & \tilde{S}_{41} \\
\tilde{S}_{12} & \tilde{S}_{22} & \tilde{S}_{32} & \tilde{S}_{42} \\
\tilde{S}_{13} & \tilde{S}_{23} & \tilde{S}_{33} & \tilde{S}_{43} \\
\tilde{S}_{14} & \tilde{S}_{24} & \tilde{S}_{34} & \tilde{S}_{44}
\end{pmatrix}
$$

$$
=
\begin{pmatrix}
S_{11}+S_{41}\underline{R}_{\downarrow4}(I-S_{44}\underline{R}_{\downarrow4})^{-1}S_{14} & S_{21}+S_{41}\underline{R}_{\downarrow4}(I-S_{44}\underline{R}_{\downarrow4})^{-1}S_{24} & S_{31}+S_{41}\underline{R}_{\downarrow4}(I-S_{44}\underline{R}_{\downarrow4})^{-1}S_{34} & S_{41}(I-\underline{R}_{\downarrow4}S_{44})^{-1}\underline{T}_{\downarrow4} \\
S_{12}+S_{42}\underline{R}_{\downarrow4}(I-S_{44}\underline{R}_{\downarrow4})^{-1}S_{14} & S_{22}+S_{42}\underline{R}_{\downarrow4}(I-S_{44}\underline{R}_{\downarrow4})^{-1}S_{24} & S_{32}+S_{42}\underline{R}_{\downarrow4}(I-S_{44}\underline{R}_{\downarrow4})^{-1}S_{34} & S_{42}(I-\underline{R}_{\downarrow4}S_{44})^{-1}\underline{T}_{\downarrow4} \\
S_{13}+S_{43}\underline{R}_{\downarrow4}(I-S_{44}\underline{R}_{\downarrow4})^{-1}S_{14} & S_{23}+S_{43}\underline{R}_{\downarrow4}(I-S_{44}\underline{R}_{\downarrow4})^{-1}S_{24} & S_{33}+S_{43}\underline{R}_{\downarrow4}(I-S_{44}\underline{R}_{\downarrow4})^{-1}S_{34} & S_{43}(I-\underline{R}_{\downarrow4}S_{44})^{-1}\underline{T}_{\downarrow4} \\
\underline{T}_{\uparrow4}(I-S_{44}\underline{R}_{\downarrow4})^{-1}S_{14} & \underline{T}_{\uparrow4}(I-S_{44}\underline{R}_{\downarrow4})^{-1}S_{24} & \underline{T}_{\uparrow4}(I-S_{44}\underline{R}_{\downarrow4})^{-1}S_{34} & \underline{R}_{\uparrow4}+\underline{T}_{\uparrow4}S_{44}(I-\underline{R}_{\downarrow4}S_{44})^{-1}\underline{T}_{\downarrow4}
\end{pmatrix}
$$

(6.35a)

$$
\begin{pmatrix}
\tilde{C}_{p,1} \\
\tilde{C}_{p,2} \\
\tilde{C}_{p,3} \\
\tilde{C}_{p,4}
\end{pmatrix}
=
\begin{pmatrix}
C_a^{(4)}(I-S_{44}\underline{R}_{\downarrow4})^{-1}S_{14} \\
C_a^{(4)}(I-S_{44}\underline{R}_{\downarrow4})^{-1}S_{24} \\
C_a^{(4)}(I-S_{44}\underline{R}_{\downarrow4})^{-1}S_{34} \\
C_b^{(4)}+C_a^{(4)}S_{44}(I-\underline{R}_{\downarrow4}S_{44})^{-1}\underline{T}_{\downarrow4}
\end{pmatrix}
$$

(6.35b)

$$
\begin{pmatrix}
\tilde{C}_{\alpha,1} & \tilde{C}_{\alpha,2} & \tilde{C}_{\alpha,3} & \tilde{C}_{\alpha,4} \\
\tilde{C}_{\beta,1} & \tilde{C}_{\beta,2} & \tilde{C}_{\beta,3} & \tilde{C}_{\beta,4}
\end{pmatrix}
$$

$$
=
\begin{pmatrix}
C_{\alpha,1}+C_{\alpha,4}\underline{R}_{\downarrow4}(I-S_{44}\underline{R}_{\downarrow4})^{-1}S_{14} & C_{\alpha,2}+C_{\alpha,4}\underline{R}_{\downarrow4}(I-S_{44}\underline{R}_{\downarrow4})^{-1}S_{24} & C_{\alpha,3}+C_{\alpha,4}\underline{R}_{\downarrow4}(I-S_{44}\underline{R}_{\downarrow4})^{-1}S_{34} & C_{\alpha,4}(I-\underline{R}_{\downarrow4}S_{44})^{-1}\underline{T}_{\downarrow4} \\
C_{\beta,1}+C_{\beta,4}\underline{R}_{\downarrow4}(I-S_{44}\underline{R}_{\downarrow4})^{-1}S_{14} & C_{\beta,2}+C_{\beta,4}\underline{R}_{\downarrow4}(I-S_{44}\underline{R}_{\downarrow4})^{-1}S_{24} & C_{\beta,3}+C_{\beta,4}\underline{R}_{\downarrow4}(I-S_{44}\underline{R}_{\downarrow4})^{-1}S_{34} & C_{\beta,4}(I-\underline{R}_{\downarrow4}S_{44})^{-1}\underline{T}_{\downarrow4}
\end{pmatrix}
$$

(6.35c)

Next, with the use of the prepared S-matrix recursion formulas, the total S-matrix of the extended four-port cross-block can be constructed via a stepwise procedure. In the first step, the S-matrix of the combined structure of the two-port block and four-port cross-block through port 1 is analyzed using Equations (6.32a) to (6.32c). The combined structure can be viewed as a four-port block with its own S-matrix and internal coupling coefficient operators. Hence, at the second step, the S-matrix formulas of the interconnection of the two-port block and the four-port cross-block through port 2 can be applied to interconnect this combined four-port block structure to a two-port block through port 2 in a straightforward manner, with no modifications. Similarly, the S-matrix formulas can be applied to interconnect the combined four-port block and a two-port block to build the extended four-port cross-block. This analysis procedure is illustrated in Figure 6.16.

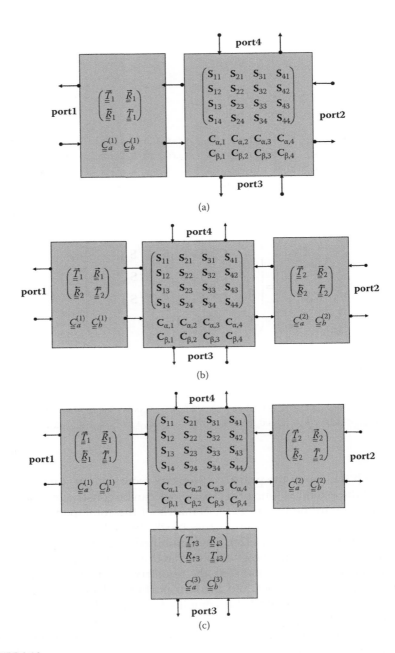

FIGURE 6.16
Building the extended four-port cross-block: (a) step 1: interconnection of a two-port block to the four-port cross-block through port 1, (b) step 2: interconnection of a two-port block to the combined four-port cross-block through port 2, (c) step 3: interconnection of a two-port block to the combined four-port cross-block through port 3, and (d) step 4: building the extended four-port cross-block by the interconnection of a two-port block to the combined four-port through port 4.

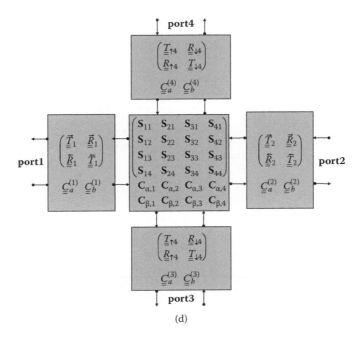

(d)

FIGURE 6.16
(Continued).

For validating the developed GSMM formulas, the electric field distributions at each step of building the extended four-port cross-block by the stated step-by-step interconnection procedure are visualized. The results of the cross-waveguide structure are shown in Figure 6.17. In Figure 6.17(a), the combined structure is illustrated. In Figure 6.17(b) to (e), the y-polarization electric field distributions at steps 1 to 4 are presented, respectively. The field visualization results are quite well matched with the results of the previous work done by the finite-difference time-domain (FDTD) method (reference [44] in Chapter 1). For comparison, additional simulation results of the T-branch waveguide structure (reference [47] in Chapter 1) and the 90° bending waveguide structure (reference [48] in Chapter 1) are presented, respectively, in Figures 6.18 and 6.19.

In summary, the developed scheme is composed of two main subtheories: (1) a local Fourier modal analysis method for analyzing internal eigenmodes of four-port cross-blocks and (2) a generalized S-matrix method for modeling crossed nanophotonic structures by interconnecting four two-port blocks and a four-port block. The established modeling and analysis of crossed nanophotonic structures is a basic element for modeling generalized large-scale nanophotonic networks. In the aspects of methodology, the local analysis scheme is efficient since the local field analysis is performed with the reasonable and practical assumption of the field localization in

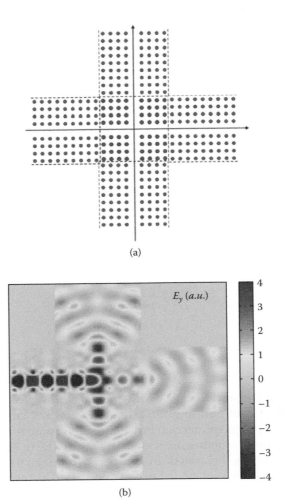

(a)

(b)

FIGURE 6.17
(a) Cross-waveguide structure and y-polarization electric field distributions at each step of building the extended four-port cross-block by the step-by-step interconnection procedure: (b) step 1, (c) step 2, (d) step 3, and (e) step 4. (From H. Kim and B. Lee, *J. Opt. Soc. Am. B*, 25(4), 2008. With permission.)

nanophotonic structures. Instead of computing the whole structure of a network, local regions occupied by functional photonic blocks are characterized by the LFMA. Eventually this local analysis scheme would provide a method for the efficient analysis of large-scale nanophotonic networks and systematic methods for the advanced design, analysis, and fabrication of nanophotonic networks.

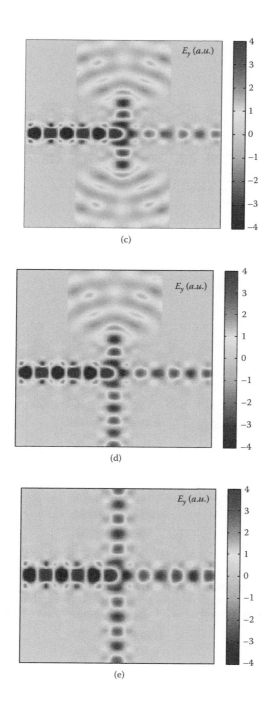

(c)

(d)

(e)

FIGURE 6.17
(Continued).

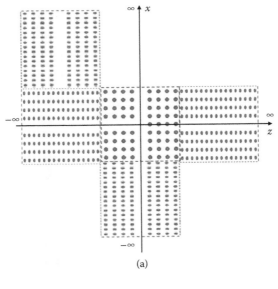

(a)

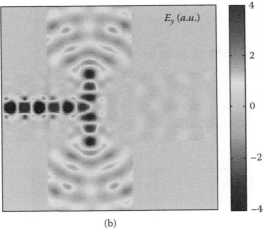

(b)

FIGURE 6.18
(a) T-branch waveguide structure and y-polarization electric field distributions at each step of
building the extended four-port cross-block by the step-by-step interconnection procedure:
(b) step 1, (c) step 2, (d) step 3, and (e) step 4. (From H. Kim and B. Lee, *J. Opt. Soc. Am. B*, 25(4),
2008. With permission.)

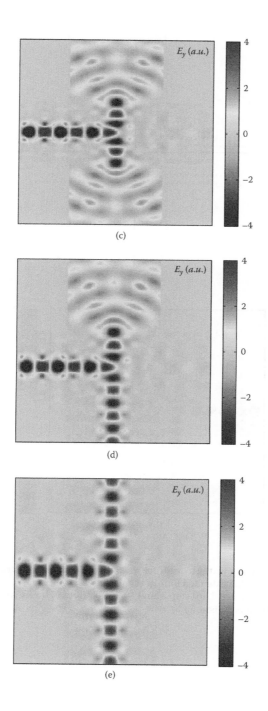

(c)

(d)

(e)

FIGURE 6.18
(Continued).

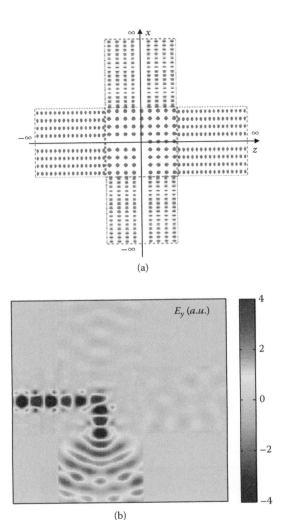

(a)

(b)

FIGURE 6.19
(a) Ninety-degree bending waveguide structure and *y*-polarization electric field distributions at each step of building the extended four-port cross-block by the step-by-step interconnection procedure: (b) step 1, (c) step 2, (d) step 3, and (e) step 4. (From H. Kim and B. Lee, *J. Opt. Soc. Am. B*, 25(4), 2008. With permission.)

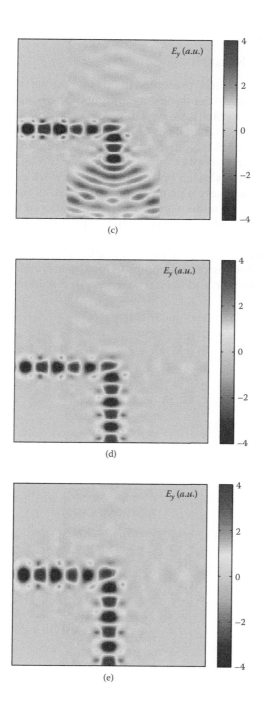

(c)

(d)

(e)

FIGURE 6.19
(Continued).

6.5 Concluding Remarks

Our hope is that the material presented in this monograph will permit readers to develop an overall, in-depth understanding of the Fourier modal method as a branch of the general linear system theory. The philosophy and concept associated with the modal analysis of linear systems are vividly reflected in the development of FMM. Numerous interesting issues dealing with mathematical modeling in nanophotonics research that can be efficiently studied with the modal analysis based on the FMM framework remain. Large-scale nanophotonic networks are a prime issue in the field of nanophotonics. This treatise attempts to provide a powerful computational and mathematical framework for advanced and practical studies related to the design, analysis, and fabrication of nanophotonics. We hope that the material presented here will contribute to advancing the newborn field of integrated nanophotonics. The field of nanophotonics and computational nanophotonics is unlimited and open to enthusiastic and keen researchers and students.

References

Chapter 1

1. S. V. Gaponenko, *Introduction to Nanophotonics* (Cambridge University Press, Cambridge, 2010).
2. P. N. Prasad, *Nanophotonics* (Wiley-Interscience, Hoboken, 2004).
3. M. Ohtsu, K. Kobayashi, T. Kawazoe, T. Yatsui, and M. Naruse, *Principles of Nanophotonics* (Series in Optics and Optoelectronics) (Taylor & Francis, Boca Raton, 2008).
4. H. Rigneault, J.-M. Lourtioz, C. Delalande, and A. Levenson, *Nanophotonics* (Wiley-ISTE, Hoboken, 2006).
5. S. A. Maier, *Plasmonics Fundamentals and Applications* (Springer, New York, 2007).
6. S. I. Bozhevolnyi, *Plasmonic Nanoguides and Circuits* (Pan Stanford, Singapore, 2008).
7. Y. Urzhumov, *From Optical Magnetism to Super-Resolution: Sub-Wavelength Electromagnetic Phenomena in Plasmonic and Polaritonic Nanostructures* (LAP Lambert Academic, Saarbrücken, 2009).
8. D. Sarid and W. Challener, *Modern Introduction to Surface Plasmons: Theory, Mathematica Modeling, and Applications* (Cambridge University Press, Cambridge, 2010).
9. J. D. Joannopoulos, S. G. Johnson, J. N. Winn, and R. D. Meade, *Photonic Crystals: Molding the Flow of Light*, 2nd ed. (Princeton University Press, Princeton, NJ, 2008).
10. J.-M. Lourtioz, H. Benisty, V. Berger, J.-M. Gerard, D. Maystre, and A. Tchelnokov, *Photonic Crystals: Towards Nanoscale Photonic Devices* (Springer, Berlin, 2008).
11. M. Skorobogatiy and J. Yang, *Fundamentals of Photonic Crystal Guiding* (Cambridge University Press, Cambridge, 2009).
12. I. A. Sukhoivanov and I. V. Guryev, *Photonic Crystals: Physics and Practical Modeling* (Springer Series in Optical Sciences) (Springer, Berlin, 2009).
13. C. Sibilia, T. M. Benson, M. Marciniak, and T. Szoplik, *Photonic Crystals: Physics and Technology* (Springer, Berlin, 2008).
14. K. Inoue and K. Ohtaka, *Photonic Crystals: Physics, Fabrication and Applications* (Springer, Berlin, 2008).
15. R. E. Slusher and B. J. Eggleton, *Nonlinear Photonic Crystals* (Springer, Berlin, 2009).
16. Y. S. Kivshar and G. Agrawal, *Optical Solutions: From Fibers to Photonic Crystals* (Academic Press, San Diego, 2003).
17. M. A. Noginov and V. A. Podolskiy, *Tutorials in Metamaterials* (CRC Press, Boca Raton, FL, 2011).
18. L. Solymar and E. Shamonina, *Waves in Metamaterials* (Oxford University Press, London, 2009).
19. W. Cai and V. Shalaev, *Optical Metamaterials: Fundamentals and Applications* (Springer, Berlin, 2009).

20. N. Engheta and R. W. Ziolkowski, *Electromagnetic Metamaterials: Physics and Engineering Explorations* (Wiley-IEEE Press, Hoboken, NJ, 2006).
21. V. M. Shalaev and A. K. Sarychev, *Electrodynamics of Metamaterials* (World Scientific, London, 2007).
22. F. Capolino, *Theory and Phenomena of Metamaterials* (Metamaterials Handbook) (CRC Press, Boca Raton, FL, 2009).
23. F. Capolino, *Applications of Metamaterials* (Metamaterials Handbook) (CRC Press, Boca Raton, FL, 2009).
24. E. J. Tremblay, *Metamaterials: Classes, Properties and Applications* (Nova Science, New York, 2010).
25. Finite difference time domain method, http://ab-initio.mit.edu/wiki/index.php/Meep
26. Finite integration method, http://www.cst.com/
27. Finite element method, http://www.comsol.com/products/rf/?tab=appareas
28. K. Busch, M. König, and J. Niegemann, Discontinuous Galerkin methods in nanophotonics, *Laser Photonics Rev.*, DOI 10.1002/lpor.201000045 (2011).
29. K. Stannigel, M. König, J. Niegemann, and K. Busch, Discontinuous Galerkin time-domain computations of metallic nanostructures, *Opt. Express* 17, 14934–14947 (2009).
30. M. G. Moharam and T. K. Gaylord, Rigorous coupled-wave analysis of planar-grating diffraction, *J. Opt. Soc. Am. A* 71, 811–818 (1981).
31. P. Lalanne, Improved formulation of the coupled-wave method for two-dimensional gratings, *J. Opt. Soc. Am. A* 14, 1592–1598 (1997).
32. H. Kim, S. Kim, I.-M. Lee, and B. Lee, Pseudo-Fourier modal analysis on dielectric slabs with arbitrary longitudinal permittivity and permeability profiles, *J. Opt. Soc. Am. A* 23, 2177–2191 (2006).
33. L. Li, Use of Fourier series in the analysis of discontinuous periodic structures, *J. Opt. Soc. Am. A* 13, 1870–1876 (1996).
34. L. Li, Fourier modal method for crossed anisotropic gratings with arbitrary permittivity and permeability tensors, *J. Opt. A Pure Appl. Opt.* 5, 345–355 (2003).
35. L. Li, Mathematical reflections on the Fourier modal method in grating theory, Chapter 4 in *Mathematical Modeling in Optical Science*, edited by G. Bao (SIAM, Philadelphia, 2001).
36. E. Popov and M. Neviére, Differential theory for diffraction gratings: a new formulation for TM polarization with rapid convergence, *Opt. Lett.* 25, 598–600 (2000).
37. E. Popov and M. Neviére, Grating theory: new equations in Fourier space leading to fast converging results for TM polarization, *J. Opt. Soc. Am. A* 17, 1773–1784 (2000).
38. L. Li, Formulation and comparison of two recursive matrix algorithms for modeling layered diffraction gratings, *J. Opt. Soc. Am. A* 13, 1024–1035 (1996).
39. H. Kim, I.-M. Lee, and B. Lee, Extended scattering matrix method for efficient full parallel implementation of rigorous coupled wave analysis, *J. Opt. Soc. Am. A* 24, 2313–2327 (2007).
40. E. Popov, M. Neviére, B. Gralak, and G. Tayeb, Staircase approximation validity for arbitrary-shaped gratings, *J. Opt. Soc. Am. A* 19, 33–42 (2002).
41. H. Kim and B. Lee, Pseudo-Fourier modal analysis of two-dimensional arbitrarily shaped grating structures, *J. Opt. Soc. Am. A* 25, 40–54 (2008).
42. J. P. Hugonin and P. Lalanne, Perfectly matched layers as nonlinear coordinate transforms: a generalized formalization, *J. Opt. Soc. Am. A* 22, 1844–1849 (2005).

43. N. Moll, R. Harbers, R. F. Mahrt, and G.-L. Bona, Integrated all-optical switch in a cross-waveguide geometry, *Appl. Phys. Lett.* 88, 171104-1–171104-3 (2006).
44. M. F. Yanik, S. Fan, M. Soliacic, and D. Joannopoulos, All-optical transistor action with bistable switching in a photonic crystal cross-waveguide geometry, *Opt. Lett.* 28, 2506–2508 (2003).
45. M. F. Yanik and S. Fan, Stopping light all-optically, *Phys. Rev. Lett.* 92, 083901-1–083901-4 (2004).
46. M. F. Yanik, H. A. Altug, J. Vuckovic, and S. Fan, Sub-micron all optical digital memory and integration of nano-scale photonic devices without isolators, *J. Lightwave Tech.* 22, 2316–2322 (2004).
47. S. Fan, S. G. Johnson, and J. D. Joannopoulos, Waveguide branches in photonic crystals, *J. Opt. Soc. Am. B* 18, 162–165 (2001).
48. A. Mekis, J. C. Chen, I. Kurland, S. Fan, P. R. Villeneuve, and J. D. Joannopoulos, High transmission through sharp bends in photonic crystal waveguides, *Phys. Rev. Lett.* 77, 3787–3790 (1996).

Chapter 2

1. M. G. Moharam, E. B. Grann, and D. A. Pommet, Formulation for stable and efficient implementation of the rigorous coupled-wave analysis of binary gratings, *J. Opt. Soc. Am. A* 12, 1068–1076 (1995).
2. M. G. Moharam and A. B. Greenwell, Efficient rigorous calculations of power flow in grating coupled surface-emitting devices, *Proc. SPIE* 5456, 57–67 (2004).
3. E. L. Tan, Note on formulation of the enhanced scattering- (transmittance-) matrix approach, *J. Opt. Soc. Am. A* 19, 1157–1161 (2002).

Chapter 3

1. H. A. Bethe, Theory of diffraction by small holes, *Phys. Rev.* 66, 163–182 (1944).
2. T. W. Ebbesen, H. J. Lezec, H. F. Ghaemi, T. Thio, and P. A. Wolff, Extraordinary optical transmission through sub-wavelength hole arrays, *Nature* 391, 667–669 (1998).
3. L. Martin-Moreno, F. J. Garcia-Vidal, H. J. Lezec, K. M. Pellerin, T. Thio, J. B. Pendry, and T. W. Ebbesen, Theory of extraordinary optical transmission through subwavelength hole arrays, *Phys. Rev. Lett.* 86, 1114–1117 (2001).
4. P. Lalanne and H. T. Liu, Microscopic theory of the extraordinary optical transmission, *Nature* 452, 728–731 (2008).
5. H. F. Schouten, N. Kuzmin, G. Dubois, T. D. Visser, G. Gbur, P. F. A. Alkemade, H. Blok, G. W. Hooft, D. Lenstra, and E. R. Eliel, Plasmon-assisted two-slit transmission: Young's experiment revisited, *Phys. Rev. Lett.* 94, 053901 (2005).
6. J. Weiner, Phase shifts and interference in surface plasmon polariton waves, *Opt. Express* 16, 950–956 (2008).

7. D. Pacifici, H. J. Lezec, H. A. Atwater, and J. Weiner, Quantitative determination of optical transmission through subwavelength slit arrays in ag films: role of surface wave interference and local coupling between adjacent slits, *Phys. Rev. B* 77, 115411 (2008).

Chapter 4

1. W. Klaus, M. Fujino, and K. Kodate, Optimization of absorbing boundaries for RCWA calculations of aperiodic structures, Diffractive Optics 2001 Budapest.
2. J. P. Berenger, A perfectly matched layer for the absorption of electromagnetic waves, *J. Comput. Phys.* 114, 185–200 (1994).
3. A. Taflove and S. C. Hagness, *Computational Electrodynamics: The Finite-Difference Time-Domain Method*, 3rd ed. (Artech House, Boston, 2005).
4. J. P. Hugonin and P. Lalanne, Perfectly matched layers as nonlinear coordinate transforms: a generalized formalization, *J. Opt. Soc. Am. A* 22, 1844–1849 (2005).
5. L. Martin-Moreno, F. J. Garcia-Vidal, H. J. Lezec, A. Degiron, and T. W. Ebbesen, Theory of highly directional emission from a single subwavelength aperture surrounded by surface corrugations, *Phys. Rev. Lett.* 90 (2003).
6. H. Caglayan, I. Bulu, and E. Ozbay, Extraordinary grating-coupled microwave transmission through a subwavelength annular aperture, *Opt. Express* 13, 1666–1671 (2005).
7. S. K. Morrison and Y. S. Kivshar, Engineering of directional emission from photonic-crystal waveguides, *Appl. Phys. Lett.* 86 (2005).
8. Z. B. Li, J. G. Tian, Z. B. Liu, W. Y. Zhou, and C. P. Zhang, Enhanced light transmission through a single subwavelength aperture in layered films consisting of metal and dielectric, *Opt. Express* 13, 9071–9077 (2005).
9. S. Kim, H. Kim, Y. Lim, and B. Lee, Off-axis directional beaming of optical field diffracted by a single subwavelength metal slit with asymmetric dielectric surface gratings, *Appl. Phys. Lett.* 90 (2007).
10. D. Choi, Y. Lim, S. Roh, I. M. Lee, J. Jung, and B. Lee, Optical beam focusing with a metal slit array arranged along a semicircular surface and its optimization with a genetic algorithm, *Appl. Opt.* 49, A30–A35 (2010).
11. H. Kim, J. Park, and B. Lee, Finite-size nondiffracting beam from a subwavelength metallic hole with concentric dielectric gratings, *Appl. Opt.* 48, G68–G72 (2009).
12. S. Kim, Y. Lim, H. Kim, J. Park, and B. Lee, Optical beam focusing by a single subwavelength metal slit surrounded by chirped dielectric surface gratings, *Appl. Phys. Lett.* 92 (2008).
13. Z. W. Liu, J. M. Steele, W. Srituravanich, Y. Pikus, C. Sun, and X. Zhang, Focusing surface plasmons with a plasmonic lens, *Nano Lett.* 5, 1726–1729 (2005).
14. H. Kim and B. Lee, Diffractive slit patterns for focusing surface plasmon polaritons, *Opt. Express* 16, 8969–8980 (2008).
15. U. Levy, G. M. Lerman, and A. Yanai, Demonstration of nanofocusing by the use of plasmonic lens illuminated with radially polarized light, *Nano Lett.* 9, 2139–2143 (2009).

16. H. Kim, J. Park, S. W. Cho, S. Y. Lee, M. Kang, and B. Lee, Synthesis and dynamic switching of surface plasmon vortices with plasmonic vortex lens, *Nano Lett.* 10, 529–536 (2010).

Chapter 5

1. Q. Cao, P. Lalanne, and J. P. Hugonin, Stable and efficient Bloch-mode computational method for one-dimensional grating waveguides, *J. Opt. Soc. Am. A* 19, 335–338 (2002).
2. J. D. Joannopoulos, R. D. Meade, and J. N. Winn, *Photonic Crystals* (Princeton University Press, Princeton, NJ, 1995).
3. J.-M. Lourtioz, H. Benisty, V. Berger, et al., *Photonic Crystals: Towards Nanoscale Photonic Devices* (Springer, New York, 1999).
4. S. G. Johnson and J. D. Joannopoulos, *Photonic Crystals: The Road from Theory to Practice* (Springer, New York, 2002).
5. K. Sakoda, *Optical Properties of Photonic Crystals* (Springer, Berlin, 2004).
6. J. D. Joannopoulos, P. R. Villeneuve, and S. H. Fan, Photonic crystals, *Solid State. Commun.* 102, 165–173 (1997).
7. M. Tokushima, H. Kosaka, A. Tomita, and H. Yamada, Lightwave propagation through a 120 degrees sharply bent single-line-defect photonic crystal waveguide, *Appl. Phys. Lett.* 76, 952–954 (2000).
8. E. Chow, S. Y. Lin, J. R. Wendt, S. G. Johnson, and J. D. Joannopoulos, Quantitative analysis of bending efficiency in photonic-crystal waveguide bends at lambda=1.55 mu m wavelengths, *Opt. Lett.* 26, 286–288 (2001).
9. Y. Akahane, T. Asano, B. S. Song, and S. Noda, High-Q photonic nanocavity in a two-dimensional photonic crystal, *Nature* 425, 944–947 (2003).
10. T. Tanabe, M. Notomi, E. Kuramochi, A. Shinya, and H. Taniyama, Trapping and delaying photons for one nanosecond in an ultrasmall high-q photonic-crystal nanocavity, *Nat. Photonics* 1, 49–52 (2007).

Chapter 6

1. H. Kim and B. Lee, Mathematical modeling of crossed nanophotonic structures with generalized scattering-matrix method and local Fourier modal analysis, *J. Opt. Soc. Am. B* 25, 518–544, 2008.

Index

For Product Safety Concerns and Information please contact our EU
representative GPSR@taylorandfrancis.com Taylor & Francis Verlag GmbH,
Kaufingerstraße 24, 80331 München, Germany

Printed and bound by CPI Group (UK) Ltd, Croydon, CR0 4YY
01/05/2025
01858516-0001